NATURALISTS AT SEA

GLYN WILLIAMS

NATURALISTS AT SEA

SCIENTIFIC TRAVELLERS FROM DAMPIER TO DARWIN

YALE UNIVERSITY PRESS
NEW HAVEN AND LONDON

Published with assistance from the Annie Burr Lewis Fund

For information about this and other Yale University Press publications, please contact:
U.S. Office: sales.press@yale.edu   www.yalebooks.com
Europe Office: sales@yaleup.co.uk   www.yalebooks.co.uk

Set in Adobe Caslon Pro by IDSUK (DataConnection) Ltd
Printed in Great Britain by TJ International Ltd, Padstow, Cornwall

Library of Congress Cataloging-in-Publication Data

Williams, Glyndwr.
   Naturalists at sea : scientific travellers from Dampier to Darwin / Glyn Williams.
      pages cm
   Includes bibliographical references.
      ISBN 978-0-300-18073-2 (cloth : alkaline paper)
1. Natural history—History. 2. Scientific expeditions—History. 3. Ocean travel—History. 4. Naturalists—History. 5. Travelers—History. 6. Naturalists—Biography.
I. Title.
   QH11.W45 2013
   508—dc23
                                                                                    2013010060

A catalogue record for this book is available from the British Library.

10 9 8 7 6 5 4 3 2 1

For Sophie Forgan

CONTENTS

List of Illustrations and Map	*ix*
Acknowledgements	*xiii*
Introduction	1
1 The 'rambling voyages' of William Dampier, Self-Taught Naturalist	7
2 'Ten years of preparation; ten hours of exploration': The Alaskan Tribulations of Georg Wilhelm Steller	32
3 'My plants, my beloved plants, have consoled me for everything': The Fortunes and Misfortunes of Philibert Commerson	54
4 'No people ever went to sea better fitted out for the purposes of Natural History': Joseph Banks and Daniel Solander	73
5 'A kind of Linnaean being': The Woes of Johann Reinhold Forster	95
6 'Curse scientists, and all science into the bargain': Cook, Vancouver and 'experimental gentlemen'	122

7 'Devilish fellows who test patience to the very limit':
Naturalists with La Pérouse and d'Entrecasteaux   150

8 'All our efforts will be focussed on natural history':
The Scientific and Political Voyage of Alejandro
Malaspina   179

9 'When a botanist first enters so remote a country
he finds himself in a new world': The Australian
Surveys of Nicolas Baudin and Matthew Flinders   201

10 'Like giving to a blind man eyes': Charles Darwin
on the *Beagle*   232

Conclusion   260

Notes   *264*
Select Bibliography   *289*
Index   *297*

ILLUSTRATIONS AND MAP

Plates

1 Engraving after drawing by Lieutenant Peircy Brett, 'A View of the Watering Place at Tenian', 1748. National Library of Australia, an10098525.

2 'Plants found in New Holland and Timor', in William Dampier, *A Voyage to New Holland*, vol. I, 1703, Tab. 4. Author's collection.

3 Herbarium specimen of *Swainsona formosa* (Sturt's desert pea). Oxford University Herbaria, Department of Plant Sciences.

4 Sven Waxell, an Aleut in his baidarka, detail of his chart of the *St Peter*, 1742. Archives of the Ministry of Marine, Petrograd.

5 Sven Waxell, drawing of a fur seal, sea lion and sea cow on his chart of Bering's voyage, 1741. Alaska State Library Historical Collections.

6 Georg Dionysus Ehret, illustration of Linnaeus's system of classification, 1736. © The Natural History Museum, London.

7 Sydney Parkinson, bougainvillea plant (*Bougainvillea spectabilis*), 1768. © The Natural History Museum, London.

8 Sydney Parkinson, Tahitian breadfruit (*Artocarpus altilis*), 1768–71 © The Natural History Museum, London.

9 Sydney Parkinson, 'Kangaru, Endeavour's River', 1770. © The Natural History Museum, London.

10 Peter Brown, 'The Blue-Bellied Parrot', in *Nouvelles illustrations de zoologie*, 1776, plate VII. © The British Library Board (1255.k.9).

11 'The Fly-Catching Macaroni', published by M. Darly, 1772. National Library of Australia, an9283268.

12 'The Simpling Macaroni', published by M. Darly, 1772. National Library of Australia, an2983270.

13 George Forster, chinstrap penguin (*Pygoscelis antarcticus*), 1772. © The Natural History Museum, London.

14 John Francis Rigaud, 'Johann Reinhold Forster and Georg Forster in Tahiti', 1780. Private collection.

15 George Forster, devilfish (*Mobula mobular*), 1774. © The Natural History Museum, London.

16 John Webber, 'A Sea Otter', (*Enhydra lutris*), 1784. Author's collection.

17 John Ellis, 'A Wired Cage for bringing over the Bread Fruit Tree...', in *Description of the Mangostan and Bread-Fruit*, 1775, plate IX. © The British Library Board (34.e.21).

18 Gaspard Duche de Vancy's drawings of receptacles for holding plants during the expedition of La Pérouse, eighteenth century. Bibliothèque Mazarine, Paris, France/Archives Charmet/The Bridgeman Art Library.

19 Nicolas Ozanne, *Massacre of De Langles*, 1797. © National Maritime Museum, Greenwich, London.

20 Francis Boott, Joseph Banks's herbarium and library at Soho Square, 1820. © The Natural History Museum, London.

21 *Eucalyptus globulus* (blue gum), in Labillardière's *Atlas pour server à la relation du voyage à le recherche de la Pérouse*, 1800. National Library of Australia, an20974042.

22 Engraving by Jacques Louis Copia, after Jean Piron, 'Sauvages du cap de Diemen préparant leur repas', 1793. National Library of Australia, an20973389.

23 José Cardero. 'Isla de Naos' (Puerto de Panama), 1790. Museo Naval, Madrid.

24 José Cardero, 'Pira y sepulcros ... en el puerto de Mulgrave', 1791. Museo Naval, Madrid.

25 Juan Ravenet, 'La Muerte de Antonio Pineda', 1792. Museo Naval, Madrid.

26 Juan Ravenet, 'La experiencia de la Gravedad en Puerto Egmont', 1794. Museo Naval, Madrid.

27 Plan of the quarter-deck of HMS *Investigator*, 1801. © National Maritime Museum, Greenwich, London.

28 C.A. Lesueur's drawing of Péron excavating Aboriginal cremation site, in Louis de Freycinet's *Atlas* accompanying *Voyage de Découvertes aux Terres Australes*, 1817, plate XVI. National Library of Australia, an7573653.

29 C.A. Lesueur, 'Mollusques et zoophytes', in *Voyage de Découvertes aux Terres Australes*, 1807, plate XXX. National Library of Australia, an7573695.

30 C.A. Lesueur, 'François Péron', frontispiece of *Éloge Historique de François Péron*, 1811. Allport Library and Museum of Fine Arts, Tasmania.

31 'La Nouvelle-Hollande mieux connue: Végétaux utiles naturalisés en France', title-page of *Voyage de Découvertes aux Terres Australes*, 1824. National Library of Australia, an20231293-v.

32 Ferdinand Bauer, sand palm (*Livistona humilis*), in *Botanical Drawings from Australia*, 1801, plate 225. © The Natural History Museum, London.

33 William Westall, 'Chasm Island, native cave painting', 1803. National Library of Australia, an4565185.

34 Matthew Flinder's chart of 'Australia or Terra Australis', sheet III. The National Archives (ADM22/436).

35 P.A. King, 'Fuegian (Yapoo Tekeenica)', frontispiece of Fitzroy's *Narrative of the Surveying Voyages of the Beagle*, vol. II, 1839. © The Natural History Museum, London.

36 John Clements Wickham, 'Remains of the Cathedral at Concepcíon, ruined by the Great Earthquake of 1835', in Fitzroy's *Narrative of the Surveying Voyages of His Majesty's*

Ships Adventure and Beagle, 1839, plate VI. Reproduced with permission from John van Wyhe, ed., *The Complete Work of Charles Darwin Online* (http://darwin-online.org.uk/).

37 'Four Species of Galapagos Finch with Different Beaks', in Charles Darwin's *Journal of Researches*, 1870. © The Natural History Museum, London.

38 George Sowerby, 'Chthamalus', in Darwin's *A monograph on the sub-class Cirripedia, with figures of all the species*, II, 1854, Plate XVIII. Reproduced with permission from John van Wyhe, ed., *The Complete Work of Charles Darwin Online* (http://darwin-online.org. uk/).

39 Darwin's microscope at Down House. © A.J. Southward, Marine Biological Association.

Map   *Page*

The Pacific Ocean.   xiv

ACKNOWLEDGEMENTS

WHEN, IN THE late 1950s, I first began research on the history of European incursions into the Pacific and the Arctic, a book covering the subject area of this volume and based on original sources would have been an almost impossible project for a single scholar. Younger colleagues are still taken aback by the fact that my doctoral research on Captain Cook's third voyage was carried out at a time when none of the main manuscript sources, including Cook's own journal, had been published in acceptable form: J.C. Beaglehole's monumental edition, with its book-length introduction and lavish annotations, was still some years away. Today the records kept on most of the voyages described here have been published in scholarly editions, and my first debt is to the editors and translators of those documents. Their painstaking work, often not fully appreciated even within the academic world, has made my task infinitely more straightforward. Apart from the incomparable Beaglehole, who added Joseph Banks's *Endeavour* journal to his comprehensive editions of Cook's journals, the list includes Raymond H. Fisher, O.W. Frost, Margritt A. Engel, Carol Urness, Etienne Taillemite, John Dunmore, Edward and Maryse Duyker, Neil Chambers, Michael E. Hoare, Nicholas Thomas, Oliver Bergh, Jennifer Newell, Harriet Guest, Michael Dettelbach, Eivor Cormack, Robert Galois, W. Kaye Lamb, Paul Brunton, Phyllis Edwards, Jacqueline Bonnemains, Jean-Marc Argentin, Martine Marin, Maria Victoria Ibañez Montoya, Dolores Higueras Rodríguez, Robert J. King, Christine Cornell, T.G. Vallance,

D.T. Moore, E.W. Groves, Anthony J. Brown, Gillian Dooley, Richard A. Pierce, Victoria Joan Moessner, Frederick Burkhard, Richard Keynes, and my co-editors on the first English translation of Alejandro Malaspina's 'Diario', Andrew David, Felipe Fernández Armesto and Carlos Novi, who, together with Donald C. Cutter and Sylvia Jamieson, taught me much about the daunting task of translating and annotating manuscript journals.

The essential corollary to the efforts of these scholars has been the publication of their work, and I must pay tribute here to the learned bodies and academic presses whose printed editions of the journals and letters of explorers and naturalists have been indispensable to my work: Allen & Unwin (Alejandro Malaspina); Australia Biological Resources Study (Robert Brown); Australian Maritime Museum with Hordern House (George Forster); British Museum, Natural History (Peter Good); Adam and Charles Black (William Dampier); Cambridge University Press (Charles Darwin); Friends of the State Library of South Australia (François Péron); Hakluyt Society (Louis Antoine de Bougainville, John Byron, James Cook, Semen Dezhnev, Johann Reinhold Forster, Jean-François de Galaup de la Pérouse, Alejandro Malaspina, George Vancouver); IK Foundation (Anders Sparrman); Imprimerie Nationale (Louis Antoine de Bougainville, Nicholas Baudin); Libraries Board of South Australia (Nicholas Baudin); Limestone Press (George Heinrich von Langsdorff, Frederic Litke); Melbourne University Press (Bruny d'Entrecasteaux); Miegunyah Press (Jacques-Julien Houtou de Labillardière); Museo Naval (Antonio Pineda Ramírez, Tadeo Haenke, Luis Neé, Alejandro Malaspina); the New York American Geographical Society (Vitus Bering); Penguin Books (Alexander von Humboldt); Pickering & Chatto (Joseph Banks); Public Library of New South Wales (Joseph Banks); Stanford University Press (Georg Wilhelm Steller); State Library of New South Wales with Hordern House (Matthew Flinders); UBC Press (James Colnett); University of Alaska Press (Hermann Ludwig von Löwenstern, Gerhard Friedrich Müller); University of Hawaii Press (Adelbert von Chamisso, Johann Reinhold Forster, George Forster); University of Minnesota Press (Philibert Commerson); Wakefield Press (Nicholas Baudin, Matthew Flinders); William Hodge & Co. (Sven Waxell); Yale University Press (James Cook). The scholarly world owes a great deal to enlightened publishers such as these.

In carrying out the research for this book I was fortunate to be living within easy reach of the superb repositories of printed and manuscript material in the British Library, the Natural History Museum and the National Maritime Museum, whose resources and staff expertise I have exploited to the full. Finally, I must record my gratitude to those colleagues who have read and commented on individual chapters: William Barr, Andrew David, Robin Inglis, Iain McCalman, Nigel Rigby, Nicholas Thomas and Carol Urness. My greatest debt is to Sophie Forgan and Alan Frost, who read the entire text in typescript, made many constructive comments and have been constant sources of encouragement. My thanks also go to Robert Shore who did a meticulous editing job on my draft manuscript, while at Yale University Press Rachael Lonsdale searched archives and libraries in tracking down the illustrations that are so important in a work of this kind. This list of acknowledgements, personal and impersonal, makes it the more necessary to stress that all errors and obscurities remain my responsibility.

Glyn Williams
West Malling, Kent
31 December 2012

THE
PACIFIC OCEAN

180°

Arctic Circle

S I B E R I A

Bering

Okhotsk

Kamchatka

Aleutian

A S I A

Sakhalin

Petropavlovsk

Ur

JAPAN

C H I N A

Macau

Philippine
Islands

Guam

P A C I F I C

Borneo

Moluccas

Singapore

New
Guinea

New Britain

Batavia

Solomon Is.

Timor

Gulf of
Carpentaria

Vanikoro

King Sound

Endeavour River

New Hebrides

Tana

Fiji

Northwest Cape

New Caledonia

Tonga
(Friendly

Shark Bay

AUSTRALIA

Cape Leeuwin

Great
Australian
Bight

Port Jackson
(Sydney)

King George
Sound

Botany Bay

Bass Strait

Poverty B

Tasmania
(Van Diemen's Land)

Dusky Sound

Queen Charlot

NEW ZEALAND

I n d i a n

O c e a n

180°

Alaska

Mt St Elias ▲   *Yakutat Bay*

et

Kodiak I.   *Lituya Bay*

gin Is.

Nootka Sound

Vancouver Island

Juan de Fuca Strait

NORTH

AMERICA

*Hawaii
(Sandwich
Islands)*

CEAN

MEXICO

Bay of Campeche

Tropic of Cancer

Acapulco ●

Panama ●

Equator

Galapagos Islands ▸

Marquesas Is.

*Society
Islands*

Tuamotu Is.

Tahiti

Easter Island

Juan Fernández ·

Callao ●   PERU

Valparaiso ●

Concepción ●

Tropic of Capricorn

Patagonia

*Tierra
del
Fuego*

Strait of Magellan ●

Cape Horn

INTRODUCTION

T HE DISCOVERY IN the late fifteenth century of a New World in America, and the opening of seaborne trade routes to the East, transformed the economic and cultural life of Europe. Despite its spectacular impact, silver was not the most important product brought back by the early adventurers and traders. Rather it was the food-producing plants of the Americas – maize, potatoes, sweet potatoes, beans and manioc – that together with Asian spices became vital elements in the diets of the growing populations of the Old World. At a different level, there was an intense curiosity about the flora and fauna of the newly-discovered regions, driven partly by interest in the strange life forms that seemed to appear at every turn, but above all by the hope that some of the host of new plants would be of commercial or medicinal value. Within days of reaching the Antilles in October 1492, Columbus reacted with 'a hyperbolic intensity, a sense of awed delight' to the sights and sounds of nature – to the luxuriant trees, the colourful fish, the flocks of songbirds. As he put it in his log-book entry of 19 October, 'I do not know where to go first; nor do my eyes tire of seeing such beautiful verdure', before regretting his inability to identify the 'many herbs and trees which will be of great value in Spain for dyes and as medicinal spices'.[1] In the 1520s Fernández Oviedo y Valdés included descriptions and drawings of Caribbean flora in his *Historia general y natural de las Indias*, published in 1535, while in Philip II's reign the pioneering study by Francisco Hernández of the natural history of New Spain described two hundred animals and birds and more than a thousand plants. In a forewarning of much that was to follow in the realm of natural-history investigations, Hernández's work was not published at the time, and his original volumes in the Escorial were destroyed by fire in 1671.[2] Even so, the activity of collectors overseas was matched by the attention paid to their specimens at home. In sixteenth-century Italy several botanical gardens were established, usually attached to a university faculty of medicine, followed by others in Austria, Germany, the Dutch Republic and Spain. In England the first botanical garden was opened at Oxford in 1621, while in Paris the Jardin du Roi was established by royal decree in 1635. As yet, the main focus was on the healing properties of plants new and old, for as Samuel Hartlib declared in 1651, 'Where any Endemicall or Natural disease reigneth, there God hath also planted a specifique for it.'[3]

Sometimes an unexpected figure would emerge with an interest in natural history. On his plundering voyage around the world in the 1570s, Francis Drake was recorded by one of his prisoners as keeping a great book 'in which he delineated birds, trees and sea-lions',[4] while a few years later John White, the artist-governor of the abortive Roanoke colony in present-day North Carolina, produced outstanding paintings of the region's flora and fauna, and of its indigenous inhabitants. In Canada, Jesuit missionaries living among 'les sauvages' sent to France each year *Relations* that gave detailed accounts of the Iroquois and Huron peoples and their natural environment. However, credit for the first systematic publication of the natural history of the New World belongs to the Dutch West India Company. Its governor in Brazil from 1638 to 1644, Count Johan Maurits of Nassau-Siegen, employed a team of scholars and artists whose researches were published at Leiden in 1648. In time, the work of Dutch collectors in Brazil was dwarfed by that of the servants of the Dutch East India Company, whose monopolistic tentacles extended across the eastern seas as far as Japan in the north and Timor in the south. Dutch ships brought back thousands of specimens, reports and drawings, some for transmission to the Company's superior officials, others for sale to private collectors looking to stock their 'Cabinets of Curiosities'. The zeal for collecting reached such proportions that in 1675 the Company was forced to take action against the 'floating botanical gardens' that consumed too much shipboard space and scarce drinking water.[5] Five botanical gardens were established in the Dutch Republic, the earliest at Leiden University before the end of the sixteenth century, as well as one in Batavia in 1650, although the Company's insistence on secrecy hindered the spread of botanical information. The main thrust of the Company's search was for new plants that might supplement existing spices as commercial products or be of therapeutic importance, but investigations of more intrinsic scholarly value were carried out by George Eberhard Rumphius in Amboina, Paul Hermann in Ceylon, Engelbert Kaempfer in Japan, and Adriaan and Simon van der Stel, father and son, at the Cape of Good Hope.

In the seventeenth century, as the French and English joined the Dutch in exploring and exploiting overseas territories, the number of exotic plants known to Europeans had outrun ways of describing and listing them, and calls grew for a new and more logical system of classification. In his *Historia plantarum generalis*, the English botanist John Ray indicated the size of the problem when he wrote of the Dutch colony at Malabar: 'Who could believe in the one province ... hardly a vast place, that there would be three hundred unique indigenous species of trees and fruit',[6] while in a fifteen-month stay in Jamaica in the 1680s the physician Hans Sloane collected eight hundred plants and other specimens. In the eighteenth century the issue became ever more urgent as the almost unknown lands of the Pacific, covering one-third of the globe's surface, became the focus of the new wave of European discovery voyages that form the subject of this book. They were prompted by a variety of motives: commercial, strategic and scientific. Governments were not in the business of sponsoring voyages for the sake of science, but were susceptible to the pressures exerted by learned societies, notably the Royal Society of London and the Académie des Sciences in Paris, and by scholarly collectors such as Nicolas Witsen in the Netherlands and Joseph Banks in England. Whether under formal orders or not, voyagers described and drew scenes and peoples from distant parts, and brought back ethnographic and natural-history items for sale or study.

A turning point in official attitudes came after the mid-eighteenth century, when Britain, France, Russia and Spain sent expeditions to the Pacific that carried on board contingents of 'experimental gentlemen'. Prominent among these were the naturalists, eager to collect flora and fauna from the lands of the great ocean. By this time the term 'naturalist' had taken on its modern meaning of one 'who is interested, or makes special study of, animals or plants', as opposed to earlier definitions of one 'who studies natural, in contrast to spiritual things' or 'follows the light of nature, as contrasted with revelation'.[7] Botany, zoology and the emerging disciplines of geology and ethnology owed much to the scholars on these voyages. Their task was helped by the invention of the simple microscope, and by the publications of Carl Linnaeus and Georges-Louis Leclerc, Comte de Buffon, who although they often disagreed were key figures in the systemisation of natural history. So great was the number of specimens brought back from the long sea voyages that the products and peoples of the Pacific Islands became better known to scholars in Europe than were those of the great continental landmasses of Asia and Africa, or even America, where collectors had to travel on foot or, at best, on horseback, without the security, scientific equipment and storage capacity of an armed vessel nearby.

This book describes the fortunes and misfortunes of a devoted if sometimes eccentric band of scholars as they ventured far from home. Their task was not an easy one, for in addition to the unaccustomed hardships of long oceanic voyages and the dangers faced on unknown shores, they were frequently at odds with the commanders of the discovery expeditions. The observations of other scientific personnel such as hydrographers and astronomers were closely linked with the shipboard business of navigation. By contrast, naturalists needed long spells ashore, and when they returned on board demanded space and care for their hauls, often at the expense of the routine running of the vessel. They were 'all at sea' in more ways than one, and arguments between them and the ships' officers were frequent, sometimes heated. The routes and landfalls of the expeditions were rarely those that the naturalists would have chosen, and shipboard differences were exacerbated by a culture gap between civilians and serving officers, between scientists and non-scientists, and by differences in age. Only William Dampier, sailing before the classic era of Pacific exploration, closed (in his own person) the gap in comprehension between seaman and naturalist. The German naturalist Georg Steller on the Russian expedition commanded by Vitus Bering complained that ten years of preparation had resulted only in ten hours of investigation on a remote Alaskan island while Captain Cook on his last voyage refused to take any civilian naturalists with him, supposedly exclaiming, 'Curse scientists, and all science into the bargain.'[8]

Even without shipboard tensions, the collections of fragile plants and other specimens were at risk from natural hazards. Not until the invention in the 1820s of the Wardian sealed glass container that regulated humidity and temperature were live plants likely to survive long sea voyages. Before the use of arsenic as a preservative, zoological specimens were at constant risk from insects, damp and rot. Nor was the transportation of specimens to Europe the end of the matter. Specimens had to be classified, drawn and preserved, and the resources to do this were not always available. The scale of the task was vast. After Malaspina's voyage in the late eighteenth century almost sixteen thousand plants and other specimens arrived at the Royal Botanic Garden in Madrid, and the expedition's senior naturalist estimated that the botanical specimens brought back by the two ships had increased the number of the world's known plants by a third. Some of the plants collected by the expedition's Bohemian naturalist Tadeo Haenke have only recently been identified in Prague. Plants sent to Joseph Banks's herbarium in his London house spilled out of their cramped premises and were moved to more spacious quarters in Kew Gardens, but his plan for a full publication of the plants gathered on Cook's *Endeavour* voyage was not realised until the 1980s. The French ships of the Baudin expedition in the early nineteenth century brought back seventy crates of live plants, comprising almost two hundred different species. It was as if a second New World had been discovered.

By the late eighteenth century botanical and zoological shipments had become a two-way process by which European plants and animals were sent overseas in the hope that they would flourish in remote regions, for the benefit both of the indigenous inhabitants and the metropolitan power. Botany remained a fashionable pastime for the leisured, the curious and the scholarly, but it was also attracting the attention of ministers and merchants. William Bligh made the link between the Pacific discovery voyages and his breadfruit mission on the *Bounty* when he claimed (with a degree of exaggeration): 'This Voyage may be reckoned the first the intention of which has been to derive benefits from those distant discoveries.'[9] Whatever the personal preoccupations of the naturalists, there were usually commercial or political motives behind their appointment. Utility as well as scientific interest was invariably present, and it was appropriate that Britain's first settlement in Australia was planned for a spot called Botany Bay.

The achievement of the seagoing naturalists was substantial, but it was not above criticism. In his years of reflection after sailing on Cook's second voyage, that gifted but troublesome scholar Johann Reinhold Forster thought that he and his fellows treated natural history 'too microscopically', spending their time 'obscurely rummaging around', counting hairs, feathers and fins.[10] It was a theme developed by the Prussian polymath Alexander von Humboldt after his land-based travels in South and Central America at the beginning of the nineteenth century. Although asserting that he was 'passionately keen on botany', Humboldt was unimpressed by the seagoing naturalists of his day. He complained that they never ventured far from their ships, that their researches were confined to islands and coasts, and that they were interested only in the mechanical tasks of collecting and listing. Above all, they had 'neglected to track the great and constant laws of nature manifested in the rapid flow of phenomena'.[11] It was a challenge that one of Humboldt's greatest admirers, Charles Darwin, was to take up on the voyage of the Beagle.

CHAPTER 1

The 'rambling voyages' of William Dampier, Self-Taught Naturalist

FOR THE MARITIME nations of Europe, the late seventeenth century was part of 'the Dark Age of Pacific historiography',[1] that long interval between the discovery voyages of the sixteenth century and the systematic explorations of the age of Cook. The immensity of the Pacific Ocean, inexact navigational instruments, the ravages of scurvy and the straitjacket of winds and currents posed huge problems to methodical exploration. After the Dutch explorations of Tasman in the 1640s the slow-moving course of Pacific exploration came to a halt. English enterprise in the ocean had been represented by the predatory voyages of Drake and Cavendish in the reign of Elizabeth I, and when interest revived in the 1670s the motives were the same as those that had prompted the Tudor adventurers: trade and plunder. The Pacific caught the English imagination not as a vast, trackless ocean but as the western rim of Spain's rich American empire. The 'South Sea' that now began to feature in English enterprise and literature was confined to the waters that lapped the shores of Chile, Peru and Mexico, the hunting ground of those rapacious marauders, the buccaneers.

The first incursions of the buccaneers were chaotic ventures as they straggled across the Panama Isthmus and relied on seizing local craft once they reached the South Sea; but during the 1680s buccaneer ships fitted out in Europe entered the Pacific through the Strait of Magellan or around Cape Horn. Neither exploration nor legitimate trade featured high on the buccaneers' list of objectives. 'Gold was the bait that tempted a Pack of Merry Boys of us',[2] one of them admitted. Violent, disputatious, anarchic, they looted and burned their way along the Pacific coast from Valdivia to Acapulco. Moving between sea and land, switching from ship to ship, often taking to small boats and frail canoes, the buccaneers led a life of hardship and danger. With all ports closed to them, they might be at sea in their overcrowded craft for months on end; food and water were often in short supply; and scurvy was an ever-present menace. They received no regular wages, but operated on the chancy basis of 'no purchase, no pay'. The crews were made up of many nationalities, but in England they held a place in popular esteem that reflected admiration for their perceived role as fighters against Spain and popery, and for the 'rags to riches' aspect of their lives. Essential to this heroising process were the buccaneers' own accounts. Given the conditions in which they travelled, it is remarkable that a handful managed to write and preserve journals that were a cut above the usual sea-logs. Among them were Basil Ringrose, who had enough classical learning to negotiate with the Spaniards in Latin, and whose narrative was published in 1685 in the second English edition of the Dutch writer Olivier Exquemelin's *History of the Bucaniers of America*; Bartholomew Sharpe, the first English seaman to round Cape Horn in an easterly direction; ship's surgeon Lionel Wafer, whose notes on the Cuna Indians of southeast Panama were of interest to anthropologists well into the twentieth century; and, above all, William Dampier, 'an exact Observer of all things in Earth, Sea and Air',[3] whose *New Voyage round the World*, published in 1697, became a classic of travel and adventure.

Born in East Coker, Somerset, in 1651, and educated at a local grammar school, Dampier made his first voyage as an eighteen-year-old to Newfoundland before joining an East Indiaman bound for Java in 1671. On his return he enlisted in the Royal Navy and served in the Third Dutch War before being hospitalised in 1673. The next year Dampier sailed to Jamaica, where he spent a few months helping with plantation management before quarrelling with the manager, who referred to Dampier as 'a self-conceited young man and one that understands little or nothing, and one that has been given to rambling, and therefore cannot settle himself to stay long in any place'.[4] We get some hint of Dampier's intention to keep a record of his activities as he prepared for his voyage to Jamaica, for he asked for paper, ink and quills as well as a pair of stout shoes and ingredients for making punch.[5] After abandoning his plantation job, Dampier sailed to the Bay of Campeachy (Campeche) in the southern part of the Gulf of Mexico where he spent most of the next three years cutting logwood, a hazardous and strenuous occupation, although the profits made from the red dye extracted from the trees were high. On the mosquito-infested marshy shores where the logwood trees grew, the cutters lived in crude huts thatched with palm leaves that offered little protection in the rainy season, when the cutters might step from their beds 'into the Water perhaps two Feet deep, and continue standing in the wet all Day, till they go to Bed again'.[6]

In 1678, Dampier returned to England with some capital, and married. He must also have brought home with him a journal about his experiences as a logwooder, which was to appear as *Voyages to Campeachy*. This was not published until 1699 as part of *Voyages and Descriptions*, the second volume of his travels, and has still not received as much attention as it deserves since it stands in the shadow of the more celebrated *New Voyage round the World* and deals with a less dramatic period of Dampier's life. Although it was almost certainly written after the *New Voyage*, and was intended to profit from that book's success, *Voyages to Campeachy* shows that Dampier was keeping detailed notes while in his twenties and before he became a buccaneer. His account described the perilous life of the logwood cutters, and the dangers they faced from alligators, snakes, hurricanes and hostile Spaniards. In June 1676 a hurricane destroyed the cutters' huts and their equipment, tore up trees and turned the shoreline into a flooded shambles. In the midst of the chaos, Dampier had enough presence of mind to save his notes, which included one of the first detailed descriptions of a hurricane. In less dramatic sections Dampier described the Bay area and its Indian inhabitants, and the wildlife: animals, fish, birds, insects and vegetation. The contents page for Chapter II included 'A Description of some Animals, Squashes, large long-tail'd Monkies, Ant-bears, Sloths, Armadillos, Tigre Catts, Snakes of three sorts, Calliwasps, Huge Spiders, Great Ants and their Nests, Rambling Ants, Humming Birds, Black-Birds, Turtle Doves' – to give only the first part of the list.

In 1679, Dampier returned to Jamaica, where he joined a force of buccaneers who the following April crossed the Isthmus of Panama. It was the beginning of Dampier's long career as a buccaneer and global traveller, and he did not return to England until 1691. Six years after his return the firm of James Knapton published an account of his travels. The title-page of the *New Voyage round the World* reveals the extent of his journeys: not a single cruise, but a series of wanderings and diversions to 'the *Isthmus* of *America*, several Coasts and islands in the *West Indies*, the Isles of *Cape Verd*, the Passage by *Terra del Fuego*, the *South Sea* Coasts of *Chili, Peru,* and *Mexico*, the Isle of *Guam*, one of the *Ladrones, Mindanao*, and other *Philippine* and *East-India* Islands near *Cambodia, China, Formosa, Luconia, Celebes,* &c. *New Holland, Sumatra, Nicobar* Isles, the *Cape* of *Good Hope,* and *Santa Hellena*'. It had taken him thirteen years to complete his interrupted circumnavigation, a fact he turned to his advantage in his account: 'one who rambles about a Country can usually give a better account of it, than a Carrier who jogs on to his Inn, without ever going out of his Road'.[7] His 'better account' was clearly intended to be different from the blood-stained narratives written by Exquemelin, Sharpe and Ringrose; rather it would be 'a mixt Relation of Places and Actions', with descriptions of the 'Soil, Rivers, Harbours, Plants, Fruits, Animals, and Inhabitants. Their Customs, Religion, Government, Trade, &c.' Adding to the book's appeal were the maps drawn by one of the leading cartographers of the day, Herman Moll. There were 'Actions' in plenty, and these probably formed the main attraction for most of his readers.

Serving on a half-dozen different ships, Dampier took part in raids on Spanish shipping and settlements along the coasts of Chile, Peru and Mexico. His account of the buccaneers' violent activities makes vivid reading, but contains little about Dampier's own participation in them. Usually, he was at pains to hint that his motives were different from those of his companions, and that he was driven by a quest for knowledge rather than for wealth. For example, he explained that in 1686 he remained on the buccaneering vessel the *Cygnet* as she left the South Sea to sail west across the Pacific to Guam and Mindanao because that was 'a way very agreeable to my Inclination' and it would 'Endulge my curiosity'.[8]

During 1687 the *Cygnet* cruised in the North China Seas before heading south through the Dutch East Indies to Timor, and farther south still to the little-known coast of New Holland (modern Australia) 'to see what that Country would afford us'.[9] In January 1688 the *Cygnet* anchored in Karrakatta Bay in King Sound on Australia's northwest coast where she remained for five weeks. From there the vessel sailed into the Indian Ocean to the Nicobar Islands, where Dampier left her. His adventures were far from over as he sailed with a few companions in an outrigger canoe to Sumatra and then on to Achin. Off the coast of Sumatra the little craft was caught in a storm that threatened to swamp it: 'The Sea was already roaring in a white Foam about us; a dark Night coming on, and no Land in sight to shelter us . . . I had a lingering View of approaching Death, and little or no hopes of escaping it; and I must confess that my Courage, which I had hitherto kept up, failed me here.'[10] Surviving this crisis, Dampier was struck down with a serious illness at Achin. There he described, but claimed never to have tried, marijuana: 'They have here a sort of plant called ganga, or band . . . It is reported of this plant, that if it is infused in any liquor, it will stupefy the brains of any person that drinks thereof; but it operates diversely, according to the constitution of the person. Some it keeps sleepy, some merry, putting them into a laughing fit, and others it makes mad.'[11] Once recovered, Dampier resumed his wanderings, spending time in Malacca (Melaka), Tonkin (Vietnam), Cambodia and Madras, as well as serving as a gunner in Sumatra. At last he was showing signs of travel-weariness – 'I began to long after my native Country, after so tedious a Ramble from it'[12] – and in January 1691 he set sail from Sumatra in an East Indiaman, and arrived home in September. With him came an enforced visitor to England: Jeoly, 'the painted [tattooed] Prince', captured on a small island near Mindanao. Once back in England, shortage of funds compelled Dampier to sell his half-share in the unfortunate Jeoly, who was exhibited at sideshows before dying of smallpox.

Little is known about Dampier's movements between his return to England and the publication of his *New Voyage* six years later. Recent research has shown that in 1694 and 1695 he was involved in an abortive project at La Coruña in northern Spain to salvage Spanish wrecks in American waters.[13] The venture collapsed when mutineers led by Henry Every (soon to become better known as the pirate 'Captain John Avery') sailed away on the flagship to embark on a career of freebooting in eastern waters. It was possibly during his months of idleness at La Coruña that Dampier began the task of revising and polishing his journal with a view to publication. Of Dampier's care in writing and safeguarding his notes there can be no doubt. Escaping from the *Cygnet* among the Nicobar Islands in May 1688, he was in a canoe that over-turned, soaking 'my Journal and some Drafts of Land of my own taking, which I much prized', and which were only saved after much drying in front of 'great fires'. During this canoe voyage Dampier also referred to his 'Pocket-book' in which he had entered navigational details before leaving the ship.[14] Three years later, he described another escapade in which 'I came by stealth from Bencooly [Benkulen], and left all my books Drafts and Instruments Cloaths bedding . . . and wages behind. I only brought with me this Journall and my painted prince.'[15] The length of time Dampier was away argues against the existence of a single journal. As he crossed the rivers and swamps of the Panama Isthmus in 1681 he described how he placed 'my Journal and other writings' in a bamboo cane plugged with wax to keep them dry,[16] and 'the journal' that Dampier brought back to England probably consisted of a number of separate logs, notebooks and loose sheets. Given the precarious circumstances of his travels, moving from ship to ship, sometimes in small boats, sometimes living a hand-to-mouth existence on land, it is remarkable that Dampier managed to obtain the necessary writing materials to keep a detailed record, and then protected it from insect ravages, enemy action and the carelessness of shipmates. Of these original manuscripts there is no trace, although the Sloane Manuscripts in the British Library contain a journal of Dampier's voyages which internal evidence shows was written after his return to England. The main text is in the hand of an unknown clerk, but it contains many additions and corrections in Dampier's handwriting. Even so, it is considerably shorter than the published account of 1697, and contains little of the information on the natural history of the places Dampier visited that made *A New Voyage* so original and valuable an account.

One of the most puzzling aspects of the authorial relationship between Dampier's *New Voyage* and the Sloane journal is their different descriptions of the Aborigines of New Holland. The *Cygnet's* anchorage of five weeks in King Sound in early 1688 was the longest known stay by Europeans on the Australian mainland.[17] Earlier Dutch landings had been for a matter of days, sometimes only hours, and often without any contact with the local inhabitants. Dampier, by contrast, saw enough of the Aborigines to devote several pages to them, and his description of them in the *New Voyage* was to live long in the European memory. Naked, black, without dwellings,

> the Inhabitants of this Country are the miserablest People in the World ... setting aside their Humane Shape, they differ little from Brutes. They are tall, strait-bodied, and thin, with small long Limbs ... Their Eyelids are always half closed, to keep the Flies out of their Eyes ... They are long-visaged, and of a very unpleasing Aspect, having no one graceful Feature in their Faces. Their Hair is black, short and curl'd like that of the Negroes, and not long and lank, like the Common Indians. The Colour of their Skins, both of their Faces, and the rest of their Body, is Coal-black, like that of the Negroes of Guinea.

They had no metal or implements; their only weapons were wooden swords and spears. They grew no crops, trapped nothing and seemed to live on small fish stranded at low tide. Their speech was unintelligible. Some of them were taken on board the ship, where they showed no curiosity about their new surroundings. Attempts were made to press them into service carrying water casks, but 'all the signs we could make were to no purpose, for they stood like Statues, without motion, but grinn'd like so many Monkeys, staring one upon another'.[18]

This unprepossessing description was to be transmitted to genera-tions of readers and scholars. In the mid-eighteenth century one of Europe's leading natural scientists, the Comte de Buffon, simply trans-lated Dampier when he came to categorise the inhabitants of New Holland. In Australian waters more than seventy years after Dampier's landing, James Cook and Joseph Banks on the *Endeavour* had the *New Voyage* to hand as they strained for their first sight of life on the shores of southeast Australia. Banks wrote, 'We stood in with the land near enough to discern 5 people who appeared through our glasses to be enormously black: so far did the prejudices which we had built on Dampiers account influence us that we fancied we could see their Colour when we could scarce distinguish whether or not they were men.'[19] In his earlier manuscript account Dampier is briefer and more accurate – for example, on the nature of the Aborigines' hair: 'They are people of good stature but very thin and leane I judge for want of food[;] they are black yet I belive their haires would be long if it was comed out but for want of Combs it is matted up like a negroes' hair.' He does not refer to them as 'the miserablest People in the World', there are no references to beastlike appearances, and the Aboriginal lifestyle is fairly described as a simple one, in which 'they are not troubled with household goods nor cloaths'.[20]

It would seem that the Sloane manuscript was Dampier's first attempt at a shortened and revised version of his journal for publication. That he had publication in mind is suggested by the enticing heading on the first page – 'The Adventures of William Dampier ... in the South Seas' – and by references in the text to 'my book'.[21] At some stage he decided, or was persuaded, to make it a more comprehensive account of his voyages, centred on his years as a buccaneer, but including natural-history observations. Persuasion seems the more likely, for Dampier later conceded that it was 'far from being a Diminution to one of my Education and Employment, to have, what I write, Revised and Corrected by Friends'.[22] Guesses at the identity of these helpers have included Jonathan Swift, who referred on the first page of *Gulliver's Travels* to 'my cousin Dampier', and Daniel Defoe, who was beginning to show an interest in the South Sea that lasted a quarter-century, and the title of whose *New Voyage round the World by a Course Never Sailed Before* (1724) was a respectful bow in Dampier's direction. A more likely candidate is Dampier's publisher, James Knapton, who can be assumed to have had an eye on the commercial appeal of a book that was a 'mixt Relation of Places and Actions'. It is also possible that Dampier's original description of the Aborigines was changed and dramatised editorially to meet public fascination with encounters with alien, subhuman creatures. (As an aside it is worth noting that a broadsheet of 1692 about the public showings of Jeoly is now filed in the British Library with advertisements for the display of giants, monsters, dwarfs and hermaphrodites.)[23]

A New Voyage was published in February 1697, and by the end of the year was in its third edition. By 1699 there was a fourth edition, and in that year Knapton published a second volume of Dampier's *Voyages and Descriptions* containing material that had been omitted from the 1697 volume. This described Dampier's activities before and after his years as a buccaneer – his early voyages to Campeche as a young man, followed by chapters on his final wanderings in Tonkin, Achin and Malacca.

Most notably, the second volume contained 'A Discourse on the Trade-Winds, Breezes, Storms, Season of the year. Tides and Currents of the Torrid Zone throughout the World', a pioneering piece of work, full of perceptive observations, which has been described as 'a classic of the pre-scientific era' and was used in the standard *Admiralty Sailing Directions* well into the twentieth century.[24] An anonymous writer in the first of England's monthly reviews, *The Works of the Learned*, was enthusiastic in his praise of the book as a whole. After summarising its contents, he concluded: 'What has cost the Captain so much Time, Fatigue, and Dangers this useful Companion will inform the Curious of, and at the same instant please even the *Sedentary Traveller*, with the variety of Descriptions, and the surprizingness of the Incidents therein contained.'[25]

Dampier's volumes met the needs of two different kinds of reader. Their account of his adventures and misadventures, often reprinted in cheaper or pirated editions, appealed to a popular audience. By contrast, their closely observed descriptions of the natural history of distant regions met scholarly demands for a more informative kind of travel narrative than the usual recitation of daily events embellished with well-worn passages from earlier accounts. In 1667, Thomas Sprat, first historian of the Royal Society, which had been founded five years earlier, hoped that 'there will scarce a ship come up the Thames, that does not make some return of experiments';[26] and the Society's *Philosophical Transactions* became a hospitable if chaotic repository of travel accounts. The Society's emphasis on the importance of observations of physical phenomena by seamen was justification, if any were needed, for Dampier's approach. In 1694 a Fellow of the Society, Tancred Robinson, wrote the introduction to *An Account of Several Late Voyages & Discoveries* in which he complained that he could not say anything about recent English voyages to the South Sea because he had not seen any journals. Ignorant of Dampier's voluminous but as yet unpublished observations, he wrote: ''Tis to be lamented, that the English Nation have not sent along with their Navigators some skilful Painters, Naturalists, and Mechanicks.'[27] Then in 1696 the Royal Society issued Dr John Woodward's *Brief Instructions for Making Observations in All Parts of the World in Order to Promote Natural History*, whose emphasis on climate, land forms and natural history matched Dampier's

New Voyage of the following year. It was altogether appropriate that Dampier, 'although a Stranger to your Person', dedicated *A New Voyage* to the then president of the Royal Society, Lord Montagu (the earl of Halifax), hoping that his book would satisfy the Society's 'zeal for the advancement of knowledge, and anything that may . . . tend to my countries advantage'. Dampier's relationship with Hans Sloane, secretary to the Society, was probably more significant. Sloane was a fashionable physician and devoted collector whose published account of his voyage to the West Indies in 1687 was a treasure-house of information for naturalists.[28] In time he became president of the Royal Society, and on his death in 1753 his collections formed the basis of the British Museum, established in June of that year by Act of Parliament (and financed by a national lottery). At about the time of the publication of the *New Voyage*, Sloane arranged for Thomas Murray to paint Dampier's portrait, and at some stage acquired a copy of Dampier's journal, together with the journals of several of his buccaneer contemporaries in the South Sea. If it wasn't James Knapton who persuaded Dampier to expand the journal by including his natural-history observations, then it seems most likely that Sloane was the person responsible.

Dampier had a practical turn of mind and, as his dedication suggested, the *New Voyage*'s natural-history descriptions were often accompanied by reflections on possible advantages for English enterprise. In this respect, if not in others, he was a forerunner of such entrepreneurial naturalists as Joseph Banks. So, while at Guam in 1686, Dampier made notes on the various uses of the coconut – drink and food; its shell made into cups and dishes, its husk turned into rope and cable – that ran to six printed pages. He explained that he had been 'the longer on this subject' because the coconut tree 'is scarce regarded in the *West-Indies*, for want of the knowledge of the benefit which it may produce. And 'tis partly for the sake of my Country-men, in our *American* Plantations, that I have spoken so largely of it.'[29] Also at Guam, Dampier gave the first published description of breadfruit and its importance as a staple food: 'The Natives of the Island use it for Bread: they gather it when full grown, while it is green and hard; then they bake it in an Oven, which scorcheth the rind and makes it black: but they scrape off the outside black Crust . . . and the inside is soft, tender and white, like the Crumb of a Penny Loaf.'[30] When the crew of Commodore Anson's *Centurion* arrived at Tinian in the Mariana Islands in 1742 they much preferred the local breadfruit to the ship's bread, and in the 'View of the Watering Place at Tiniam' drawn by Lieutenant Peircy Brett the breadfruit tree in the foreground dominates the scene (Pl. 1). In Dampier's description and Brett's drawing there is a premonition of images that were to shape Europe's vision of the South Seas – Tahiti, Bligh and the *Bounty*.

Dampier was at pains to avoid the fanciful exaggerations of many earlier travel accounts, insisting, 'I have been exactly and strictly careful to give only True Relations and Descriptions of Things.'[31] At Guam he was fascinated by the islanders' great outrigger canoes or proas, and sailed on one to check its speed. He used a logline to determine that it was making twelve miles an hour but thought that, if pushed, it could achieve twice that speed. Dampier was an unschooled, instinctive observer, describing rather than classifying, and adding personal touches that held his reader's interest. These characteristics were already evident in his early notes from the Bay of Campeche, published in 1699. The following is his description of the giant anteater:

It lays its Nose down flat on the Ground, close by the Path that the Ants travel in, (whereof there are many in this Country) and then puts out its Tongue athwart the Path: the Ants passing forwards and backwards continually, when they come to the Tongue, make a Stop, and in two or three Minutes time it will be covered all over with Ants; which she perceiving, draws in her Tongue, and then eats them; and after puts it out again to trapan more. They smell very strong of Ants, and taste much stronger, for I have eaten of them.

Dampier's description of the hummingbird is a classic of its kind, written in beguiling, non-technical terms:

The Humming Bird is a pretty little feather'd Creature, no bigger than a small Needle, and his Legs and Feet in proportion to his body. This Creature does not wave his Wings like other Birds when it flies, but keeps them in a continual quick Motion like Bees and other Insects, and like them makes a continual humming Noise as it flies. It is very quick in motion, and haunts about Flowers and Fruit, like a Bee gathering Honey, making many near addresses to its delightful Objects, by visiting them on all sides, and yet still keeps in motion, sometimes on one side, sometimes on the other; as often rebounding a foot or two back on a sudden, and as quickly returns again, keeping thus about one Flower five or six minutes, or more.

Less appealing to the timorous reader was the description of Campeche's spiders, 'of a prodigious size, some near as big as a Man's Fist'. These had 'two Teeth, or rather Horns an Inch and a half, or two Inches long . . . their small end sharp as a Thorn'. Undeterred, Dampier's companions put these stabbing weapons to good use: 'Some wear them in their Tobacco pouches to pick their Pipes. Others preserve them for Tooth-Pickers, especially such as were troubled with the Toothache; for by report they will expel that pain, tho' I cannot justifie it of my own Knowledge.' Dampier was rarely at a loss for a good story. A long description of alligators was followed by an account of an encounter between one and an Irish companion of Dampier's.

Going to the Pond in the Night [he] stumbled over an Alligator that lay in the Path: the Alligator seized him by the Knee; at which the Man cries out, *Help! Help!* His consorts, not knowing what the matter was, ran all away from their Huts, supposing that he was fallen into the clutches of some Spaniards. But poor *Daniel*, not finding any assistance, waited until the Beast opened his Jaws to take better hold; because it is usual for the Alligator to do so; and then snatch'd away his Knee, and slipt the But-end of his Gun in the room of it . . . he was in a deplorable condition, and not able to stand on his Feet, his Knee was so torn with the Alligators Teeth. His Gun was found the next day ten or twelve Paces from the Place where he was seized, with two large Holes made in the But-end of it, one on each side, near an Inch deep; for I saw the Gun afterwards.[32]

While on the buccaneer vessel the *Bachelor's Delight* in June 1684, Dampier spent twelve days on the mysterious Galapagos Islands, whose location and even existence the Spaniards were at pains to keep secret. Four years earlier a captured Spanish officer had told the buccaneer captain Bartholomew Sharpe: 'had we gone to the Islands of *Galapagos*, as we were once determined to do, we had met on that Voyage with many Calms, and such Currents, that many ships have by them been lost, and never heard of to this day.'[33] The sailing master of the *Bachelor's Delight*, Ambrose Cowley, heard even more off-putting stories from the

Spanish prisoners on board: 'The Spaniards laugh at us telling us they were Inchanted Islands ... and that they were but Shadowes and not real Islands.'[34] The Spanish attitude was understandable. With its most easterly island almost six hundred miles out in the Pacific, the uninhabited Galapagos group offered buccaneers security from Spanish search vessels – and, as Dampier noted, a plentiful food supply in the form of giant tortoises: 'The Land-turtle here are so numerous, that 5 or 600 Men might subsist on them alone for several Months, without any other sort of Provision.'

Dampier wrote several thousand words describing the '4 sorts of Sea-turtle, *viz.* the Trunk-turtle, the Loggerhead, the Hawks-bill, and the Green-turtle'. They had distinctly different appearances:

> The Trunk turtle is commonly bigger than the other, their Backs are higher and rounder, and their Flesh rank and not wholsome. The Loggerhead is so call'd, because it hath a great Head, much bigger than the other sorts; their flesh is likewise very rank, and seldom eaten but in case of Necessity ... The Hawks-bill Turtle is the least kind, they are so call'd because their Mouths are long and small, somewhat resembling the Bill of a Hawk: on the Backs of these Hawks-bill Turtle grows that shell which is so much esteem'd for making Cabinets, Combs, and other things ... The green Turtle are so called, because their shell is greener than any other. It is very thin and clear, and better clouded than the Hawks-bill; but 'tis used only for inlays, being extraordinary thin. These Turtles are generally larger than the Hawks-bill; one will weigh 2 or 3 hundred pound.[35]

More significantly, Dampier called the green turtle 'a sort of bastard', 'for their shell is thicker than other green Turtle in the *West* or *East-Indies*, and their flesh is not so sweet'. This remark suggested that there were differences within species depending on their place of habitation, an observation given additional emphasis when, 150 years later, Charles Darwin visited the Galapagos with the *New Voyage* in hand. Darwin made an important addition to Dampier's description of the islands' tortoises, for he reported that 'the tortoises coming from different islands in the archipelago were slightly different in form', and that locals 'could at once tell from which island any one was brought'.[36] Since all but a day of Dampier's visit was spent on one island, and in his time there were no human inhabitants who might pass on knowledge of the archipelago's wildlife, it can hardly be expected that he would have noticed this; and Darwin admitted that he himself did not realise the implications of these differences until after he had left the islands.

Dampier's *New Voyage* was intended for the general reader rather than for specialists in any form of natural history. Although a perceptive observer, Dampier did not claim specialist authority in any field except, perhaps, that of meteorology. The last page of his manuscript journal also shows that he was criticised, or thought he was about to be criticised, for daring to write an account of voyages in which he played only a subordinate role. Its erratic spelling and lack of punctuation also suggest that his more polished book had undergone extensive editorial work:

> It may be Demanded by som why I call these voyages and discoverys myne seeing I was neither master nor mate of any of the ships; to such demands I answer that I might have been master of the first I went out in if I would have accepted it for it was known to most men that were in the seas, that I kept a Journall and all that knew me well did Ever judge my accounts were kept as Exact as any mans besides most if not all that kept Journalls either Loozed them before they got to Europe or Else are not returned nor Ever likely to com home therefore I judge that mine is the more entire I having still perused my writing therefore I think I may most justly Challenge a Right to these Discoverys than any other man ... yet such is the opinion of most men that nothing pleaseth them but what comes from the highest hand though from men of the meanest Capacities.[37]

Dampier's assertion was a declaration of the importance of the observer on discovery voyages, while the actual text of the *New Voyage* set the pattern for future travel accounts by providing more than a mere narrative of events. His complaints about the deficiencies of 'the highest hand' were soon to strike an incongruous note for the publication of his book made Dampier a minor celebrity, and in August 1698 as part of his new social life Dampier dined with Samuel Pepys. Also present was that other famous diarist, John Evelyn, who was impressed by Dampier's 'very extraordinary' adventures and 'very profitable' observations, and thought him 'a more modest man than one would imagine by the relation of the crew he had consorted with'.[38] As important as these social contacts was Lord Montagu's introduction of Dampier to the First Lord of the Admiralty, the earl of Orford, who asked Dampier 'to make a proposal of some voyage wherein I might be serviceable to my Nation'. Dampier responded with a plan for a voyage to 'ye remoter parts of the *East India Islands* and the Neighbouring Coast of *Terra Australis*' on the grounds that it was 'reasonable to conceive that so great a part of the World is not without very valuable commodities'; and during 1698 the Admiralty accepted the plan. The detached observer who had made a virtue out of his refusal to take a position of authority on his voyages was appointed a captain in the Royal Navy. A hint that the *New Voyage* had brought him little in the way of financial gain, however, came with his request to the Admiralty for an advance of £100 because of 'the Lowness of my present circumstances'.[39]

Dampier's instructions laid down that he was to make careful observations, collect specimens and bring back 'some of the Natives, provided they shall be willing to come along'. A rather superfluous instruction ordered him to keep an 'Exact Journall', followed by an intriguing reminder that on his return it was to be handed in to the Admiralty 'and to no other'. To help him, he was provided with the services of 'a Person skill'd in drawing', an anticipation of later expeditions when artists such as Parkinson and Hodges added much to Europe's knowledge of the Pacific. Having agreed to the idea of an official discovery expedition – a rarity in itself – the Admiralty then seems to have lost interest. Dampier was provided with only one vessel, the *Roebuck*, rather than the two he had requested; moreover, the 290-ton vessel was in a state of disrepair, and only two of her crew had 'crossed the Line' before. Whatever his merits as a navigator and author, Dampier had no experience of command. At odds with his first lieutenant, a regular naval officer who regarded his captain as 'an Old Pyrating Dog', Dampier was in constant fear of mutiny, 'being forced to keep my self all the way upon my Guard, and to lie with my Officers, such as I could trust, and with small Arms upon the Quarter-Deck; it scarce being safe for me to lie in my Cabbin, by Reasons of the Discontents among my Men'.[40]

Having assaulted his lieutenant, put him in irons and sent him back to England, Dampier managed to nurse his vessel across the southern reaches of the Atlantic and Indian Oceans as far as the west coast of New Holland,

and on to Timor and New Guinea, before reaching his major discovery of
'Nova Britannia'. Dampier then considered turning south to explore the
unknown east coast of Australia, but the condition of both ship and crew
was so poor that he decided to head for home. That the decision was a
wise one was shown a year later when the bottom fell out of the *Roebuck*
in mid-Atlantic on her homeward voyage, and the crew were fortunate
to save their lives by scrambling ashore on Ascension Island. In the confusion, many of Dampier's books and papers were lost. Worse followed, for
when he returned to England in the summer of 1701, Dampier faced not
only the customary court martial for the loss of his ship, but a second
court martial arising from his treatment of his lieutenant on the outward
leg of the voyage. The court found him 'guilty of very Hard and cruel
usage', deducted all his pay for the voyage and declared him to be 'not a
Fitt person to be Employ'd as comdr of any of her Maty ships'.[41] The
period of enforced idleness that followed at least gave Dampier time to
prepare his account of the voyage for publication, 'notwithstanding the
Objections which have been raised against me by prejudiced Persons', and
the first volume of *A Voyage to New-Holland*, published by James and John
Knapton, appeared in 1703. That it was dedicated to the earl of Pembroke,
Lord President of the Council and a former First Lord of the Admiralty,
was a sign that Dampier's disgrace had been quickly overlooked following
the outbreak of the War of the Spanish Succession, and in January 1703
he was appointed to the command of a two hundred-ton privateer, the
St George, bound for the South Sea. A further mark of official favour came
in April when he was ushered into Queen Anne's presence by her consort,
Prince George of Denmark, to kiss hands.[42] Clearly, the allure of Dampier's
knowledge of Spanish American waters far outweighed his occasional
misdemeanours.

Dampier did not return from the voyage until 1707, which explains
why the second volume, *A Continuation of a Voyage to New-Holland*, was
not published until 1709, the author 'being obliged to prepare for another
Voyage, sooner than I at first expected'. The title-page of the first volume
was almost a replica of that of the *New Voyage*. It began with a list of the
countries visited, including 'Their Inhabitants, Manners, Customs,
Trade &c. Their Harbours, Soil, Beasts, Fish &c. Trees, Plants, Fruits
&c.' Then came an addition, for the presence of an artist on board the
Roebuck enabled the inclusion of 'divers Birds, Fishes and Plants . . .

Curiously Ingraven on Copper Plates'. Also included were maps, again drawn by Herman Moll. In other ways *A Voyage to New-Holland* lacked the popular appeal of the earlier book. It began with some curiously defensive remarks: 'It has almost always been the Fate of those who have made new Discoveries, to be disesteemed and slightly spoken of, by such as either have had no true Relish and Value for the Things themselves that are discovered, or have had some Prejudice against the Persons by whom the Discoveries were made.'[43] The book had none of the swashbuckling adventures of Dampier and his associates in their bucca-neering days; for long periods the search for water rather than for Spanish treasure dominated the narrative. The first volume gave a detailed description of the Cape Verde Islands and of Bahia in Brazil, from where Dampier took the *Roebuck* across the southern oceans to the west coast of New Holland to begin exploration proper. With him he had a manuscript copy of a Dutch map showing Abel Tasman's voyages in Australian waters. In August 1699 the *Roebuck* reached the west coast of Australia and anchored in a bay visited by a Dutch expedition under Willem de Vlamingh only two years earlier. Located in latitude 25°20′S. longitude 113°30′E., well to the south of Dampier's landing place in 1688, the area is still known by his name of Shark's (now Shark) Bay.[44]

After a week's search failed to find water, Dampier sailed north to North West Cape and then followed the coast through a chain of islands later named Dampier Archipelago. Still looking for water, Dampier continued in a northeast direction until he landed again, probably at Lagrange Bay, south of the modern town of Broome. Inland, all that could be seen were sand hills and some coarse grass, and for the first time on the voyage a number of Aborigines, 'tall black naked Men'.[45] In a running skirmish with them, one was shot and wounded, while a seaman was struck in the face by a wooden spear which at first he feared was poisoned. Swift's *Gulliver's Travels* has more than an echo of this incident as its hero reaches the southwest point of New Holland by canoe, and encounters a group of its inhabitants:

They were stark naked, men, women, and children, round a fire, as I could discover by the smoke. One of them spied me, and gave notice to the rest; five of them advanced towards me, leaving the women and children at the fire. I made what haste I could to the shore, and getting into my canoe, shoved off: the savages observing me retreat, ran after me; and before I could get far enough into the sea, discharged an arrow, which wounded me deeply on the inside of my left knee (I shall carry the mark to my grave). I apprehended the arrow might be poisoned, and paddling out of the reach of their darts (being a calm day) I made a shift to suck the wound, and dress it as well as I could.[46]

Dampier's meeting with the Aborigines on this occasion was too brief and violent for him to add much to the description he had already given in the *New Voyage*. The only significant addition concerned the body markings of 'a kind of Prince or Captain [who] was painted (which none of the rest were at all) with a Circle of white Paste or Pigment (a sort of Lime, as we thought) about his eyes, and a white streak down his Nose from his Forehead to the tip of it. And his Breast and some part of his Arms were also made white with the same paint.' Otherwise, Dampier contented himself with the observation that these 'New-Hollanders' seemed similar to the ones he had met in 1688, forty or fifty leagues to the northeast, with 'the most unpleasant Looks and the worst Features of any people that I ever saw'.[47] Here, with scurvy affecting some of the crew, and with little hope of finding food or water, Dampier bore away for Timor. Although commander of a Royal Navy ship, he had made no attempt to take possession of the land at any of the spots where he went ashore – perhaps because of his pessimistic appraisal of the region, or possibly because of his assumption of prior Dutch claims (although the Dutch had not taken possession of any part of the Australian mainland).

Disappointing though he found New Holland, Dampier took pains to describe it in as much detail as he could. His account was illustrated by several pages of drawings of fish, birds and coastal profiles, presumably by the unknown artist on board whose sketches were never more than workmanlike. Dampier also collected plants, and against all the odds managed to bring home some dried specimens, twenty-three of which can still be seen in the Sherardian Herbarium, Oxford.[48] Many were collected during the *Roebuck*'s stay at Rottnest Island in Shark Bay, where Dampier described how

[m]ost of the Trees and Shrubs had at this Time [August] either Blossoms or Berries on them. The Blossoms of the different Sort of Trees were

of several Colours, as red, white, yellow, &c. but mostly blue. And these generally smelt very sweet and fragrant, as did some also of the rest. There were also beside some Plants, Herbs, and tall Flowers, some very small Flowers, growing on the Ground, that were sweet and beautiful, and for the most part unlike any I have seen elsewhere.[49]

The drawings of the plants in the published account have more detail and are of a higher quality than the book's other illustrations; they were probably made in England by a professional artist from the specimens that Dampier brought back with him (Pl. 2).[50] The latter are of prime importance for they represent the first known collecting effort by any European on Australian soil,[51] and most are in remarkably good condition, showing that they were pressed soon after collection. As Dampier explained, 'I brought home with me . . . a good Number of Plants, dried between the leaves of Books.'[52] Notable was 'Sturt's desert pea', for a time known as *Clianthus dampieri* in recognition of Dampier's priority in recording and collecting it (Pl. 3). He was clearly aware of the significance of his collections, for in the preface to his book he explained that the plants he had brought back were 'in the Hands of the Ingenious Dr. Woodward', the author of the Royal Society's *Brief Instructions for Making Observations and Collections* of 1696. Woodward lent some of the specimens to John Ray, the most important English botanist of the age, who had introduced the concept of species into the classification of plants. Ray defined species as a group of individuals sharing characteristics that would be repeated in their offspring. In his system plants fell into three main categories: those without flowers, those with one seed leaf (*Monocotyledons*) and those with two (*Dicotyledons*). It was a 'natural' system of classification that would be largely replaced in the mid-eighteenth century by the simpler 'artificial' system of Linnaeus.[53] Ray described in Latin the eleven plants brought back by Dampier from 'Nova Hollandia' (together with four from Brazil, two from New Guinea and one from Timor) in the supplement to the third volume of his ground-breaking work *Historia plantarum generalis*, published in 1704. And Ray was almost certainly responsible for the detailed notes in English, complete with Latin names, on eighteen plants listed and illustrated at the end of the first volume of *A Voyage to New-Holland*. However unlearned Dampier was, his wide-ranging collecting activities would have appealed to Ray, who in 1691 had advised: 'let it not suffice to be book-learned ... but let us ourselves examine things as we have opportunity, and converse with Nature as well as with books.'[54] Some of Dampier's plant specimens were also passed on to Leonard Plukenet, botanist to Queen Anne, and he included illustrations of them in the 1705 edition of his *Amaltheum botanicum*. In this way Australian plants unknown in Europe found their way into early English botanical works – however, because Dampier's specimens predated the introduction of modern plant nomenclature, today's names are those of later collectors.

For Dampier, the plants, together with some 'strange and beautiful Shells', were the only redeeming feature of New Holland. As already mentioned, his references to the Aborigines reinforced the disparaging account of them given in the *New Voyage*, and his descriptions of other forms of life did little to dispel the overwhelming sense of backwardness, even monstrosity. A glimpse of dingoes left a distinctly unfavourable impression: 'two or three Beasts like hungry Wolves, lean like so many Skeletons, being nothing but Skin and Bones.'[55] Even the innocuous bobtail lizard was described in a way that made it seem uniquely repulsive:

> A sort of Guano's, of the same Shape and Size with other Guano's, describ'd but differing from them in 3 remarkable Particulars: for these had a larger and uglier Head, and had no Tail. And at the Rump, instead of the Tail there, they had a Stump of a Tail, which appear'd like another Head; but not really such, being without Mouth or Eyes: Yet this Creature seem'd by this Means to have a Head at each End ... They were speckled black and yellow like Toads, and had Scales or Knobs on their Backs like those of Crocodiles ... Their livers are also spotted black and yellow. And the Body when opened hath a very unsavory Smell. I did never see such ugly Creatures any where but here.[56]

On shore, bush flies made life unpleasant, while the only mammals Dampier sighted were 'a sort of Raccoons, different from those of the West Indies, chiefly as to their Legs; for these have very short fore Legs, but go jumping upon them as others do'. It was the first description of the hopping movement later associated with kangaroos, although he made no mention of the marsupial's pouch. The creature was probably a banded hare-wallaby, which afforded, Dampier added, 'very good

Meat'.[57] One oddity that was to puzzle naturalists for many years was Dampier's discovery in the stomach of a shark caught in Shark Bay of what he described as 'the Head and Boans of an Hippopotamus', complete with jaw, hairy lips and teeth.[58]

Dampier's original intention of sailing south from his landing place in New Holland before turning east along its southern coast was abandoned; instead he sailed northeast to Timor and New Guinea, and then kept farther east into waters never before visited by European vessels, where he discovered the large island that he named 'Nova Britannia' (later found to be two islands, New Britain and New Ireland). For the first time, Dampier began to scatter the names of royalty, noble patrons and admirals across the chart. Contact with the shore was fleeting, and as he bore away Dampier could only report that 'this Island may afford as many rich Commodities as any in the World; and the Natives may be easily brought to Commerce, though I could not pretend to it under my present Circumstances'.[59]

Dampier's main contribution on his two visits to New Holland was to make its western shores a real, if unattractive, landfall rather than a wavering line on the map. Although he added little to the earlier Dutch discoveries, his journals were published soon after his return. In contrast, no full description of Willem de Vlamingh's voyage was published until 1753, and the watercolour views of the coast by the expedition's artist disappeared from view until 1970.[60] Nor were the natural-history specimens collected by the Dutch crew at the request of Nicolas Witsen much regarded. On the expedition's return to Holland the officers presented Witsen with 'a little box containing shells, fruits, plants etc.' gathered on Australia's west coast, but he concluded that they were of 'little value'.[61] Dampier's description of the scrubby coastal area and its inhabitants, the drawings of its flora and fauna, Moll's maps and the coastal profiles, gave solidity to what had been a phantom presence. As we have seen, so vivid were Dampier's word-paintings of the Aborigines, 'the miserablest People in the World', and their arid, fly-blown habitat, that seventy years later and a continent's span away Joseph Banks looked at the Aborigines of Botany Bay through his eyes.

In terms of commercial importance, the English voyages to the Pacific were far surpassed by French ventures, which took advantage of Spanish naval weakness and wartime shortages in Chile and Peru. Between 1698 and 1725 no fewer than 168 ships sailed from France for the South Sea, but their voyages were not matched by published accounts of the kind that accompanied the English voyages. This may have been a consequence of the relatively mundane nature of their trading activities, or the result of a reluctance to publicise ventures that were diplomatically sensitive. Although only four English expeditions reached the South Sea in the first twenty years of the eighteenth century (Dampier's twice, Rogers's and Shelvocke's), they resulted in six books as well as several pamphlets and other published pieces. In contrast, a mere three books appeared describing the much more numerous French voyages in the same period: Froget's account (1702) of the de Gennes voyage, which never got farther than the Strait of Magellan; three volumes of scientific observations along the Atlantic and Pacific coasts of South America by the mathematician and botanist Louis Feuillée, whose name was given to the genus of plant *Feuillea*; and A.-F. Frézier's *Voyage de la Mer du Sud* (1716), which alone was translated into English.

Frézier was a capable mathematician and engineer, and his account of his voyage of 1714–16 contained detailed charts, town plans and coastal profiles of places along Spanish America's Pacific seaboard as far north as Callao. The book's English edition noted that, as well as navigational and hydrographic observations, it contained 'a Description of the Animals, Plants, Fruits, Metals, and whatever the Earth produces of Curious, in the richest Colonies of the World'; it also included four plates of Chilean and Peruvian plants, accompanied by detailed if non-technical descriptions. In places, as Frézier's narrative takes a step or two away from events, passages grow reminiscent of Dampier's *New Voyage*. In the arid hinterland of the small Peruvian port of Arica, Frézier described the successful cultivation of the Guinea pepper – 'so very hot and biting, that there is no enduring of it, unless well used to it' – in a passage that could have been written by the English voyager:

> When the seed is sprouted, and fit to be transplanted, the Plants are set winding, that is, not in a strait Line, but like an S, to the end that the Disposition of the Furrows, which convey the Water to them, may carry it gently to the Foot of the Plants; then they lay about each Plant or *Guinea* Pepper as much *Guana*, or Birds Dung as will lie in the Hollow of

a Man's Hand. When it is in Blossom, they add a little more; and lastly, when the Fruit is form'd, they add a good Handful, always taking care to water it, because it never rains in that Country.[62]

Dampier himself was to make further voyages, but as captain and navigator rather than author. He was the commander of a privateering expedition in which his ship, the *St George*, together with the *Cinque Ports*, sailed for the South Sea in September 1703. It was a disastrous venture in which both ships were lost, and little was gained in the way of prize money. As captain, Dampier must have kept a log, but the only full-length account of the voyage was published by the mate of the *St George*, William Funnell, who accused Dampier of cowardice, ineptitude and much else.[63] If Funnell and another hostile witness, midshipman John Welbe, are to be trusted, Dampier was at his worst when engaged against enemy ships. Off Juan Fernandez he was accused by Funnell of breaking off action early against a French ship, while Welbe accused him of failing to give the crew either encouragement or commands as he positioned himself unheroically on the quarterdeck 'behind a good Barricado, which he had order'd to be made of Beds, Rugs, Pillows, Blankets, &c.'[64] After this debacle the ships sailed towards the mainland, where they attacked Santa Maria, across the bay from Panama. Warned in advance, the inhabitants had fled, taking their valuables with them, but it was less this than the refusal to give the attackers their customary issue of brandy that infuriated Welbe as he remembered Dampier's insouciant explanation of his decision: 'If we take the Town, they will get Brandy enough, but if we don't take the Town, I shall want it my self.'[65] Most galling of all was the failure to take the Manila galleon as she approached Acapulco on the last leg of her long voyage. Dampier refused to board the galleon, and as her cannon began to batter the *St George*, he bore away. According to Welbe, his only comment was: 'Well, Gentlemen, I will not say, as Johnny Armstrong said, I'll lay me down and bleed a while; but I will lay me down and sleep a while' – which he proceeded to do with such effect that he did not wake until the next morning.[66] Dampier defended himself in a short, angry pamphlet, *Captain Dampier's Vindication*, in which he insisted that his crew were 'drunk and bewitched', and failed to obey his orders. Apart from these few pages, he published nothing about the voyage. The expedition's lack of success, and its final disintegration, confirmed the evidence of the *Roebuck* voyage – whatever his skills as a navigator and observer, Dampier lacked the qualities of an effective captain.

There was one more voyage to come, the privateering expedition of Woodes Rogers, on which Dampier sailed as 'Pilot for the South Seas' and was promised a sixteenth share of the expedition's profits. Remarkably, it was his third circumnavigation. By now almost sixty years old, he cut an increasingly forlorn figure on the voyage. Asked to guide the ships to the Galapagos Islands, he relied on his erroneous statement in *A New Voyage* that the nearest islands lay little more than three hundred miles off the mainland; the ships had to sail several hundred farther miles west before he would admit his mistake. On another occasion, as the ships approached the Mexican coast, Rogers conceded that 'Capt. Dampier has been here also,' before adding: 'but it's a long Time ago.' And there was surely some irony in Rogers's congratulations when the expedition arrived safely at an island 'which Capt. Dampier, I do believe, can remember he was at'.[67]

At home, Dampier's name was still one to conjure with. When the privateers reached the Dutch coast in 1711 on their return voyage, news of their arrival sent by an agent in Amsterdam to Robert Harley, Lord Treasurer and effective head of the government in London, began with the simple message: 'Dampier is alive.'[68] Rogers and Edward Cooke, captain of the expedition's second ship, published rival accounts of the voyage, but Dampier remained silent. If he kept a journal he seems to have made no effort to publish it, even though the voyage was marked by a series of colourful incidents, including the rescue of Alexander Selkirk, that strange figure 'cloth'd in Goat-Skins' who served as part of the inspiration for Robinson Crusoe, from Juan Fernandez, and the capture of a Spanish treasure-ship off the Californian coast. The official account by Rogers, *A Cruising Voyage round the World*, sold steadily, and was translated into French and German. It gave a commanding officer's view of the voyage, much of it taken up with routine descriptions of the places visited, based on existing printed sources. Of the first-hand natural-history observations that formed so striking an element in Dampier's books, there were very few. Two passages declare Rogers's lack of competence (and perhaps interest) in the subject. Off the coast of Brazil in November 1708, he refers readers

who might want to know more about the natural history of the region
to the account of the Dutch traveller Jan Nieuhof, which had recently
been reprinted in *A Collection of Voyages and Travels*, published by
Awnsham Churchill and John Churchill. 'The Descriptions of such
things are not my Province,' he continues, 'but I thought it convenient
to give this Hint for the Diversion of such Readers as may relish it
better than a Mariner's bare Journal.' Then in August 1709 on the coast
of Peru he notes: 'There's great Variety of Plants and Trees peculiar to
these hot Climates ... but it being out of my Road to describe such
things, I refer 'em to such whose Talents lie that way.'[69]

After his return from the voyage Dampier evidently declined in
health; he died in 1715 in his mid-sixties, 'diseased and weak in body
but of sound and perfect mind'.[70] There is no record of his place of
burial, and his will does not mention his wife, presumably also dead by
this time. Only his journals remained. They were republished as part of
a four-volume edition of voyages in 1729, and extracts from them
featured in collections of travels for the rest of the century. The appeal
of Dampier's writings was widespread. Scholars, publishers, seamen and
merchants, and casual readers after a good yarn were all attracted. For
some, such accounts amounted to little more than an enlargement of
the Grand Tour by proxy. As Defoe explained, his 'Compleat English
Gentleman' 'may go round the globe with Dampier and Rogers';[71] but
for others, Dampier's books became the standard model for voyagers
sailing to distant parts of the world.

CHAPTER 2

'Ten years of preparation; ten hours of exploration'

The Alaskan Tribulations of Georg Wilhelm Steller

IN THE EARLY eighteenth century the lands and waters of the North Pacific were for Europeans among the least-known parts of the inhabited globe. Sailing from Mexico along the coast of Lower California, Spanish ships in 1603 had reached as far north as Cape Blanco in latitude 43°N., but then turned back. A century later and five thousand miles distant the next known point of land was the northern peninsula of Kamchatka on the eastern fringes of Asia, reached by Russians in 1706. The physical contrast between the two peninsulas, the one hot and arid, the other snow-covered and fogbound for much of the year, was an indication of the immensity of the task ahead as navigators tried to close the gap. Kamchatka was the far limit of a Russian empire that stretched from the Baltic to Siberia, and what lay across the sea to the east was unknown. In 1648 a Russian trader, Semen Deshnev, had sailed round the easternmost extremity of Siberia (now Cape Deshneva) in a flimsy craft, but knowledge of his voyage was fragmentary.[1] When Peter the Great visited West European capitals in 1716 and 1717 he was unable to give specific answers when questioned about the extent of the Asian continent. On this and an earlier visit, the tsar had been impressed by the work carried out by the learned societies of Europe: the Royal Society of London, the Académie des Sciences of Paris and the Berlin Academy. In 1724 he sent a proposal for a Russian Academy of Sciences to the Senate, and one was established at St Petersburg in December 1725, soon after Peter's death. It had sixteen founding members, none of them Russian. Shortly before his death, the tsar also appointed Danish-born Vitus Bering, who had joined the Russian navy in 1704 and risen steadily through its ranks, to lead an expedition east from Kamchatka.[2]

In recent years scholars have argued intensively but inconclusively about the reasons behind Bering's first voyage, usually known as the First Kamchatka Expedition. Several motives have been put forward: the charting of the Asian coast north of Kamchatka; the exploration of lands rumoured to have been found by Dutch seamen in the ocean southeast of Kamchatka; and the discovery of the relationship between Asia and America. After a colossal trek of six thousand miles from the Russian capital of St Petersburg on the Baltic across Siberia that took them almost three years, Bering and his party reached the east coast of Kamchatka. There he built a small vessel, the *St Gabriel*, and in the summer of 1728 sailed north through the strait dividing Asia and America that now bears his name. Bering then turned back in his estimated latitude of 67°24´N. without sighting the opposite coast, so leaving the way open for differing interpretations of his voyage. Four years later Mikhail Gvosdev sighted the Alaskan coast while on a voyage in Bering's old ship, but was uncertain whether it was the American mainland. He referred to it cautiously as *bolshaya zemlya* (the 'big land' or 'large country').[3]

In 1731, after discussions between the Admiralty College, the Academy of Sciences and the Senate, Bering was entrusted with another discovery expedition, one of the most ambitious undertaken by any government in the eighteenth century. The Second Kamchatka Expedition or Great Northern Expedition involved a whole series of ventures: surveys along the rivers and coasts of Siberia, voyages south through the Kuril Islands to Japan, and the main expedition under Bering, which was to survey the uncharted American coast as far south as California. He was to build two brigs at Okhotsk, a small port founded by the Russians in 1647 which faced Kamchatka across the Sea of Okhotsk, and was then to winter at Avacha Bay on the southeast coast of Kamchatka in readiness for his voyage of discovery into the open ocean. For ten years Fleet Captain Bering and an army of associates, scientists and officials laboured at their task, struggling against obstacles of distance and climate, financial stringency and local opposition. Across thousands of miles of difficult terrain, frozen in winter, swampy and flooded in summer, everything that could not be obtained locally – from shipbuilding materials to weapons and ammunition – had to be carried by sledge, raft or packhorse. The number of porters and other carriers employed ran into the thousands, and the costs were huge, equalling one-sixth of the total state income in the last year of Peter I's reign.[4] So laborious was the overland route that at one time serious consideration was given to the possibility of supplying the expedition at Okhotsk by sending ships from Russia round the world by way of Cape Horn.

Among those appointed to the expedition was the French astronomer Louis Delisle de la Croyère, whose half-brother, the geographer Joseph-Nicolas Delisle, had worked at the St Petersburg Academy of Sciences since 1726. At the request of the Russian government Delisle produced a map that greatly influenced Bering's route, for it indicated the existence of large and potentially wealthy countries – Yezo, Company Land and Gama Land – in the northern seas between Asia and America.[5] Also appointed to the expedition were two German scholars from the St Petersburg Academy: Gerhard Friedrich Müller and Johann Georg Gmelin. Müller's task was 'to describe the history of those peoples' encountered; Gmelin's to investigate the natural history of the regions crossed by the expedition. The latter task was regarded in severely practical terms, as the only part of the instructions given to Bering by the Senate in 1732 that bears any relation to natural history shows. This was a standard clause similar to that issued to all Russian discovery expeditions of the period: 'In these voyages search should be made for good harbors and for forests where timber for shipbuilding is to be had. Let mineralogists with a guard go ashore and prospect. If precious metals are found in some place under Russian jurisdiction, the commander of Okhotsk and the principal officers elsewhere should be notified, and they shall send ships, miners, workmen, instruments, machinery and provisions and begin working the mines.'[6]

In the event, both Müller and Gmelin withdrew from the expedition during the winter of 1736–7, although the former was later to publish a history of Bering's voyages. Gmelin was replaced as mineralogist and botanist by Georg Wilhelm Steller, a young German-born scholar who was to play an unexpected and, in the end, critical role in the story of

Bering's last voyage. Steller was educated at the University of Halle, several of whose faculty had strong links with Russia, and in 1736 his 'insatiable desire to visit foreign lands'[7] led him to take up a position in Russia, where at the age of twenty-seven he joined the staff of the archbishop of Novgorod as a physician. The next year the St Petersburg Academy of Sciences appointed him as an adjunct (the rank below professor) in natural history with orders to join Martin Spanberg's ship which was to sail to Japan as part of the Great Northern Expedition. His long trek across Siberia revealed something of the nature of the man. Gmelin and Müller had travelled in style, accompanied by dozens of porters, servants and cooks, but Steller's journey was a far less elaborate affair. As Gmelin noted: 'He had reduced [his housekeeping] to the least possible compass. His drinking cup for beer was the same as his cup for mead and whiskey. Wine he dispensed with entirely. He had only one dish out of which he ate and in which was served all his food. For this he needed no chef. He cooked everything himself . . . It was no hardship for him to go hungry and thirsty a whole day if he was able to accomplish something advantageous to science.'[8]

In August 1740, Steller joined Bering at Okhotsk, where the building of two ships, the *St Peter* and the *St Paul*, had just been completed. There Bering used his seniority and the lure of unknown lands to persuade Steller to sail with him rather than accompany Spanberg because, in Gmelin's absence, he had 'the necessary skill in searching for and assaying metals and minerals'. This was to be Steller's primary responsibility, but he told Bering that he would also 'make various observations on the voyage concerning the natural history, peoples, conditions of the land, etc.'[9] On the voyage, although the *St Peter* had an assistant surgeon, Matthias Betge, Steller acted as personal physician to Bering, whose cabin he shared and who was becoming increasingly infirm. After the voyage Steller completed 'a short, impartial, and true account of the voyage and what happened to me on it',[10] although it was not published until many years later, and then only in edited form.

The *St Peter* and the *St Paul* were sturdy vessels, ninety feet in length, and of just over two hundred tons displacement. In September they sailed from Okhotsk to the superb natural harbour of Avacha Bay on the southeast coast of Kamchatka, where they spent the winter in the newly established port of St Peter and St Paul (modern Petropavlovsk).

The following June the *St Peter*, commanded by Bering, and the *St Paul*, commanded by Alexsei Chirikov, left harbour and headed southeast across the North Pacific towards latitude 45°N. where the lands shown on Delisle's map were thought to lie. It was an unwise decision, for in searching for the imaginary Gama Land the expedition lost so much time that the chances of the ships crossing the ocean, reaching the still-unknown coast of America and returning in a single season were much reduced. Bering's second-in-command on the *St Peter*, the Swedish-born Sven Waxell, wrote a narrative of the expedition in the 1750s in which he was fiercely critical of the sailing directions drawn up by Delisle and his colleagues at the St Petersburg Academy, who 'obtained all their knowledge from visions ... my blood still boils whenever I think of the scandalous deception of which we were the victims'.[11] After becoming separated from Bering, Chirikov in the *St Paul* sailed north-east until he sighted the American coast in latitude 55°21′N. (near present-day Sitka in Alaska), but soon after lost both his boats and their crews. Unable to land, Chirikov turned for home and, after a difficult passage through the Aleutian chain of islands during which La Croyère and five others died from scurvy, reached Avacha Bay in October 1741.

For events on the *St Peter* we have several sources, although no personal record seems to have been kept by Bering himself: ship's logs by the fleet master, Sofron Khitrov, and the assistant navigator, Kharlam Yushin, Sven Waxell's narrative (written, or at least completed, more than a dozen years after the voyage) and Steller's personal journal. The latter's experience as a landsman, physician and scientist on board the *St Peter* was not a happy one. Despite sharing a cabin with Bering, he remained an outsider on the ship, with no place in the naval command hierarchy; and his priorities were very different from Bering's, as the events of the voyage were to show. Whereas Steller was excited by the prospect of being the first European naturalist to investigate the unknown lands of northwest America, Bering was primarily concerned with the safety of the ship and her crew. To Steller, a shore looming out of the mist promised opportunities for investigations into the new land's inhabitants, flora and fauna that would bring him publicity and fame, but to Bering and his officers it threatened navigational hazards and potentially hostile natives. For them, charting the new coasts and returning home with that information were the priorities. And there were problems arising from the differences between scholarly and service backgrounds. As O.W. Frost, the modern editor of Steller's journal, points out: 'he was no doubt accustomed to open discussion of issues and reasoned discourse based on available knowledge. Sven Waxell, Bering's lieutenant and routinely in charge of the ship, was, however, not inclined to recognise any expert testimony outside the naval chain of command.'[12]

Nine days after leaving Avacha Bay on 4 June 1741, with no sight of the lands marked on Delisle's map, Bering and Chirikov decided to turn northeast. The separation of the two ships a week later, in foggy weather, was the first of several disasters that overtook the expedition. By this time Bering, almost sixty years old, ill and exhausted from his responsibilities and the hardships of the long trek to Kamchatka, was rarely seen on deck. He is scarcely mentioned in Waxell's account of the voyage, while Steller maintained: 'Even at this early time, a scheme was begun to tell the Captain-Commander, who remained constantly in his cabin, no more than was deemed expedient.' Decisions of importance were made by the council of officers – standard practice in the Russian navy – and Bering's was only one of several voices. In his journal Steller raged about the refusal of Waxell and others to accept his conviction (an entirely mistaken one) that they 'were running along land' to the south. His comments indicate something of his contemptuous attitude towards the ship's officers, and explain why he was not the most popular of shipboard companions: 'Most of the officers had lived ten years in Siberia, each as he wished, and had acquired and maintained rank and honor from the ignorant rabble ... by force of habit [they] had deluded themselves into believing themselves highly insulted if anyone should say anything they did not know.'[13]

By 14 July half the ship's supply of water had been consumed, and it was agreed that if land were not reached within a week the ship would return to Avacha Bay. Then, on 17 July, a shoreline was sighted to the north and soon the towering peak of a great mountain (Mount St Elias) was visible above the clouds. This Alaskan landfall was one of the defining moments of world geography, although this was clearer in retrospect than at the time. Amid the crew's mutual congratulations, Bering remained unmoved, simply shrugging his shoulders as he looked at the distant land. Later, in his cabin, he complained to Steller about the hysterical reaction of the 'pregnant windbags' on board, and he worried about the accidents that could befall a ship so far from home.[14] Three days later the *St Peter* anchored off the oceanic coast of Kayak Island, and two boats were sent ashore. One was to find water; the other, under Sofron Khitrov, was to explore more generally. After a heated argument with Bering, Steller was allowed to accompany Khitrov, but with the assistance of only one person, the Cossack Thomas Lepekhin. Realising that his time away from the ship might be limited, Steller hurried along the shore and soon found evidence of recent occupation by the unseen inhabitants: the remains of a meal cooked over hot stones, a wooden hammer and an arrow. Deeper in the woods he came across a cellar dug in the ground. It was roofed with bark and contained baskets filled with smoked fish, straps made from seaweed, rolls of dried larch or spruce, and some arrows. Modern ethnologists have identified the cellar as a storage pit dug by the Chugach Eskimos of the region.[15] Lepekhin took smoked fish and other items back to the ship, together with a request from Steller that Bering should send him two or three more men to help him explore the area.

Meanwhile the naturalist walked several miles along the beach, collecting plants as he went, before finding his way blocked by steep cliffs. From the top of a hill Steller saw smoke less than a mile distant, and decided to return to the landing place to ask for the use of the small yawl and several men so that he could get farther along the shore. His solitary venture on a totally unknown island, armed only with a knife for digging up plants, showed professional commitment and personal courage of the highest order, but Bering's response to his request was a curt ultimatum that unless he were back on board within an hour he would be left behind. When Khitrov returned to the ship he reported that he had found a possible anchoring place and a small hut whose inhabitants, he thought, had only recently fled. Bering sent ashore some iron items, beads and tobacco to be left in the cellar Steller had found, then weighed anchor before dawn on the morning of 21 July, ignoring Waxell's advice that they should stay longer to take on more water (thirty-five barrels of fresh water had been brought on board, twenty remained empty). Steller expressed his frustration in an exclamation in his journal that for later critics summed up the missed opportunity of the second Bering expedition: 'We have come only to take American water to Asia.'[16] Bering's most recent biographer has noted that a disappointed Steller did not appreciate that Bering 'knew that the ship was their lifeline and that every precaution should be taken to preserve it'.[17] He would have been even less inclined to take risks if he had known about Chirikov's loss of his two boats and fifteen men farther east along the Alaskan coast at about the same time. In his journal Steller returned to the charge: 'The only reason no landing was attempted on the mainland was a sluggish stubbornness and cowardly fear of a handful of unarmed and even more terrified savages . . . and a cowardly longing for home . . . The time spent for investigation bore an arithmetical relationship: the preparation for this ultimate purpose lasted ten years; twenty hours were devoted to the matter itself.'[18]

The journals show that the expedition's officers agreed that they had reached America, and for Steller views of the land from the ship and his brief excursion ashore were enough to suggest a land of promise:

> America is of noticeably better character than the most extreme northeast part of Asia, although the land toward the sea, whether viewed close up or from afar, everywhere consists of amazingly high mountain ranges, most of whose peaks are covered with perpetual snow. Yet these mountains, compared with those of Asia, have a much better nature and character . . . [they] are firm, covered above the rock mantle not with moss but with good soil; therefore up to the highest peaks, they are densely overgrown with the most beautiful trees . . . here at 60 degrees latitude the very beach itself is right from the waterline studded with the most beautiful forests.

He had found no minerals, for his walk along the beach had been on sand and grey rock, and most of the berries he had noticed were familiar: blueberries, cloudberries and crowberries. But there was one exception, the salmonberry, 'doubtless of superb flavour', to which he gave a significant name, *Rubus americanus* (now *Rubus spectabilis*). Steller carefully dug out several bushes in the hope that they could be sent to St Petersburg, but Bering refused his request because of lack of room or, as Steller put it, 'since I myself as protester now took up too much space'. The only land animals Steller noticed were foxes, but he saw more than ten new species of birds, including a jay that he thought he had seen illustrated in a recent history of the plants and birds of Carolina by an author whose name he could not remember (Mark Catesby, whose *Natural History of Carolina* was published in 1731). 'This bird alone sufficiently convinced me that we were really in America,' he wrote; in fact, it was a different species from Catesby's jay and was later named after Steller (*Cyanocitta stelleri*).[19]

From Kayak Island the *St Peter* sailed southwest along the Alaska Peninsula. It was a risky business, with the ship several times running without warning into shallow, rock-strewn waters. On 10 August a meeting of the ship's council decided that since the weather was closing in and they were in uncharted waters they should return to Avacha Bay, still fifteen hundred nautical miles distant, without attempting further exploration. The resultant document was signed as usual by the officers and also by petty officers down to the boatswain's mate (but not by Steller, who as a supernumerary was not a member of the ship's council). A key factor in the decision was the report by the assistant surgeon, Matthias Betge, that five men were unfit for any duty, and that sixteen others were 'badly affected' by scurvy.[20] With water running short again, at the end of August the *St Peter* anchored off Nagai Island in the outer Shumagin chain. Here Steller once again found himself at odds with Waxell and the other ship's officers, this time over the best watering place. Waxell was now clearly in charge of more than the routine operations of the ship, although he occasionally consulted Bering, ill in bed.[21] On or near the island, Steller sighted many birds new to him, but of more practical use was his finding of scurvy grass, dock and other 'magnificent antiscorbutic plants' which he gathered for his use and Bering's. It was helpful that after his arrival in Kamchatka in 1740 he had paid particular attention to plants used by the native Kamchadals to ward off scurvy. In his journal Steller noted disapprovingly: 'the medicine chest had been meagrely and miserably furnished, filled with the most useless medicines, almost nothing but plasters, ointments, oils, and other surgical supplies needed for four or five hundred men with wounds from great battles, but with nothing whatever needed on a sea voyage where scurvy and asthma are the chief complaints'.[22] At first the officers scorned Steller's use of greens to cure scurvy, and refused to send help for him as he gathered scurvy grass, dock, cress and berries. But they changed their attitude when they saw how the fresh greens helped

Bering, 'so bedridden with scurvy that he had already lost the use of his limbs – so far within eight days that he was able to get out of bed and on deck to feel as vigorous as he had been at the beginning of the voyage'.[23]

There was nothing surprising about the officers' initial scepticism in response to Steller's treatment. Long the scourge of the seas, scurvy was a mystery to the medical profession, its causes unknown and its treatment uncertain. The opinions of ships' surgeons seem not to have been sought by physicians on shore, whose explanations ranged from a melancholic humour to exposure to sea air. Much pernicious nonsense had clouded the matter, notably that scurvy and idleness went hand in hand. Expeditions to northern regions were particularly vulnerable, and only a few years before Bering's second voyage a Russian party taken by Peter Lassenius down the Lena River to the Arctic Ocean had lost thirty-eight out of its forty-six men to scurvy. At the same time that Bering was at sea, the circumnavigation by a British naval squadron under Commodore George Anson had suffered huge losses from the disease. Out of nineteen hundred men who sailed from England in 1740, almost fourteen hundred died during the voyage, most from scurvy.

After Anson's voyage the naval surgeon James Lind carried out experiments with different treatments that provided evidence of the antiscorbutic properties of lemon juice (long used on East Indiamen), although he had no more explanation of why it should be effective than had any of his predecessors, and another fifty years passed before lemon juice became regular issue on Royal Navy vessels. Only with the discovery of the existence of vitamins in the early twentieth century was scurvy properly diagnosed as resulting from a deficiency of vitamin C or ascorbic acid. This was shown to be an essential element in the body tissues, and could only be obtained through the food supply. Greens, milk and citrus fruit all contain vitamin C, as does fresh meat to a limited degree. After a few weeks without an intake, the amount of ascorbic acid in the body falls below measurable levels, and scurvy can appear at any time.[24] By the time of Betge's report on 10 August the crew of the *St Peter* had been at sea for ten weeks, with little in the way of a balanced diet, and scurvy was bound to follow. Chirikov's crew was also affected. Six men died and Chirikov lay helpless in his cabin, preparing for death; had the *St Paul* not made harbour at Avacha Bay on 12 October, the death toll would have been much higher.[25]

On 3 September, while the *St Peter* was anchored off Bird Island in the outer Shumagin Islands, Bering's men encountered their first 'Americans' (Aleuts) when two sealskin kayaks with one paddler in each were sighted coming towards them. It was an unanticipated meeting for, as Steller wrote, 'We did not expect any trace of human beings on this miserable island twenty miles away from the mainland.'[26] The sighting was followed by a period of mutual incomprehension, with neither side able to understand the shouts of the other. The seamen beckoned the two men to come closer to the ship, while they in turn pointed to the shore where their companions were watching and calling. After painting his face with earth, one of the paddlers threw towards the ship a stick of spruce wood (Pl. 4). Painted red and with two falcon wings bound to it, the offering puzzled Steller, who wondered whether it was a sacrifice or a gesture of friendship. Eventually, the lieutenant took a boat on shore, accompanied by Steller, nine seaman and soldiers, and the ship's Chukchi interpreter. Heavy surf prevented the boat from landing, but two of the seamen and the interpreter waded ashore where they were presented with pieces of blubber, although they were careful not to lose sight of their boat during the ceremony. An elderly Aleut came out to the boat in his kayak, but refused gifts and spat out a beakerful of spirits he was offered. After an hour Waxell decided it was time to return to the ship, but the Aleuts only released the interpreter when soldiers fired muskets over their heads. The din of the shots echoing off the cliffs had its effect. As Steller wrote: 'when they heard it, they all looked so stunned that, as if struck by thunder, they all fell to the ground and let go of everything in their hands.'[27] Until this moment the encounter had been peaceable if fraught, but Waxell noted that his men 'would have much preferred to have fired at the savages themselves'.[28] The next morning seven kayaks approached the ship and two came up to the gangway, where their occupants exchanged gifts before rising seas forced them back to shore. Waxell suggested to Bering that he should seize the men, but was given written orders forbidding him from using any force – a sign that Bering, despite his frailty, was still in command.

Both Waxell and Steller summed up their impressions of this historic encounter, the first between Europeans and the native inhabitants of the northwest coast of America. Waxell left a succinct description of their craft:

> In the middle of the kayak is a raised part like a wooden bowl and in the centre of this a hole large enough for a man to get the lower part of his body through it and down into the kayak. Round this hole is fastened a sealskin bag which in its turn is fastened round the body with a long thong. Once seated in the kayak and thus fastened in it, not a drop of water can find its way inside. The natives are accustomed to these craft from youth and are perfectly able to maintain their balance – the whole secret of sailing them – even in very rough weather.[29]

Steller was also impressed by the paddlers' skill, and gave a detailed description of the Aleuts' physical appearance and their clothing, some elements of which – their sealskin boots and pants, and the hats made of tree bark that shaded their eyes – he thought resembled that worn by the Kamchadals. In an attempt to converse with them, Waxell turned to a Huron vocabulary printed in Baron Lahontan's celebrated *Nouveaux Voyages dans l'Amérique septentrionale* of 1703. He claimed to have used it successfully, testing the Aleuts with words for water, wood and blubber. In a conclusion that impressed no one apart from himself, Waxell wrote: 'These and several other questions I put to them so as to learn whether or not they really were Americans, and as they answered all my queries to my satisfaction, I was completely convinced that we were in America.'[30] Steller, by contrast, was more interested in the relationship between the natives he had briefly encountered and the Kamchadals and Chukchi to the west, and correctly concluded that he had 'found a clear indication that the Americans originated in Asia'.[31]

As the *St Peter* skirted the Aleutian Islands on the return voyage to Kamchatka, still a thousand miles distant, conditions on board worsened and a dozen men found themselves on the sick list.[32] With the vessel struggling against headwinds, 'Waves struck like shot out of a cannon,' Steller wrote, and he added that the ship's veteran senior navigator could not remember in his fifty years' experience a storm as violent. 'Every moment we expected the shattering of our ship . . . No one could remain at his station, but we were drifting under God's terrible power wherever the enraged heavens wanted to take us. Half the men lay sick and weak, and the other half was healthy out of necessity but thoroughly crazed and maddened by the terrifying movements of the sea and ship.'[33] 'The old navigators on board had seldom seen such violent storms, while Yushin's log – usually a model of restraint – referred to a 'terrific storm' (7 October), 'a terrific gale' (9 October) and 'frightful squalls' (10 October).[34] Waxell's account described the sad state of the crew: 'They were attacked by scurvy so violently that most of them were unable to move either their hands or feet, let alone use them ... a day seldom passed without our having to throw the corpse of one of our men overboard.' By mid-October, he wrote,

> so many of our people were ill that I had, so to speak, no one to steer the ship ... when it came to a man's turn at the helm, he was dragged to it by two other of the invalids who were still able to walk a little, and set down at the wheel. There he had to sit and steer as well as he could, and when he could sit no more, he had to be replaced by another in no better case than he. I myself was scarcely able to move about the deck without holding on to something ... Our ship was like a piece of dead wood, with none to direct it; we had to drift hither and thither at the whim of the wind and waves.[35]

Steller was one of only four reasonably fit men in the last stages of the voyage, and noted that, when asked for help, 'I gave as best as I could with bare hands' – before adding, pointedly, 'even though it was not in my job description'.[36]

On 4 November land was sighted and, hoping that it was Avacha Bay, even 'the half-dead crawled out to see it'; but, writing later, Waxell pointed out that cloudy weather had prevented them taking a latitude observation 'for a long time'.[37] Yushin's log shows that only one observation for latitude had been possible from 15 October to 3 November (on 25 October), and when the clearing weather allowed them to take an observation on 4 November they were found to be at least a hundred miles north of Avacha.[38] A meeting of the council decided that in view, as Waxell put it, 'of our helpless condition in which we were more like a wreck than a ship', they should head for the land, about six miles distance.[39] In heavy seas the small bower anchor parted and the *St Peter* crashed over rocks before being brought up short with its sheet anchor in calm water a few hundred yards from the shore. Gradually the sick were taken on land and placed in sand pits dug in the beach: 'Many died on deck as soon as they were brought out into the air; others died in the boat before they ever reached soil, and others died the moment they were set ashore.'[40] Characteristically, as soon as Steller arrived, he searched for plants, which, together with some ptarmigan shot by the draughtsman Friedrich Plenisner, he sent on board the ship for Bering. There was fresh water nearby, but no trees of any kind. For the moment at least food was plentiful: sea otter, fox, seal and ptarmigan. Where they had landed was not clear to Bering and his officers. It was presumably some part of Kamchatka but whether mainland or island was not discovered until December, when a reconnaissance party climbed hills inland and saw open sea to the west. The news was the more depressing because at the end of November gales had driven the ship ashore, where she sank into the sand and filled with water. Waxell put the situation bluntly:

> We now saw ourselves threatened with certain destruction, being on an unknown desert island without a ship or timber with which to build a new one, and at the same time with little or no provisions. Our people were very sick and we had no medicines or drugs of any description. Nor were we even decently quartered, but lodged, so to speak, under the open sky. The whole ground was covered with snow and we quailed at the thought of the long winter and the fierce cold that would descend upon us here, where there was no sign of fuel.[41]

During November, December and early January more than thirty of the crew died from scurvy and exposure, and the survivors were so weak that the living and the dead lay next to each other in their scooped-out hollows, the corpses half-eaten by foxes. Bering died on 8 December, 'more from hunger, cold, thirst, vermin, and grief than from a disease', according to Steller.[42] Waxell gave more details 'of the wretched state of the Captain-Commander as he lay in his agony. The half of his body was already buried in the ground while he still lived, for he said to us: "The deeper in the ground I lie, the warmer I am; the part of my body that lies above the ground suffers from the cold." '[43] According to Waxell, Bering was buried 'tied fast to a plank and thrust down into the ground', but an excavation that followed the discovery of his grave by a Russian-Danish scientific team in 1991 revealed, surprisingly, that he was buried in a wooden coffin.[44] The wood probably came from the wreck of the *St Peter*. None of the other five graves excavated showed any sign of a coffin, but that containing Bering's remains indicates a more formal burial than the perfunctory one described by Waxell. The officers named the island in memory of their dead commander.

The forty-six survivors spent the winter in five underground dwellings in imitation of the Kamchadals, but since there was no wood available the shelters of Bering's men were simply hollows scooped out of the sand, covered with sails. As Waxell pointed out: 'from the sea came thick mists and dampness that caused the sails to rot, until they were no longer able to withstand the violence of the storms, but were blown away on the first gust, leaving us lying under the open sky.'[45] Arguments and disputes continued, sometimes over apparently trivial matters. Waxell allowed the men – against regulations – to play cards as a help to pass the time. Steller criticised 'the dissolute gambling' and accused the officers of using their superior skills to win money and sea-otter pelts from the men.[46] Fleet Master Khitrov, blamed by many of the crew for their troubles, begged Steller to allow him to join his little group, but was turned away. Waxell was also unpopular, but Steller helped him for fear that if he died Khitrov would take command. Waxell at one stage appeared terminally ill with scurvy. He was further weakened by the fact that he had to share his meagre ration of food with his twelve-year-old son, who had been taken on the voyage but was not officially a member of the ship's crew and so was allocated no allowance. With Bering dead, and Waxell and Khitrov incapacitated for much of the time, Steller became the effective leader of the survivors, although his was leadership by example rather than in terms of formal command. It is intriguing that, out of the seven hundred sea-otter pelts reported to have been brought back by the crew to Avacha Bay, two hundred belonged to Steller. Many of these, it has been suggested, were given to him in gratitude for his services.[47] Waxell later described his condition and that of his shipmates who were also suffering from scurvy.

> The disease first showed itself in a heaviness and weariness in all our limbs, such that we were all the time wanting to sleep ... We became

more and more depressed mentally ... the least movement made them unable to get their breath. This was soon followed by stiffness in the limbs, swelling of the feet and legs, while the face became yellow. The whole mouth, and especially the gums, bled and the teeth became loose. All this can happen in the course of eight days, unless you try to fight against it in time, and having got to this state it is already all up with the patient.[48]

Steller had helped sufferers on the voyage by giving them scurvy grass and dock, but in winter no plants were to be found on Bering Island. Some rye-flour had been rescued from the wreck, but the crew's main source of food was the flesh of the sea otter. Waxell wrote: 'even if you can perhaps endure the smell of sea-otter meat, it is extremely hard and as tough as sole-leather and full of sinews, so that however much you chew at it, you have to swallow it in large lumps'.[49] However unpalatable the meat, for Bering's men it was a life-saver, for it was a source of vitamin C. As the sea otters moved out of danger, the hunting parties had to travel farther in search of their prey and risked being caught in violent snowstorms. Steller described how after one storm a hunting party failed to return to the snowed-in camp before nightfall: 'The next day we worked for several hours before we were able to dig ourselves out of our dwelling; at the very time the entrance had luckily been cleared, three of our men arrived, senseless and speechless, and so stiff from the cold that, like immovable machines, they could hardly move their feet. The assistant surgeon, totally blind, unable to see, walked behind the others.'[50]

As fresh meat became increasingly difficult to find, the men eyed with mounting desperation the giant sea cows or northern manatee (*Hydrodamalis gigas*), thirty feet long and twenty feet around, grazing on kelp in the shallow water. For some weeks they tried to kill one of these giant creatures. Iron hooks were thrust into its hide and attempts made to drag it ashore, but its strength and bulk were such that it escaped to deep water, carrying the hooks and ropes with it. After several failures, a new method was tried. The yawl was repaired, and a crude harpoon was fastened to a long rope, with its end held by forty men on the shore. 'As soon as the harpooner had struck one of them, the men on shore started to pull it to the beach while those in the yawl rowed towards it and by their agitation exhausted it even more. As soon as it had been somewhat enfeebled, the men in the yawl thrust large knives and bayonets in all parts of its body until, quite weak through the large quantities of blood gushing high like a fountain from its wounds, it was pulled ashore at high tide and made fast.'[51] The flesh of one of these huge sea mammals was enough to feed the entire crew for two weeks; its meat tasted like beef, and its fat 'tastes so sweet and desirous that we lost all desire for butter'.[52]

By March most of the snow had gone, and Waxell – unusually – acknowledged the part Steller played in the recovery of the men's health as the naturalist encouraged them to collect herbs and roots.[53] As conditions improved, Steller continued his investigations into the wildlife of the island. Among the birds he observed were an eider, later named *Anas stelleri* in honour of its discoverer, and the flightless, spectacled cormorant (*Phalacrocorax perspicillatus*), known only on Bering Island, where it was exterminated in the nineteenth century. In June, Steller made a special journey to the south side of the island, where he spent six days in a driftwood shelter observing the appearance and habits of the little-known northern fur seal (*Callorhinus ursinus*): 'If I were asked to state how many I have seen on Bering Island, I can say without lying that it is impossible to make a computation. They are innumerable.'[54]

At the beginning of April work had begun on dismantling the hulk of the *St Peter* and building a smaller vessel to take the survivors back to Avacha Bay. It was a laborious task, made more difficult by the lack of tools and skilled carpenters. Waxell explained how 'for a keel we used the old ship's mainmast which we sawed off three feet above the deck, not having either the tools or the strength to get it right out. The remaining stump of the mainmast had to serve as the new vessel's prow; the stern-post we made from a capstan which we had on the old ship.'[55] By early August the work was finished, and on the 13th the new *St Peter* weighed anchor with its crew of forty-six. The small, forty-foot craft sat so low in the water that most of the items brought on board – from bedding to cannonballs – had to be thrown overboard. Steller had hoped to take on board the skeleton and skin of a young sea cow, but was refused space, and in the end the only remnant that reached St Petersburg was a pair of the sea cow's horny palatal plates.[56] After two weeks the vessel reached Avacha Bay, where the crew discovered that they had been given up for lost: the belongings they had left behind the previous year had been sold. It was a tribute to the skill of Waxell and his sailors that in 1752 their homemade vessel was still being used to transport goods from Okhotsk to Kamchatka.

Steller did not long survive the Bering expedition. Soon after returning to Avacha Bay, he continued his travels across Kamchatka, while during the winter he worked on his journal and natural-history notes. His sympathy for the native Kamchadals led to problems that almost certainly shortened his life. His release of Kamchadal prisoners accused of rebellion against the Russian authorities brought two spells of detention at Irkutsk, and delayed his return to St Petersburg.[57] On his final journey Steller died of a fever at Tiumen, Siberia, in November 1746, with much of his work incomplete and none of it published. This can partly be explained by his untimely death, partly by the decision of the Russian government that there would be no official follow-up expedition to Bering's and that its findings should remain secret. Occasional rumours and reports of the voyage reached Western Europe but they were brief and inaccurate. Among the early reports was one published in the *Gentleman's Magazine* in London in October 1743. It claimed that Bering was shipwrecked on an island on his *outward* voyage, thus denying him credit for his Alaskan discoveries. It went on to describe how 'Mr. Stoller [sic], with the assistance of some of his companions, found Means to build, out of the ruins of their great Ship, a little shallop in which he himself and 19 others, after running through a thousand perilous Adventures, arrived at Kamschatka'. It took eighteen months for Steller in Irkutsk to receive this news item from his brother, whereupon he wrote back angrily to him: 'I should like very much to know who has been making me out a sailor and a windbag. My desire is to fill gaps in the realm of science, not vacant space in the newspapers.'[58]

Russian official attitudes changed after the return to France in 1747 of Joseph-Nicolas Delisle, who angered the Russian government by taking away confidential maps and reports. These he showed to his uncle, Philippe Buache, the leading French geographer of the day, and in 1750 Delisle presented to the Paris Académie des Sciences a paper on Bering's voyages, accompanying it with a map constructed by Buache.

As far as the second voyage was concerned, Delisle repeated the fiction that Bering's ship was wrecked soon after leaving Kamchatka, and he devoted most of his attention to what he claimed was the more significant voyage of Chirikov and his own half-brother, Louis de la Croyère, in the *St Paul*. The memoir and map were published in 1752 amid a swirl of publicity and controversy, not least in Russia, where the president of the St Petersburg Academy of Sciences commissioned the Academy's secretary, Gerhard Friedrich Müller, to write a pamphlet refuting Delisle's 'evil representations'.[59] This took the form of a *Lettre d'un officier de la marine russienne*, published in 1753, with an English edition the next year. The pamphlet was the prelude to Müller's full-length account of Bering's two expeditions, *Voyages from Asia to America*. This was first published in German in 1758, and then translated into English in 1761 (with an improved second edition in 1764). First in the pamphlet and then at greater length in the book, Müller described the explorations of both Bering and Chirikov on the Second Kamchatka Expedition as well as those of the subordinate expeditions connected with the project. He interviewed some of the survivors from the *St Peter*, and paid glowing tribute to Steller's role during the shipwreck on Bering Island: 'Men could not lose heart because they had Steller with them. Steller was a doctor who at the same time administered to the spirit. He cheered everyone with his lively and agreeable company.'[60]

Steller's own journal, an invaluable if partial account of Bering's second voyage, was not published until late in the century. Its description of Bering Island appeared in 1781, while the main part of the journal was issued in 1793. Both instalments were published under the aegis of Peter Simon Pallas, an outstanding natural historian and president of the St Petersburg Academy of Sciences. Unfortunately Pallas edited, censored and amplified Steller's journal, which was not published in its original form (in English translation) until 1988.[61] Five years after Steller's death his detailed descriptions of the four sea mammals – sea otters, fur seals, sea lions and sea cows – that he had observed while on Bering's voyage were published under the title of *De bestiis marinis* ('Beasts of the Sea'). Of lasting value was his record of the sea cow, for by 1768 this huge, placid creature had been exterminated by Russian fur traders, and Steller was the only naturalist to have seen one alive. The largest member of the *Sirenia* order, it was related to the manatee of the shallow waters of the Caribbean, the West African coast and the Amazon delta. It was evidence of the breadth of Steller's reading that he was able to cite sixteenth-century Spanish accounts of the creature before turning to William Dampier: 'Among all who have written about the manatee, none has produced a more complete and painstaking description than the very inquisitive and industrious Captain Dampier.'[62] In his account Sven Waxell described the animal whose meat restored the crew to health on Bering Island – 'I can in very truth say that none of us properly recovered until we began eating them' – but he hoped that a fuller description would be found among Steller's papers, for he was 'a great botanist and anatomist'.[63] Steller began his account by noting that every day for ten months he had observed the behaviour and habits of this remarkable creature:

> These animals love shallow and sandy places along the seashore ... They keep their half-grown and young in front of them when pasturing, and are very careful to guard them in the rear and on the sides when travelling, always keeping them in the middle of the herd. With the rising tide they come in so close to the shore that not only did I on many occasions prod them with a pole or a spear, but sometimes even stroked their back with my hand ... These gluttonous animals eat incessantly, and because of their enormous voracity keep their heads always under water with but slight concern for their life and security ... While browsing they move slowly forward, one foot after the other, and in this manner half swim, half walk like cattle or sheep grazing ... In the spring they mate like human beings, particularly towards evening when the sea is calm.[64]

Steller was not content with watching and describing the sea mammals of Bering Island. During his six-day stay on the south coast of the island overlooking a huge colony of fur seals, he not only recorded their appearance and habits, but dissected a male fur seal, making thirty-one different measurements to add to his thirteen-page description. Then in July he turned to the altogether more difficult task of dissecting a huge female sea cow, weighing, he thought, eight thousand pounds. Its stomach was 'of amazing size, 6 feet long, 5 feet wide, and so stuffed with food and seaweed that four strong men with a rope attached could scarcely move it from its place and with great effort drag it out'.[65]

Adding to Steller's difficulties were 'packs of the most despicable Arctic foxes [that] were tearing with their vile teeth and stealing everything from under my very hands, carrying off my papers, books and inkstand while I examined the animal and ripping it while I was writing'.[66] A member of the crew, almost certainly the young draughtsman and artist Friedrich Plemisner, drew six sketches of the sea mammals, including two of the sea cow, and these were sent to the Academy by Steller in 1743 but never arrived. Fortunately, copies of the sketches appeared on two of Waxell's charts (Pl. 5). They are the only ones known to have been drawn by (or at least under the direction of) someone who had actually seen the mammal, and settled, for example, the question of whether or not the creature had a forked tail.[67]

Some indication of Steller's energy as a natural historian is shown by the miscellaneous sets of papers he left behind, twenty-five in all, ranging from lists of minerals to descriptions of insects and various vocabularies.[68] However, it is clear that many of his notes have been lost, or remain unidentified. The specimens of fishes he sent to St Petersburg from Siberia included at least thirty new species, but they did not become known to ichthyologists until the third volume of Pallas's *Zoographia* was published in 1826, and by then other naturalists had claimed much of the credit for their discovery. Steller's field notes on the Kamchadals were incorporated into the pioneering account of Kamchatka by Stepan P. Krasheninnikov published in Russian in 1755, but because of the death of its author before publication his debt to Steller was never properly acknowledged. More general problems remain in assessing Steller's work, for he died seven years before Linnaeus in his *Species plantarum* introduced the scientific world to his binomial system of classifying plants.[69] Even though Gmelin in his *Flora Sibirica*, published in four volumes between 1747 and 1769, included many of the new species described by Steller, he referred to them by outdated pre-Linnaean names: as a result some were claimed by a later generation of botanists, unaware of the work of their German predecessor. On the other hand, the unattributed nature of much of the documentation in the Russian archives relating to natural history in this period has led to some claims on Steller's behalf that cannot be substantiated. Among the manuscripts thought to be by Steller is *Catalogus plantarum intra sex horas*.[70] Although not in Steller's handwriting, it has long been thought to be a record of the plants either collected or noted by Steller on Kayak Island. It contains no fewer than 143 entries, representing forty-nine modern plant families. A problem is that many of the plants are not known on Kayak Island, or indeed in Alaska generally, and it has been suggested that the unknown compiler of the catalogue included plants collected by Steller in Siberia as well as those found by other botanists. Given that Steller spent only six hours on Kayak Island, and had to scramble over troublesome terrain, it is hard to believe that he collected, or even noted, the number of plants listed in the catalogue. Without the actual specimens, much must remain conjectural about both the contents and the origin of the catalogue.

Steller, who died at the age of thirty-seven, accomplished much while in the service of the Russian government. As an ethnologist he carried out investigations into the lifestyle of the Kamchadals during the early period of Russian rule over Kamchatka. He used that knowledge to good effect during the struggle for survival of the *St Peter*'s crew on Bering Island, where his efforts played a vital part in keeping men alive. As a naturalist he gathered a remarkable amount of information in dangerous conditions and with little in the way of official encouragement. Many of his notes have been lost or remain unidentified, but his name will long be remembered by association with some of his most intriguing discoveries – above all, the mighty mammal known as Steller's sea cow. Linnaeus at least had no doubts about Steller's achievement. Writing to Gmelin before he knew of Steller's death, the great Swedish botanist suggested that a new plant species should be named after the naturalist, 'who has discovered so many new plants during so many years of most laborious travel . . . Everybody in the botanical world who knows plants loves Mr. Steller.'[71]

CHAPTER 3

'My plants, my beloved plants, have consoled me for everything'

The Fortunes and Misfortunes of Philibert de Commerson

A FTER THE ENDING of the global Seven Years' War in 1763, Britain and France experienced a Pacific craze in which new national heroes emerged in the shape of naval explorers and itinerant scientists. Expeditions set off into the unknown expanses of the Pacific Ocean, to return after three years laden with artefacts and specimens, and with their crews eager to publish descriptions, charts and views of the wondrous places visited and peoples seen. The first discovery voyages of the new era of oceanic exploration were set in train by the British Admiralty. In 1764, Commodore John Byron sailed with instructions to carry out explorations in both the South and North Pacific, while two years later Captain Samuel Wallis and Lieutenant Philip Carteret also left for the Pacific. Byron's instructions set the tone for the British expeditions: 'nothing can redound more to the honor of this Nation as a Maritime Power, and to the Advancement of the Trade and Navigation thereof, than to make Discoveries of Countries hitherto unknown.'[1] There were no civilian scientists, artists or observers on the British ships, for the Admiralty was more interested in tangible returns for British trade than in scholarly investigation. By contrast, the French expedition under Louis-Antoine de Bougainville that left Europe a few months after Wallis was a more brilliant affair. The nobility was well represented among the officers, while Bougainville himself was a figure of European renown. Aristocrat, soldier, diplomat, well read in the philosophical writers of the day, he fulfilled the ideals of the Enlightenment.

And among the complements of his ships, the *Boudeuse* and the *Etoile*, were the astronomer Pierre-Antoine Véron and the naturalist Philibert de Commerson.

For reasons not necessarily connected with his professional duties, Commerson was to become one of the most celebrated (and controversial) members of the Bougainville expedition. Son of a lawyer, he had studied medicine, but his great passion was collecting plants. It was a mark of Commerson's growing reputation that in 1754 the Swedish naturalist Carl Linnaeus, professor at Uppsala University, whose great work *Species plantarum* had been published the previous year, asked him to investigate the marine plants, fishes and shells of the Mediterranean. Linnaeus had developed a distinctive way of classifying and naming plants which, although not universally accepted, was soon adopted by most naturalists. Until this time, botanists had identified plants with lengthy descriptions involving the appearance, colour and shape of their leaves and flowers. But as travellers returned with more and more specimens from distant lands, this method became unmanageable, and the search began for an alternative. In 1718 the botanist Sébastien Vaillant described the sexual function of the stamen and pistil in plants, and in the 1730s Linnaeus built on this identification to introduce an entire system based on a plant's sexual characteristics. He established twenty-three classes based on the features of the pollen-bearing stamens or male parts (and a final, twenty-fourth class of plants such as mosses with no visible flowers), which were then divided into orders based on their pistils or female parts (Pl. 6). Each plant was named with two words in Latin, the first denoting its genus (a group of plants with common characteristics), the second its species (distinctive features that differentiate plants within the same genus). This was followed by the name or initial of the person who discovered or named the plant. Linnaeus's binomial system was first set out in detail in his *Species plantarum*, whose two volumes of 1,200 pages included all the then-known plants. He listed five thousand species, placed in 1,098 genera, but this number was to increase in dramatic fashion as naturalists scoured the world for new specimens.

Linnaeus attempted a similar task for mammals in his *Systema naturae*: its first edition, of 1735, devoted only two pages to mammals, but in its tenth edition, of 1758, the lists of binomials for the animal world covered 824 octavo pages. They included mammals (with *Homo sapiens* leading the way among the primates), birds, amphibians, fish, insects and invertebrates, and became the starting point for modern zoological nomenclature.[2] As one scholar has put it, 'Linnaeus' fame in the eighteenth century rested in the democratizing accessibility of his achievement. The virtue of his classificatory system consisted neither in its faithfulness to the natural order (it was patently artificial), nor in its inherent logic. Rather, its workaday usefulness appealed to both learned men and novices.' Linnaeus established, in effect, a 'filing cabinet of nature',[3] and his publications stimulated an informed interest in botany from both professionals and amateurs. His sexual classification method might, in W.T. Stearn's words, be a stop-gap system, but it was simple to use. The Dutch botanist Johan Gronovius the Younger said of the lists in *Systema naturae*, 'I think these tables so eminently useful that everyone ought to have them hanging up in his study, like maps.'[4]

Not all accepted Linnaeus's taxonomic system. Some deplored the way in which all previous botanical names suddenly became out of date; as Johann Amman, professor of botany at St Petersburg, remarked, the changes would lead to a situation 'worse than the confusion of Babel'.[5] Amman also objected to the way in which plants that differed in every way except for the number of their stamens and pistils were linked together in the same class. Others were repelled by the insistence on the sexuality of plants, regarding it as 'a sort of veiled pornography'.[6] In 1768 the first edition of the *Encyclopaedia Britannica* complained: 'Obscenity is the very basis of the Linnean system.'[7] Foremost among its critics was the eminent French naturalist the Comte de Buffon, director of the Jardin du Roi (later the Jardin des Plantes) in Paris. He pointed out that Linnaeus linked plants that differed markedly in appearance, and argued in favour of a system of classification that gave priority to the *uses* made of plants. Another alternative was offered by Michel Adanson, who came back from six years in Senegal with a huge collection of plants, and whose *Familles naturelles des plantes* in 1763 advocated a system based on the similarity or otherwise of individual organs. The most influential of all the attempts to provide a natural system of classification came late in the century with the publication of Antoine-Laurent de Jussieu's *Genera plantarum* of 1789, which was to attract many adherents. Despite criticism and the emergence of rival systems, Linnaeus remained the dominant force in eighteenth-century botany. His books were translated into all the major European languages, plagiarised, summarised and endlessly reprinted. In terms of the sales of works on natural history, only Buffon's immense *Histoire naturelle*, the first volume of which appeared in 1749 and the forty-fourth (posthumously) in 1804, could rival him.

Characteristic of Linnaeus's proselytising zeal was the way in which he included in his best-known theoretical treatise, the *Philosophia botanica* of 1751, practical instructions on how to organise botanical expeditions, collect specimens and establish herbaria. Another example of his influence came in his encouragement to his students to venture on long-distance voyages and travels to investigate the natural history of faraway regions, and to bring back specimens. His 'apostles', as he called them, included Pehr Kalm, who travelled through much of colonial North America; Johan Peter Falck, who served on a Russian expedition that reached as far as western Siberia; and Carl Peter Thunberg, who, as a surgeon for the Dutch East Company, visited Japan. Among his other students, Daniel Solander sailed on Captain Cook's first Pacific voyage, and Anders Sparrman on his second. From his university quarters at Uppsala, Linnaeus spun a spider's web of global routes for his students to follow, and over the decades nineteen of them left Sweden for distant parts, although the casualty rate among them was heavy. As early as 1737, when discussing earlier botanical travellers, Linnaeus had written, 'Good God! When I observe the fate of botanists, upon my word, I doubt whether to call them sane or mad in their devotion to plants.'[8] His instructions to the travellers show that Linnaeus was moved by more than scientific curiosity. They were exhorted to bring back seeds, bulbs, plants and trees that might be grown in Sweden and so help the country's economy. To Swedish naturalists sailing on foreign expeditions he sometimes suggested that they might smuggle specimens back to their homeland. Linnaeus especially favoured the cultivation in Sweden of tea bushes or seeds, and mulberry trees for silk production, but problems of passage and an unhelpful climate brought frustration and failure. In his relating of botany to the national economy, Linnaeus was the forerunner of Joseph Banks and other naturalist-entrepreneurs of the second half of the century. His successors would also have approved of the stern instruction that Linnaeus gave his travelling naturalists: 'The voyage should not be frittered away with gossip, chats, songs, fairy tales, jokes, playing, and vanities.'[9] Several of the naturalists published accounts of their adventures and investigations, and, as Mary Louise Pratt has put it, 'Travel and travel writing would never be the same again ... Alongside the frontier figures of the seafarer, the conqueror, the captive, the diplomat, there began to appear everywhere the benign, decidedly literate figure of the "herborizer", armed with nothing more than a collector's bag, a notebook, and some specimen bottles.'[10]

After completing his work for Linnaeus and returning to his duties as a doctor, Commerson lavished time and money on establishing a botanical garden in his birthplace of Châtillon-les-Dombes. Each year he spent his summers botanising in the Alps and the Pyrenees, sometimes going fifteen or twenty days with little sleep, staying with shepherds, and living on bread and cheese. His brother-in-law later estimated that Commerson had sold property worth 5,000 livres to finance the gathering of plants for his garden.[11] In spending, often extravagantly, on his botanic garden and herbarium, Commerson was following the enthusiasms of many in French society from the royal family downwards, as botanical books, botanical societies and botanical research flourished to an unprecedented degree.[12] In 1764 he moved to Paris where he became known to some of the country's leading botanists. Two years later, having met and impressed Bougainville, he was appointed to the forthcoming discovery expedition as 'Médicin naturaliste et botaniste du Roi' after he sent to the minister of the marine a seventeen-page programme of his proposed activities on the voyage. He intended to report, among much else, on birds, fish, quadrupeds, mammals and insects; collect plants, minerals, shells and fossils; and make geological and meteorological observations.[13] It was a programme as ambitious in scope as it was challenging to realise in practice.

Before sailing, Commerson received from Charles de Brosses a copy of his *Histoire des navigations aux Terres Australes*, with a request to make marginal notes of any inaccuracies.[14] Understandably, he was flattered to be entrusted with such a task by the leading authority on Pacific voyages, and moreover one who was a friend of the great Buffon. De Brosses stressed the importance of natural-history artists who could draw and paint specimens collected on discovery voyages but, surprisingly, none was taken on the Bougainville expedition. Together with Véron, Commerson sailed, not with Bougainville in the newly built frigate the *Boudeuse*, but in the workaday *Etoile*, a former emigrant vessel. He was given a generous annual stipend of 2,000 livres, more than a captain's salary,[15] and when he joined his ship in December 1766 he was accompanied by a servant, Jean Baret. It was a sign of the respect accorded to Commerson that the captain of the *Etoile*, Chesnard de la Giraudais, gave up his cabin to him so that he had room in it both for his materials and his servant.

For the voyage, Bougainville was given two quite different tasks. The first was to hand over to Spain the disputed Falkland Islands (the Malouines for the French, the Malvinas for the Spaniards). He was then to enter the Pacific Ocean and examine the lands lying in the vast expanse between the western coast of America and the East Indies. 'Knowledge of these islands or continent being very slight,' he was told, 'it is of interest to improve it. Furthermore, as no European nation has any establishment or claim over these lands, it can only be in France's interest to survey them and take possession of them.'[16] Then, in a clause that helps to explain Commerson's appointment, Bougainville was ordered to 'examine the soils, trees and main productions; he will bring back samples and drawings of everything he may consider worthy of attention'.

The *Boudeuse* and *Etoile* sailed separately from France, with Bougainville first calling at the Falklands to formalise the handover of the islands to Spain. In June 1767 the ships joined forces at Rio de Janeiro. The *Etoile* had experienced stormy weather on her voyage into the South Atlantic, and Commerson had suffered from continuous seasickness He had clearly intended to keep a journal of the voyage, but gave up the task even before he reached Montevideo, although a large number of notes in his hand from later periods in the voyage has survived. Commerson made no bones about the fact that he was unhappy both with conditions on board and with most of his companions. He complained that the private trade goods taken on board by La Giraudais and other officers had resulted in intolerable overcrowding, and that they thought it 'an original sin for me to occupy fifteen or twenty cubic feet in a ship that seemed to those unbridled traders to have been solely intended to make them wealthy'.[17] Commerson's journal and notes reveal an unpleasant, sometimes vindictive, attitude. The *Etoile*'s second-in-command, Jean-Louis Caro, was stupid, another officer, Pierre Landais, persecuted him incessantly, while on one occasion he accused the surgeon, François Vivez, of trying to poison him. In summary, the ship was 'that hellish den where hatred, insubordination, bad faith, brigandage, cruelty and all kinds of disorders reign'.[18] As the modern editor of Bougainville's voyage puts it, for all his scientific enthusiasm, Commerson displayed a darker side to his character.[19] Among those with whom Commerson remained on good terms was the quiet, good-natured astronomer Véron, and two young volunteers – Pierre Duclos-Guyot (son of the second-in-command of the *Boudeuse*), who helped Commerson with his journal in the early stages of the voyage, and Charles-Félix-Pierre Fesche. The journals of both men have annotations by Commerson.

His relations with Bougainville seem to have been amicable except for the unexplained order before the ships left Rio that he was to be confined to his cabin for a month. The two men were sometimes together on shore excursions: for example, in the River Plate region, where they inspected bones reputed to be of giants; frustratingly they proved to be from quadrupeds – perhaps mammoths, Bougainville thought. In Patagonia, he noted approvingly that the naturalist gathered a good collection of plants, while a few days later, in the Strait of Magellan, Bougainville wrote that Commerson 'is adding to his botanical treasures and daily finds new plants'.[20] During the expedition's stay in Rio de Janeiro, Jean Baret came across a climbing plant that blossomed with brilliant white flowers inside the red leaves of the bark. Commerson named it after his commanding officer – *Bougainvillea spectabilis* – and specimens that he collected eventually reached France, although not until after his death (Pl. 7). It was first published by Antoine-Laurent de Jussieu in his *Genera plantarum*, and in time it became one of the most ubiquitous and spectacular of all exotic shrubs imported into Europe, enthused over by many who had no knowledge of the man after whom it was named. It was a sign of the generally good relations between the two men that Bougainville offered Commerson (and Véron) accommodation on board the *Boudeuse*. Commerson declined on the grounds that he had more room for his books, instruments and twenty or more storage chests on the *Etoile*,[21] but events later in the voyage showed that there may have been another motive for Commerson's reluctance to move into close proximity with Bougainville. At Buenos Aires, Commerson was able to turn his medical skills to advantage. Charging the equivalent of 150 livres for a consultation, he wrote that if he had stayed in the city he would have made a fortune in less than three years.[22]

The ships' passage through the Strait of Magellan in early 1768 took almost two months. The weather was stormy most of the time, and the occasional encounters with the semi-naked Fuegan inhabitants depressing. On 11 January, Bougainville summed up his reactions. 'Frightful night, deplorable day, rain, squalls, violent WNW winds. What a sequence of bad weather ... One cannot live in this horrible climate which is equally shunned by quadrupeds, birds, and fish and where only a handful of savages live, whose wretchedness has been increased by their dealings with us.'[23] The French had already come across evidence that their discovery voyage had been anticipated by a British one, for on the north coast of the Strait they found sail-cloth with the inscription 'Chatham 1766', and a message dated 1767 carved into a tree. They had been left by the crew of Carteret's *Swallow*, consort vessel to Wallis's *Dolphin* on a circumnavigation during which the two ships became separated. Since Commerson was no longer keeping a personal record, we have to rely on the journals of others to give some idea of his activities. The most revealing is that kept by the surgeon Vivez, who disliked Commerson and claimed that almost from the beginning of the voyage he suspected that the botanist's personal servant, Baret, was a woman. His journal was completed after the voyage so cannot be trusted as an immediate record of events, but it seems credible that some at least of the crew were convinced that – contrary to official regulations, and alarming to those of a superstitious nature – there was a woman in disguise on board. Vivez described Baret as 'small of stature, short and plump, wide-hipped, shoulders in keeping, a prominent chest, a small round head, a freckled complexion, a gentle and clear voice'. Baret put in long, arduous hours botanising with Commerson, who was hindered by a bad leg. As Vivez noted, Baret 'worked like a black. During our period of call at the River Plate, she went to collect plants in the plain, in the mountains two or three leagues away carrying a musket, a game-bag, food supply and paper for the plants.' In the

Strait of Magellan, Vivez wrote, 'these exertions doubled, spending entire days in the forest with snow, rain and ice to seek plants or along the seashore for shells'.[24]

Once in the Pacific Ocean, the ships sailed northwest and then west for two months, passing the Tuamotus but not landing. By the beginning of April, the crews were in poor shape. On the *Etoile*, Vivez noted in his journal that there were 'twenty clear cases of scurvy and the rest of the crew weakened and spiritless, having lived for four months on nothing but salt meat, a bottle of stinking and rotting water, brandy rationed, one meal only with wine, the biscuit beginning to go bad, the refreshments for the sick very scarce, and our food in the officers' mess not much more appealing'.[25] At last, in early April, high land was sighted about fifteen miles distant, and Bougainville wrote:

> During the afternoon we stood in for the land. The whole coast rises in an amphitheatre with deep gullies and high mountains. Part of the land seems to be cultivated, the rest wooded. Along the sea, at the foot of the high country, runs a band of low land, covered with trees and habitations and as a whole this island presents a charming aspect. Over a hundred canoes, of various sizes, but all with outriggers, came around the ships. Several came aboard with demonstrations of friendship, all carrying tree branches, symbols of peace.[26]

They had arrived at Hitia'a on the east coast of Tahiti, the island for long to be regarded by Europeans as the geographical and emotional centre of Polynesia. By chance, Wallis had arrived at Tahiti in June of the previous year, although the French had no knowledge of this – assuming that they were the first European visitors, they gave their own names to the island and the surrounding archipelago. Bougainville stayed only nine days, but during that time his reactions and those of Commerson played a crucial role in establishing the romantic image of Tahiti that, although not unchallenged, has lasted until the present.

The next day the ships found a gap in the reef and anchored near the shore (in what proved to be a dangerous, coral-strewn anchorage where Bougainville was to lose several anchors). Again, canoes came out to trade, and among the exchanges was one famously described by Bougainville in his published account of the voyage. A young girl who came on board went onto the quarterdeck and stood over a hatchway which was open to give air to those who were heaving at the capstan below. 'The girl carelessly dropt a cloth, which covered her, and appeared to the eyes of all beholders, such as Venus showed herself to the Phrygian shepherd, having, indeed, the celestial form of that goddess. Both sailors and soldiers endeavoured to come to the hatch-way; and the capstan was never hove with more alacrity.'[27] This incident has stood as the epitome of the sensuous image of Tahiti, but it probably did not happen quite in the way described by Bougainville. His manuscript daily journal has a much briefer and matter-of-fact entry: 'A young and fine-looking young girl came in one of the canoes, almost naked, who showed her vulva in exchange for small nails.'[28] Other journals describe provocative displays by different Tahitian women who dropped their clothes, while Vivez described an incident on the day the *Boudeuse* left the island that has some similarities to Bougainville's story of the young woman whose nude form greeted the crew on their arrival: 'Bougainville. . . opened two portholes opposite the capstan on the starboard side and had all the others closed. In front of the portholes there were three canoe-loads of women to whom he threw a few pearls and signalled to them to display themselves with all their charms . . . so that our men working the capstan having seen them, driven by agreeable curiosity, pushed with all their might at the capstan in order to pass in front of the open port.' It may well be, as John Dunmore has suggested, that Bougainville fused elements from several different incidents to construct his beguiling story of the Tahitian Venus.[29]

To the French, sailors and officers alike, Tahiti seemed an enchanted island. It was the height of the Season of Plenty and fresh food was abundant – pigs, breadfruit, plantains, coconuts, bananas – while bare-breasted young women, their bodies sweet-smelling from scented oils, flocked around the strangers to the accompaniment of singing and gentle flute music. Bougainville in particular was overwhelmed by the place and its people:

> I cannot leave this fortunate island without praising it once more. Nature has placed it in the finest climate in the world, embellished it with the most attractive scenery, enriched it with all her gifts, filled it with handsome, tall and well-built inhabitants. She herself has dictated its laws,

they follow them in peace and make up what may be the happiest society in the globe. Lawmakers and philosophers, come and see here all that your imagination has not been able to dream up . . . as long as I live I shall celebrate the happy island of Cythere. It is the true Utopia.[30]

Only a few brief notes on Tahiti by Commerson have survived, and they are relatively restrained by comparison with Bougainville's outpourings. He stressed the island's fertility and praised its tranquillity: '[Its inhabitants] seemed to be ruled by a senior chief whom they respect more than fear and then by heads of families . . . Peace and unity seem to reign unbroken among them.' Tahitian men were tall and well built (and skilled thieves), but Commerson was more interested in the appearance of the women, who 'stand comparison with the finest European brunettes except they are less white. They have large eyes, blue or black and level, black eyebrows, a coquettish and seductive glance but bold . . . a fine bosom, nice plump hands and even finer arms.' With the exception of their feet and legs, 'all their body is exquisitely proportioned'.[31] Later in the voyage Commerson would have more, much more, to say about Tahiti and its charms.

For the elders and chiefs, the arrival of the strangers presented both dangers and advantages. Unknown to the French, only a few months earlier the cannon of Wallis's *Dolphin* had slaughtered many warriors in a one-sided sea battle in Matavai Bay, less than a dozen miles away. Understandably, the islanders were thus wary of the firearms that the new arrivals carried, but this did not prevent them from laying down strict conditions for the visitors. When Bougainville presented the chief with eighteen stones to signify that he needed to stay that long to take on board food and water, nine of the stones were returned to him, and in the end the French remained little longer than a week.[32]

Of Commerson's botanical activities we have few details, but the visit to Tahiti resulted in the exposure of Jean Baret as a woman. Bougainville noted that rumours were circulating in both ships: 'His build, his caution in never changing his clothes or carrying out any natural function in the presence of anyone else, the sound of his voice, his beardless chin, and several other indications had given rise to this suspicion.'[33] Vivez's journal has the most intriguing version of what happened when the expedition reached Tahiti, though given his antipathy to Commerson it may not be altogether reliable. The first Tahitian to board the ships was Ahutoru, who returned with the expedition to France and became the first Polynesian to visit Europe. On the *Etoile* he noticed Baret, 'to whom he at once gestured from the bench where he was sitting, making proposals that were unequivocal shouting *Ayenne* which means girl in the local tongue ... Nothing more was needed to confirm to the entire crew the nature of his sex and to convince the reader that her master looked disconcerted.' Vivez then moved on triumphantly to the climax of his story, describing how the next day when Baret went ashore with Commerson, she was surrounded by islanders yelling 'Ayenne' and attempting to carry her off. Only the intervention of an armed officer saved the day. After that, Baret carried two pistols for protection, but seems to have left them behind on a later shore trip on the Melanesian island of New Ireland when, as Vivez described it, 'after having gone botanizing, her master left her ashore to look for shells and the servants who were there drying the washing took advantage of the moment and found in her the concha veneris, the precious shell they had been seeking for so long. This examination greatly mortified her but she became more at her ease, no longer compelled to restrain herself or to stuff herself with cloths.'[34]

Earlier, at Tahiti, when called before Bougainville, who would have been aware of a royal ordinance of 1689 forbidding women on board the king's ships, Baret had concocted a story that absolved Commerson of any blame in the deception. Bougainville's journal entry of 29 May 1768 related how, the day before, Baret, with tears in her eyes, had told him that, determined to sail round the world, she had deceived Commerson into employing her by appearing before him in men's clothing as the ships were about to sail from Rochefort. Full of admiration for her determination, Bougainville decided to take no action, for, as he added drily, 'Her example will hardly be contagious.' The true story is somewhat different. While a young woman in her early twenties, Jeanne Baret had been employed by the Commersons as a household servant in 1762, and after the death of his wife became Commerson's mistress. It is also possible that despite her youth she made a living as a 'herb woman', whose knowledge of the medicinal value of local plants would have been much in demand; this too might help to explain her relationship with Commerson. In December 1766 master and servant arrived together at Rochefort, Jeanne dressed as a man. Whatever their sexual relationship, Baret was an invaluable help to Commerson, who was handicapped by a lame leg in his botanising activities. Bougainville noted how Baret accompanied Commerson 'in all his botanising, [and] carried weapons, food, plant notebooks with a courage and strength which had earned for him from our botanist the title of his beast of burden'. She was no beauty: 'neither ugly nor pretty,' Bougainville wrote, while the less sympathetic Vivez claimed she was 'fairly ugly and un-attractive'. Others shared Bougainville's admiration for Jeanne Baret. A wealthy young aristocratic passenger on the *Boudeuse*, Charles-Othon d'Orange et de Nassau-Siegen, praised her 'for such a bold undertaking, forsaking the peaceful occupations of her sex, she had dared to face the strains, the dangers and all the happenings that morally one can expect in such a navigation. The adventure can, I believe, be included in the history of famous girls.'[35]

This established story of the unmasking of Jeanne Baret, based on the journals of Bougainville and Vivez, has been challenged by her most recent biographer, Glynis Ridley, who believes that both men had reason to conceal the truth. Ridley argues that Bougainville's journal entry of 29 May, describing Baret's confession to him, was fabricated at a later date to conceal the embarrassing fact that the forcible 'examination' on the New Ireland beach that 'greatly mortified her', as described by Vivez, was actually a gang rape of a defenceless woman by a group of Frenchmen from the ship, perhaps including Vivez himself. Ridley also believes that Bougainville knew of Commerson's deception about Baret's identity as early as the expedition's stay at Rio de Janeiro, and that this accounts for his curious decision to confine Commerson to his cabin for a month. Much remains speculation, and efforts to reconcile inconsistencies in the record are hindered by the fact that the journal of La Giraudais, captain of the *Etoile*, who would have seen Commerson and Baret at close quarters throughout the voyage, is missing.[36]

The visit to Tahiti was the emotional climax of the voyage. The journal-keepers were ecstatic in their reactions to an island that Bougainville claimed for France and named Nouvelle-Cythère after Aphrodite's blessed Aegean birthplace of Cythera. Inevitably, the rest of the voyage was more low-key, though some important discoveries were made as the ships sailed west from Tahiti, through the Samoan group and the New Hebrides (Vanuatu), before turning away from the Great Barrier Reef and heading towards the Solomon Islands and New Guinea. Although Bougainville failed to make the connection, his sighting of the Solomon Islands was the first by a European expedition since their discovery by Alvaro de Mendaña two hundred years earlier. Landings were infrequent, and Commerson had few opportunities for botanising. During one scuffle in the New Hebrides he complained that, although he took the same risks as everyone else in the boat party, the officer in charge, Pierre Landais, kept all the weapons captured from the islanders (they would be saleable trophies on the return to Europe). Off New Britain, the *Etoile* was almost wrecked when a coral reef was mistaken for a shoal of fish. Commerson's version of events shows contempt for the officers on watch as well as affording a glimpse of his own preoccupations at the time:

> You should have seen these renowned seafarers in this dangerous situation go pale, at a loss and not knowing which way to turn. The captain [La Giraudais] was the first to lose his head and one can truthfully say that it was the forecastle that saved the quarter-deck, that is to say for those who are not sailors, that the tail ruled the head, if there was a head. As for me, I found the spectacle altogether so singular, I was so engrossed in sorting out who was right about the so-called shoal of fish (and one must agree that the sheets of white coral that passed so quickly under the keel were somewhat like one) that the peril had gone before I realized its extent.[37]

After calling at Batavia (Jakarta), the expedition sailed for the French colony of Isle de France (Mauritius), where Bougainville found a solution to an awkward problem when, at the request of the intendant or civil administrator, Pierre Poivre, he gave permission to Commerson and Baret to leave the expedition. Since departing from Tahiti, Commerson had written an article praising the island and its inhabitants, and this reached France in the spring of 1769 when the ships returned to Rochefort to complete the first French circumnavigation. They brought with them a collection, which was presumably Commerson's, of plants, birds, fishes, shells 'and other curious rarities; some naturalists have even come down from Paris to Rochefort to see them'.[38] Whatever the interest of scientists in the expedition's botanical and zoological specimens, this paled before the enthusiastic reception by a wider public of Commerson's reflections on Tahiti, sent to his friend the astronomer Lefrançais de Lalande at the Paris Académie des Sciences. In November 1769 these were published in the *Mercure de France* as 'Post-Scriptum sur l'isle de la Nouvelle-Cythère ou Tayti'. It was not the first description of Tahiti to appear in print. Earlier that summer *Relation de la découverte . . . d'une isle qu'il a nommé La Nouvelle Cythère* was printed in Paris as a newsletter. The piece was, in the words of its modern editor, 'remarkably objective in its account of Tahiti and its people . . . the island paradise is replete with human sacrifice, warfare, and slavery, and its social order is hierarchical, its political organization tyrannical'.[39] The author is unknown, but he was almost certainly on board the *Boudeuse*, where Ahutoru was informing Bougainville and others about less attractive aspects of life on Tahiti, ranging from infanticide to civil war.

Conversations with Ahutoru were complicated at the best of times, but they led Bougainville to rethink his original views of Tahiti. On arrival, Bougainville described how

> I thought I was transported into the garden of Eden, we crossed a turf, covered with fine fruit-trees, and intersected by little rivulets, which kept up a pleasant coolness in the air ... A numerous people there enjoy the blessings which nature showers liberally down upon them. We found companies of men and women sitting under the shade of their fruit-trees; they all greeted us with signs of friendship ... everywhere we found hospitality, ease, innocent joy, and every appearance of happiness among them.[40]

His later, revised opinions were set out in the second edition of his narrative of the voyage, published in 1772: 'I was mistaken; the distinction of ranks is very great at Taiti and the disproportion very tyrannical. The kings and grandees have power of life and death over their servants and slaves, and I am inclined to think they have the same barbarous prerogatives with regard to the common people.' His final assessment of the expedition's visit was a measured one: 'Our touching at Taiti has been productive of good, and of disagreeable consequences; dangers and

THE FORTUNES AND MISFORTUNES OF PHILIBERT COMMERSON 69

alarms followed all our steps to the very last moments of our stay; yet we considered this country as a friend, which we must love with all his faults.'[41]

Sailing on the *Etoile*, Commerson had no opportunity to listen to and learn from Ahutoru, and his 'Post-Scriptum' represented a rapturous assessment of the paradisiacal society he thought he had seen in Tahiti. The island seemed to provide the followers of Jean-Jacques Rousseau with all the evidence they required about the merits of primitive societies, untainted by the vices of civilisation. Some indication of the tract's tone can be gained from a characteristic passage in which Commerson referred to Tahiti as the true Utopia whose people were 'born under the most beautiful of skies, fed on the fruits of a land that is fertile and requires no cultivation, ruled by the fathers of families rather than by kings, they know no other Gods than Love. Every day is dedicated to it, the entire island is its temple, every woman is its altar, every man its priest.'[42] Following hard on the arrival of Ahutoru in Paris, the appearance of Commerson's piece produced an effusive reaction in the French literary world. There, free love, Tahitian style, was held to be superior to the social constraints of France, and the simple honesty of the islanders was compared favourably with the corruption of French political life. Unaware of Wallis's prior visit to Tahiti, Commerson unwittingly provided his own example of European rivalries when he wrote that he was not allowed to disclose the latitude and longitude of his blessed island.

Meanwhile, as the piece was published, its author was far distant in the Indian Ocean, carrying out his official scientific duties on the orders of Pierre Poivre. A keen botanist, Poivre had spent years in his youth travelling through the Eastern Seas in the hope of collecting nutmeg and clove seedlings for transplanting to Isle de France. His efforts had failed, but the arrival of Bougainville's ship with Commerson on board offered an unexpected opportunity to renew the quest. Commerson's official task was to report on the natural resources of Isle de France, Bourbon (Réunion) and Madagascar, paying special attention to plants that might have medicinal value, but secret instructions ordered him to cast his net wider and identify spice plants in the Dutch-controlled Moluccas that might be brought to Mauritius.[43] It was the beginning of a busy and arduous four years for Commerson and Baret, who were helped by two draughtsmen provided by Poivre: Paul-Philippe Sauguin de Jossigny and Poivre's godson, Pierre Sonnerat. Together Jossigny and Sonnerat began to catalogue and illustrate the specimens collected by Commerson during his time on the *Etoile*, while he was investigating the natural products of Isle de France and the neighbouring island of Bourbon. In 1770, Commerson ventured farther afield when he and Jossigny sailed to Madagascar, where the French had established a foothold at Fort Dauphin on the south coast of the island. It was there that Commerson acknowledged Baret's help by naming a plant genus, *Baretia*, after her.[44] Commerson's visit was curtailed by ill health, but even so his party brought back to Isle de France a collection described as including 'more than a thousand unknown plants, together with lapis luzuli and other precious stones'. Madagascar was, he told a colleague, a veritable treasure house of nature.[45]

Much of 1771 was spent botanising on Bourbon, where Commerson found time and energy to join a party that made the first recorded ascent of the volcano 'Le Grand Brûlé'. Commerson's return to Isle de France saw a further decline in his health, and the loss of his patron when Poivre was recalled to France. There was an air of sad resignation about one of his last letters when he wrote: 'My plants, my beloved plants, have consoled me for everything.'[46] Commerson died on 13 March 1773, aged forty-five, only a few weeks after his last botanical excursion and before the news could reach him from Paris that he had been elected to the prestigious Académie des Sciences. While admitting Commerson's abilities as a botanist, the acting intendant had no high opinion of him as a person, reporting to the ministry in Paris that he was widely regarded as 'an evil man capable of the blackest ingratitude'.[47] After Commerson's death Jeanne Baret opened a tavern on Isle de France, and in 1774 married a former army sergeant with whom she returned to France, thus becoming the first woman known to have completed a circumnavigation. This remarkable woman appears for the last time in the official records in 1785 when she is described in a letter signed by the minister of the marine as 'one who met the burdens and dangers of her work with the utmost courage', and was awarded a pension of 200 livres.[48]

The fate of Commerson's collections was wretchedly complicated. There had already been losses on the voyage, when specimens sent back to France from Brazil and La Plata failed to arrive. After his death there were squabbles over ownership between the administration on Isle de France and Commerson's executors. He seems to have had a premonition of challenges to come, for when he sent plants and seeds to his patron Louis-Guillaume Lemonnier at the Jardin du Roi in Paris he accompanied them with a plea, 'Have my priority of date conserved for those things that are truly new', for 'in the Republic of Letters, as in the beehive, the lazy, unoccupied drones depend solely upon the active and industrious bees'.[49] Matters were made more uncertain by the fact that the specimens and Commerson's notes were in no sort of order. A medical colleague on Isle de France was asked to look after and list the collection, but reported that he had only been able to protect it against accidents and thefts, and suggested that Jossigny should be given the task of arranging and cataloguing. Before Jossigny was able to make much progress on this, the decision was taken to ship the whole collection, in thirty-four cases, to France. There it arrived at the Jardin du Roi, where Buffon held sway over a much enlarged establishment and received from Jossigny many of his natural-history drawings of Commerson's specimens.

It was estimated that the plants sent to Paris included three thousand species new to European science, and in the following years, in a piecemeal and haphazard way, some of Commerson's collection began to filter through into the scholarly world, though often without proper attributions. Buffon was mainly interested in Commerson's notes on birds, and used some of them in his *Histoire des oiseaux*, but did little with the rest of the collection. Sonnerat's superb paintings of birds perhaps owed more to Commerson than he acknowledged.[50] Much the same kind of dispersal happened to Commerson's fish specimens, discovered in Buffon's attic after his death, when they were handed over to Bernard-Germain-Etienne Lacépède at the Cabinet d'Histoire Naturelle, to appear with due acknowledgement in the third volume of his *Histoire naturelle des poissons*. Antoine-Laurent de Jussieu, whose hugely influential *Genera plantarum* appeared in 1789, published thirty-seven of Commerson's plant specimens with their attributions, including the beautiful *Bougainvillea*. The anatomist Georges Cuvier recorded his astonishment that 'one man should have been able to do so much in so short a time in a tropical climate',[51] but no comprehensive description exists of Commerson's work either on Bougainville's voyage or during his years on Isle de France. The botanist is best remembered for aspects of his life that were marginal to his considerable scientific achievements – his ill-advised eulogy of Tahiti and his relationship with Jeanne Baret. As John Dunmore has observed, Commerson preceded Captain Cook's naturalists in the Pacific and 'had his work been published, or even substantially outlined in Bougainville's *Voyage*, it would have ensured a distinguished place both for him and for the whole expedition in the annals of science.'[52]

CHAPTER 4

'No people ever went to sea better fitted out for the purposes of Natural History'
Joseph Banks and Daniel Solander

JAMES COOK'S FIRST Pacific expedition was one of the most improbable of all of Europe's great discovery voyages. Commanded by a junior lieutenant sailing in a converted collier, it located the position of Polynesian island groups, charted the coasts of New Zealand, established the eastern littoral of the Australian continent and confirmed the existence of Torres Strait. So wide-reaching were its achievements that it is easy to forget that in origin it was not a voyage of geographical discovery at all, but a scientific expedition sponsored by the Royal Society with the sole aim of observing the transit of Venus in 1769 from the newly located Pacific island of Tahiti. As such, it was different in intention from the two earlier Pacific voyages of George III's reign, those of Commodore John Byron and Captain Samuel Wallis, which had clear strategic and commercial objectives and carried no scientists on board. In its proposed form, the expedition attracted the attention of Joseph Banks, a wealthy young landowner with a passion for natural history. It was evidence both of his enthusiasm for botany and of his social status that when, as an undergraduate at Oxford, Banks discovered that the Sherardian professor of botany had no interest in teaching the subject he paid a naturalist from Cambridge to come and give lectures. But Banks was no study-bound scholar. After leaving Oxford (without taking a degree), he took passage across the Atlantic on HMS *Niger* in 1766, and spent several months botanising on the coasts of Newfoundland and Labrador.[1] On the return voyage he was given a foretaste of the problems of safeguarding plant collections on board ship when heavy seas smashed into his cabin, destroyed his collection of seeds and live plants, and left only his dried plant specimens.

Elected a Fellow of the Royal Society at the age of twenty-three shortly after his return from the voyage, Banks next contemplated a visit to Uppsala to meet Linnaeus, but then heard of an even more alluring project – a voyage round the world. The Royal Society had already arranged for a civilian astronomer, Charles Green, to go on the voyage. Working with Lieutenant James Cook, the commander of a converted collier (renamed the *Endeavour*) who had just spent five summers surveying the coasts of Newfoundland, Green would be responsible for the all-important observations at Tahiti. In April 1768, the same month that Cook was appointed to the *Endeavour*, Banks wrote to a correspondent to explain that by joining the expedition he would have 'a finer opportunity for the Exercise of my Poor Abilities than Ever man before had', since in the South Seas 'almost Every Production of Nature is here very different from what we see at this End of the Globe'.[2] There the matter seemed to rest until, in June, the Council of the Royal Society informed the Admiralty that Joseph Banks was 'desirous' of joining the voyage. He would not be alone, for he had arranged to be accompanied by no fewer than eight assistants. They included the botanist Daniel Solander, a favoured student of Linnaeus who was working at the newly established British Museum, Herman Diedrich Spöring, another Linnaeus pupil who was to help Banks with secretarial duties, and the artists Sydney Parkinson and Alexander Buchan. In addition, there were four 'servants', who would help with the more menial tasks of collecting and preserving the natural history specimens. With the party went some twenty tons of luggage and equipment, including twenty storage chests, fourteen cases, some of which were partitioned to hold glass-stoppered specimen bottles (of which there may have been two hundred or more), and a range of instruments including telescopes and microscopes.[3]

The taking of two artists, one to draw natural-history specimens, the other to record landscapes and persons, was one of the most important of Banks's contributions to the voyage. De Brosses's recommendation that artists should accompany naturalists on discovery voyages had been taken seriously. The presence of Parkinson and Buchan suggests that from the beginning Banks had in mind an illustrated publication of the scientific results of the voyage.[4] For his part, Solander, in his application to the Trustees of the British Museum for leave of absence, hoped that his presence on the voyage would be 'of great utility to the British Museum, in collecting Natural Curiosities ... from Countries that perhaps never before were investigated by any curious men'.[5] Collecting would not be Solander's only value to the expedition; equally helpful would be his mastery of the Linnaean system of classification and nomenclature. As Richard Pulteney wrote some years later in his *Progress of Botany*, '[Solander's] perfect acquaintance with the whole scheme enabled him to explain its minutest parts, and elucidate all those obscurities with which, on a superficial view, it was thought to be enveloped.'[6] Added to this, Solander was not only able 'to name and classify but to formulate concise diagnoses based on essential features with the correct technology; to know which characters should be used to define genera, and which to define species'.[7]

The costs of this retinue, to be accommodated somehow on an already overcrowded vessel, were borne by Banks, described by the Royal Society in its letter to the Admiralty as, first, 'a Gentleman of large fortune', and only secondly as 'well versed in natural history'.[8] For a time the project was in the balance, for the First Lord of the Admiralty, Edward Hawke, curtly informed Banks that although he was 'very welcome to go ... we cannot find room for people skilld in Botany & drawers of Plants'. Only the quiet intervention of Philip Stephens, secretary to the Admiralty, and Banks's offer to cover the costs of his party saved the day. After that, Banks wrote, he kept well away from Hawke.[9] His only regret as he busied himself with a venture that would far outdo the conventional Grand Tour of Europe – 'Every blockhead does that; my Grand Tour shall be one round the whole Globe'[10] – was that it denied him 'an opportunity of Paying a visit to our Master Linnaeus & Profiting by his Lectures before he dies'.[11] Linnaeus had to be satisfied with news of the voyage from the London merchant and naturalist John Ellis, who informed him: 'No people ever went to sea better fitted out for the purpose of Natural History, nor more elegantly. They have got a fine library of Natural History; they have all sorts of machines for catching and preserving insects; all kinds of nets, trawls, drags and hooks for coral fishing; they even have a curious contrivance of a telescope, by which, put into the water, you can see the bottom to a great depth.'[12]

Before the *Endeavour* sailed, Cook was given additional 'secret' instructions. After the stay at Tahiti he was to sail in search of land that some of Wallis's men thought they had seen to the south during the *Dolphin*'s visit to the island in June 1767, and that they hoped might be part of the great southern continent, *Terra Australis Incognita*. These orders drastically enlarged the original intention of the voyage, and opened the way to Cook's future as an explorer. Just as the Admiralty could not have anticipated this, so the Royal Society could not have foreseen the importance of the close working relationship that developed between Banks and Cook. The additional instructions made clear the duties expected of Banks and his team, as Cook was ordered to 'observe the Nature of the Soil, and the Products thereof' and to bring home 'such Specimens of the Seeds of the Trees, Fruits and Grains as you may be able to collect'.[13] Wallis (although not Byron) had been given a similar order, but not the skilled helpers necessary to make the task feasible. In more ways than one, then, the form that the *Endeavour* voyage took was unexpected and fortuitous. A few days before sailing Banks once more revealed his excitement at the prospect ahead when he told a correspondent that 'the South Sea at least has never been visited by any man of Science'.[14]

From the first days of the voyage Banks was assiduous in making natural-history observations, and as early as his second journal entry he was concerned with the Linnaean classification of dolphins spotted from the ship. On the *Endeavour* he must have cut a conspicuous figure in his civilian dress: shooting down birds from the quarterdeck at one moment, casting nets and lines overboard to catch strange fish the next. His curiosity seemed limitless, his energy boundless. The personal cabins of Banks and Solander – luxuries on a small vessel – were only six feet square, little more than cupboards, but they and their associates were allowed the use of the captain's more spacious 'great cabin' at the stern of the vessel for their scientific work. Solander explained how they read in the morning, and then from mid-afternoon until dark sat 'by the great table with our draughtsman [Parkinson] opposite and showed him in what way to make his drawings, and ourselves made rapid descriptions of all the details of natural history, while our specimens were still fresh'.[15] The outward voyage offered Banks and Solander ample opportunity to catch and examine 'the inhabitants of the deep', and they found it 'a very incouraging circumstance to hope that so large a feild [*sic*] of natural history has remained almost untrod ... and that we may be able ... to add considerable Light to the science which we so eagerly pursue'.[16] Once ashore, the two men spent their time collecting specimens. Their single-mindedness is illustrated in one of Banks's journal entries grumbling that at Madeira they had to remain on board ship for a ceremonial visit by the governor of the island, 'so that unsought honour lost us very near the whole day'.[17] Even so, they collected 'above 300 Species of Plants 200 of Insects & about twenty of fishes many of all three kinds such as had not before been described'.[18] Since Solander, unlike Banks, kept no journal, it is easy to underestimate his role in the expedition's scientific activities; but there is no doubt that for Banks, the enthusiastic amateur, it was a marked advantage to have as his companion a botanist trained in the new system of classification by Linnaeus himself. Of Solander's Latin descriptions of the plants collected, the nineteenth-century botanist Joseph Hooker wrote that they 'have never been surpassed for fullness, terseness and accuracy'.[19]

There was no ceremonial welcome, time-consuming or otherwise, at Rio de Janeiro, where the viceroy, suspicious about the objectives and status of the expedition, refused to allow the naturalists ashore. In a letter home to a friend, Banks made his feelings clear in characteristic fashion: 'You have heard of the French man laying swaddled in linen between two of his Mistresses both naked using Every possible means to excite desire but you have never heard of a tantalizd wretch who has born [*sic*] his situation with less patience than I have done mine I have cursd swore ravd stampd & wrote memorials to no purpose in the world they only laugh at me.'[20] Through surreptitious landings and bribes to local inhabitants to bring on board plants under the pretence that they were for mealtime salads, Banks and Solander managed 'by fair means and foul' to obtain about three hundred specimens. Parkinson sketched thirty-seven plants, including the attractive *Bougainvillea spectabilis*; but overall the stay of almost three weeks was one of frustration for the naturalists. It was with feelings of relief that they finally left 'these illiterate impolite gentry' at the beginning of December 1768.

During the long weeks at sea an unexpected bonus was the interest shown by some of the sailors in catching and identifying fish. On the boat excursions, Solander wrote, 'they could remember what we had shewn them, and consequently, could look out for new ones'. That some of the sailors proved 'good philosophers' and 'very useful hands' was in its own way a tribute to Banks and Solander as their mentors, and perhaps to the depth of the naturalists' pockets.[21] During calm weather in tropical waters Banks and his helpers scooped up huge numbers of molluscs, and caught new species of crabs and jellyfish. Solander's biographer remarks that 'these were pioneering nights in the history of marine biology', and quotes a letter to the Royal Society from Solander to the effect that 'we have been very fortunate in finding a great many Sea Productions, that I hope will be better cleared up by us, than they have been by any one before', before the author added, in an aside that showed the importance of collaboration with the artists, 'especially as Mr Banks's People have had an opportunity of drawing them when fresh and alive'.[22]

In mid-January the *Endeavour* reached the Bay of Good Success in Tierra del Fuego. Here Banks's determination to venture inland led to disaster. With weather 'vastly fine much like a sunshiny day in May', Banks and his companions climbed high in search of alpine plants. As the weather closed in, with frequent flurries of snow, Buchan had an epileptic fit (and was to die at Tahiti three months later), and without adequate clothing or provisions the party spent a miserable night during which Banks's two black servants froze to death. It was typical of Banks that when he managed to reach the *Endeavour* the next morning, and the other shivering survivors were sent to bed for rest and warmth, he took a boat out to catch more fish specimens.[23]

During this early stage of the voyage Cook appeared sceptical of the worth of Banks's scientific activities. On 15 January 1769 he noted that, while the *Endeavour* was in a potentially risky situation in the Strait of Le Maire, Banks nevertheless was 'very desireous' of being put ashore, but when he returned on board he brought with him only 'several Plants Flowers &ca most of them unknown in Europe and in that alone consisted their whole Value'.[24] However, when Banks went ashore five days later, he found and described two plants whose importance Cook would have appreciated – scurvy grass (*Cardamine glacialis*) and wild celery (*Apium prostratum*). Because of their known antiscorbutic properties, these were immediately put in the crew's soup. In the same journal entry Banks seemed almost to be answering Cook's criticism of 15 January when he recorded his delight at tramping across this bleakest of all landscapes: 'Probably No botanist has ever enjoyed more pleasure in the contemplation of his Favourite pursuit than Dr Solander and myself among these plants so intirely different from any before describd that we are never tird with wondering at the infinite variety of Creation.'[25] Despite all the hardships and the barren terrain, they managed to collect 104 species of flowering plants as well as thirty-four mosses.

On the long haul from Cape Horn across the South Pacific, Banks began to suffer from scurvy despite following Cook's instructions to the crew to drink a pint of malt wort each day. The steps he took to arrest its progress formed, in retrospect, one of the most intriguing episodes in the whole voyage. As the *Endeavour* neared its destination of the Society Islands in April 1769, Banks noticed, first, an inflammation of his throat, followed by a swelling of his gums and the appearance of pimples on the inside of his mouth which threatened to become ulcers. He had on board a supply of lemon juice recommended by Nathaniel Hulme, author of *A Propasal for Preventing the Scury* (1768), and he took six ounces a day: 'The effect of this was surprising, in less than a week my gums became as firm as ever and at this time I am troubled with nothing but a few pimples on my face.' A letter to Banks from a merchant, Richard Samuel, written before the expedition left England, shows Banks's interest in the subject, and perhaps his worries about what lay ahead. Samuel told Banks, 'I have enquired of several Sea Faring persons for a remedy against the Scurvy – none could inform me though most agreed they have received the greatest benefit from lemon juice.'[26] With these informed recommendations before him, Banks took a substantial supply of lemon and orange juice, and used it to good effect. Given that Banks and Cook were living at close quarters, and that they read each other's journals, it seems odd, to say the least, that Cook appears to have ignored Banks's demonstration of the efficacy of lemon juice. He continued to put his faith in malt wort, sauerkraut, fresh food and a clean, well-aired ship, and was dismissive of lemon juice as an antiscorbutic. In a letter to Sir John Pringle, president of the Royal Society, after his second voyage, Cook wrote: 'the dearness of the rob [concentrate] of lemons and oranges will hinder them from being furnished in large quantities, but I do not think this so necessary; for, though they may assist other things, I have no great opinion of them alone.'[27]

On 13 April the *Endeavour* reached Tahiti, and the nature of Banks's journal changed. He was still collecting, but the evidence provided by notebooks kept on the voyage (now in the Natural History Museum in London) indicates that this was now primarily Solander's task, helped by Spöring and Parkinson. The latter painted no fewer than 113 plants from the Society Islands, as well as sketching another thirteen. Banks described how the artist's work was hampered by dense swarms of flies, 'for they not only covered his subjects so that no part of its surface could be seen, but even ate the colour off the paper as fast as he could lay it on'. This remark was printed in the published account of the voyage, incorporating the journals of both Cook and Banks, edited by Dr John Hawkesworth, but without the next passage in Banks's journal which described how, when a flytrap filled with a mixture of tar and molasses failed to deter the insects and was left outside, one of the islanders took 'some of the mixture into his hand, I saw and was curious to know for wht use it was intended, the gentleman had a large sore upon his back-side to which this clammy liniament was applyd, but with wht success I never took the pains to enquire'.[28] Whatever the practical challenges, the combination of botanists collecting plants and of artists painting them on the spot set a new standard for naturalists in the Pacific. The team's collections offer an invaluable picture of the plant life of Tahiti just before the full impact of the European arrival, and furnish impor-tant support for the thesis of the peopling of Polynesia from southeast Asia. Banks and Solander paid particular attention to the island's culti-vated plants, describing breadfruit, bananas, yams and sweet potatoes among others. Parkinson's comprehensive fourteen-page list was headed 'Plants of Use for Food, Medicine, &c in Otaheite', and included long descriptions of the palm and the breadfruit tree.[29] Banks's eulogy of breadfruit became celebrated: 'In the article of food these happy people may almost be said to be exempt from the curse of our forefather; scarcely can it be said that they earn their bread with the sweat of their brow when their cheifest [*sic*] sustenance Bread fruit is procured with no more trouble than that of climbing a tree and pulling it down.'[30]

Less poetically, Solander described it as 'one of the most usefull [*sic*] vegetables in the world'.[31]

Above all, Banks was fascinated by the islanders, and his journal is an ethnological treasure-house of observations on their appearance and customs. His interest was personal (not to say sexual) as well as scholarly. On only his second day in Tahiti he wrote: 'I espied among the common croud a very pretty girl with a fire in her eyes that I had not before seen in the country.'[32] More than any other member of the ship's company he spent time ashore, mixing with the islanders as he made a valiant attempt to learn their language, joining in their festivities and ceremonies, finding willing sexual partners among their young women and recording at length his experiences. He was, his editor has written, 'the founder of Pacific ethnology',[33] and his observations on the 'Manners & customs of S. Sea Islands' run to more than fifty pages in the modern edition of his journal. Both in the length of time he spent in Tahiti and in his conscientiousness in recording events and impressions, Banks's stay outdid in significance that of Commerson on Bougainville's voyage. On Banks's excursions ashore, Cook was often his companion, and each benefited from the insights of the other, as their journals show with their mutual borrowings and acknowledgements. Solander too, closer in age to Cook than Banks, was another important influence. He compiled a Tahitian vocabulary which included the names that the islanders had given to members of the ship's company: 'Tute' for Cook would soon be known throughout Polynesia. Towards the end of the voyage Cook recorded his appreciation of the work of both Banks and Solander when he reported to the Admiralty that their 'many Valuable discoveries ... in Natural History and other things usefull to the learn'd World cannot fail of contributing very much to the Success of the Voyage'.[34] The sentence was as much a tribute to Cook as to the naturalists.

The mutual understanding between Banks and Cook was the more unexpected, not only because of their different characters and backgrounds – the one a wealthy landowner, young and exuberant, the other an austere naval officer risen from humble origins and fifteen years older – but because both were more used to giving than receiving commands. In what turned out to be one of the most important moments on the voyage, Cook deferred to Banks, if reluctantly, when he

agreed to the naturalist's request on Tahiti that they should take on
board for the rest of the voyage Tupaia, an imposing warrior, navigator
and high priest from nearby Raiatea. Banks chuckled about his minor
triumph when he noted of Tupaia, 'I do not know why I may not keep
him [in England] as a curiosity as some of my neighbours do lions and
tygers'.[35] But in less patronising moments he, like Cook, appreciated
Tupaia's extraordinary skills as a navigator, chartmaker and mediator.

The necessary astronomical observations at Tahiti were made – that
they were disappointing was no fault of Green and Cook – and the
Endeavour put to sea again, first charting more of the Society Islands,
and then heading south to search for the unknown continent. At sea
Banks once more became the intent botanist:

> Now do I wish our freinds in England could by the assistance of some
> magical spying glass take a peep at our situation: Dr Solander setts at the
> Cabbin table describing, myself at my Bureau Journalizing, between us
> hangs a large bunch of sea weed, upon the table lays the wood and barna-
> cles; they would see that notwithstanding our different occupations our
> lips move very often, and without being conjurors might guess that we
> were talking about what we should see upon the land [New Zealand]
> which there is no doubt we shall see very soon.[36]

By now Banks and his team had developed a standard routine for
describing and drawing the specimens brought on board. As far as
possible the work was done while the specimens were still fresh; the
plants were then pressed and dried between sheets of papers. For this
purpose Banks and Solander used a supply of unbound sheets of Milton's
Paradise Lost which they had bought in bulk from a London printer.
Little did they imagine how appropriate that title would seem to future
generations dismayed by the impact of European firearms and venereal
diseases on Polynesia.

When the *Endeavour* reached the unknown shores of New Zealand
in October 1769 the advance party that went ashore on the second day
at Tuuranga-nui harbour to meet the 'Indians' included Cook, Banks,
Solander, Green, the surgeon Monkhouse and Tupaia.[37] The composi-
tion of the party showed that negotiation and investigation were the
main objectives, but all were armed, and when a Maori warrior seized

Green's sword Banks opened fire, though only with small shot, designed to sting and wound rather than kill. He was followed by Monkhouse, whose musket was loaded with ball and who shot the man dead. Later in the day there were further confrontations in which four Maori were killed and others wounded. In their journals both Cook and Banks agonised over the affair. Cook claimed self-defence, writing that he could not 'stand still and suffer either my self or those that were with me to be knocked on the head'.[38] Banks was clearly shaken by the deaths: 'Thus ended the most disagreeable day My life has yet seen, black be the mark for it and heaven send that such may never return to embitter future reflection.'[39]

The events at Poverty Bay, as Cook misleadingly named this first landing place, set the pattern for the expedition's survey of the New Zealand coastline. Tupaia proved an invaluable mediator, for he found 'that the language of the people was so like his own that he could tolerably well understand them and they him', but there was little of the easy-going relationship that had characterised the stay in Tahiti, and few of the extended excursions that had been possible in the Society Islands. As the *Endeavour* sailed from Poverty Bay, Banks noted that he had 'not above 40 species of Plants in our boxes, which is not to be wonderd at as we were so little ashore and always upon the same spot', although Solander's list of New Zealand plants indicates that in fact sixty-one were collected in the bay.[40] The botanising continued, not always in favourable conditions. At the Bay of Islands, in drenching rain, 'Boats went ashore ... I do not know what tempted Dr Solander and myself to go there where we almost knew nothing was to be got but wet skins.'[41] Even when the local inhabitants were friendly, 'the immense thickness of the woods which are almost rendered impassable by climbing plants intangling in every way has not a little retarded us'. The farthest inland Cook and Banks reached was a dozen miles or so up the River Thames, but they landed only once. 'The banks of the river', Banks wrote, 'were completely cloathed with the finest timber my Eyes ever beheld', and he had seen enough to recommend the area as the best site for future settlement.[42] At Anaura Bay, when Banks, Solander and their helpers returned to the beach after a full day's botanising, they found that the ship's boats were too busy carrying water casks to take them on board. Instead, they persuaded Maori nearby to ferry them out to the ship by canoe, but were so awkward that their flimsy craft overturned in the surf, leaving them 'well sousd'. A second attempt was more successful, and Banks recorded his appreciation of 'our Indian freinds [sic] who would the second time undertake to carry off such Clumsy fellows'.[43] Sometimes they were accompanied by Spöring, 'a grave thinking man', who spent his time sketching; but the main work of drawing the natural-history specimens was left to the hard-working Parkinson on the ship. Tolaga Bay (Uawa) provided plenty of material for him, for Banks collected there 158 plant specimens, many of them previously unknown, while at Mercury Bay (Te Whanganui-o-Hei) the botanists found 214 new plants, a not surprising haul, Banks wrote, 'in a countrey so totaly new'.[44] He was always on the lookout for 'useful' plants, and he paid particular attention to New Zealand flax, *Phormium tenax* – 'of a strength so superior to hemp as scarce to bear a comparison with it . . . So useful a plant would doub[t]less be a great acquisition to England, especialy as one might hope that it would thrive there with little trouble.'[45]

Banks again spent much time observing and noting the customs of the local inhabitants. On one occasion, as he and Solander were returning to the ship with 'our treasure of plants, birds &c.', they were stopped by an old man anxious to demonstrate his warlike skills. With his lance he furiously attacked a stick, representing his opponent, and having run his mimic enemy through the body, belaboured him with blows from a stone club, 'any one of which would probably have split most sculls; from hence I should be led to conclude that they give no quarter'.[46] Both Cook and Banks were struck by the courage and warlike demeanour of the Maori. As the survey continued, the journal-keepers became obsessed by the question of cannibalism, and after some weeks of disbelief finally found conclusive evidence of it at Queen Charlotte Sound (Totara-nui), referred to in many of their journals as Cannibal Harbour.

By March 1770, Cook had completed his magnificent survey of New Zealand, proving that its twin islands were not part of any continental landmass. From there the *Endeavour* sailed towards that region of mystery, the uncharted eastern parts of New Holland. As the ship crossed the Tasman Sea, Banks continued his observations. His description of the Portuguese man-of-war is a good example of his eye for detail:

The body of it Consists of a bladder on the upper side of which is fixd a
kind of Sail which he erects or depresses at pleasure . . . Under the bladder
hang down two kinds of strings, one smooth and transparent which are
harmless, the other full of small round knobs . . . With these latter however
he does his mischief, stinging or burning as it is calld if touched by any
substance: they immediately exert millions of exceedingly fine white
threads about a line in length which peirce the skin and adhere to it giving
very acute pain.[47]

The coast of southeast Australia was sighted on 19 April, but it
would be nine days before Cook managed to land. Nothing seen so far
had quite prepared the discoverers for New Holland and its Aboriginal
inhabitants. At the spot soon to be known as Botany Bay, the first
contact was made with the Gweagal clan of the Tharawa tribe, two of
whom, armed only with spears and throwing-sticks, bravely resisted the
landing. It took the firing of several muskets loaded with small shot
before they retreated, and Cook, Banks, Solander and Tupaia were able
to investigate a half-dozen bark huts, a huddle of frightened children
and a few flimsy canoes. The log of Richard Pickersgill, master's mate
on the *Endeavour*, seemed to sum up the crew's general impression:
'The people have nothing to cover themselves, but go quite naked, men
and women, and, in short, are the most wretched set I ever beheld or
heard of.'[48] At the end of their cruise along Australia's eastern shores
Banks would give a fuller and more sympathetic account of the
Aborigines, but in the meantime he was busy collecting. In the days
following the first landing so many new plants were found that in his
journal Cook changed his name for the place from the unappealing
Sting Ray Harbour to Botany Bay, by way of a compliment to the
Endeavour's naturalists. He could not have anticipated how ominous
the name would sound to future generations that knew it primarily as a
convict settlement.

During the stay Banks gave some indication of the labour involved
in drying and preserving their specimens:

Our collection of Plants was now grown so immensely large that it was
necessary that some extraordinary care should be taken of them least they
should spoil in the books. I therefore devoted this day to that business and

carried all the drying paper, near 200 Quires of which the larger part was full, ashore and spreading them upon a sail in the sun, kept them exposd the whole day, often turning them and sometimes turning the Quires in which were plants inside out. By this means they came on board at night in very good condition.[49]

For six days after sailing from Botany Bay on 6 May they were hard at work coping with their haul, and not until the evening of 12 May was Banks able to write that 'we finishd Drawing the plants got in the last harbour, which had been kept fresh till this time by means of tin chests and wet cloths'.[50] The tin chests referred to by Banks seem to have been glass containers encased in iron and partially filled with water that he and Solander took ashore with them on their collecting expeditions.[51] In two weeks Parkinson had made no fewer than ninety-four drawings, although not all would have been of complete plants. As he struggled to cope with the number of specimens brought on board by the botanists, Parkinson at first made complete coloured drawings of live plants of special interest, but for others he had time only for a quick sketch before the plant died. He was aware that once at sea there would be no question of using the precious drinking water on board the ship to keep the plants alive for any length of time. Parkinson's initial sketch, together with his notes and a drawing of the plant's individual flowers or fruit, enabled him to complete his painting at a later date.[52]

The voyage north along the (Queensland) coast was marked by a near-disastrous encounter with the Great Barrier Reef. As the crew struggled to float the *Endeavour* off the jagged coral, Banks wrote, with some justice, that 'fear of Death now stard us in the face'.[53] The fortunate discovery of a harbour near the mouth of the Endeavour River enabled repairs to be made, while Banks tried to save plants stored in the bread room that had been soaked when the sea flooded in. Once that was done, he and Solander resumed their botanising activities, collecting more than two hundred plant species. Most intriguing were the sightings of a grey animal 'that instead of Going upon all fours ... went only upon two legs, making vast bounds'.[54] On 14 July one was killed, and Banks gave it the Aboriginal name of 'kanguru'. In his journal he struggled to describe the strange creature: 'To compare it to any European animal would be impossible as it has not the least resemblance of any one I have seen. Its fore legs are extremely short and of no use to it in walking, its hind again as disproportionately long; with these it hops 7 or 8 feet at each hop.'[55] The kangaroo was quickly cooked, and one is reminded of Banks's boast in later years: 'I believe I have eaten my way into the Animal Kingdom farther than any other man.'[56]

As the *Endeavour* put to sea again Banks wrote a long description of the region (named New South Wales by Cook), its products and its inhabitants. Particularly striking were his comments on the Aborigines. During the six-week stay at the Endeavour River, he had seen a good deal of the men (though not the women) from the Guugu-Yimidhirr tribe, and although contacts were intermittent and sometimes fractious he gradually developed an appreciation of the harmony between these people and their natural environment. He had moved a long way from Dampier's strictures and even from his own original assessment of the inhabitants of Botany Bay – 'hairy, naked &c ... rank cowards' – when, in August 1770, he described what he perceived to be the Aboriginal life-style. He was not uncritical: they had neither clothing nor dwellings, and seemed ignorant of cultivation. But the fact that they carried 'their worldly treasures' in a small bag led to a surprising conclusion: 'Thus live these I had almost said happy people, content with little nay almost nothing, far enough removed from the anxieties attending upon riches, or even the possession of what we Europeans call common necessaries.' Even more unexpected was Cook's emphatic agreement with this as he related the Aboriginal lifestyle to a benign climate and an adequate supply of food, and concluded, 'in reality they are far more happier than we Europeans.'[57]

After passing through the Torres Strait, the voyage home was a miserable anticlimax. Not a single member of the crew died from scurvy, but almost a third perished from dysentery and other diseases picked up during a stay at Batavia (Jakarta). Banks, Solander and Spöring became seriously ill, and Tupaia, Monkhouse and the invaluable Parkinson all died. Despite his illness, Banks provided a good description of Batavia in his journal, while Cook sent a copy of his journal home to the Admiralty by way of a Dutch ship. Whatever happened to the *Endeavour*, at least the discoveries made by the expedition would not be lost. When the ship finally reached England in July 1771, Cook handed over his complete journals, together with charts and illustrations, to the Admiralty, and with a letter that was overmodest in hoping that they would 'convey a Tolerable knowledge of the places they are intended to illustrate, & that the discoveries we have made, tho' not great, will Apologize for the length of the Voyage'.[58]

To the modern reader, conscious of Cook's achievements on his first Pacific voyage in terms of geography, astronomy, shipboard health and much else, there is something puzzling and disconcerting about the lack of public interest accorded to him personally on the *Endeavour*'s return. In the London newspapers of July, the voyage belonged to 'Mr Banks and Dr Solander'. It was they who had sailed round the world, had 'met with great hardships, and were often in danger of being shipwrecked', and, apart from the stay at St George's Island (Tahiti), had 'touched at near forty other undiscovered Islands ... and have brought over with them above a thousand different Species of Plants, none of which were ever known in Europe before'. As a final accolade, the *Public Advertiser* of 7 August noted that 'Mr. Banks and Dr. Solander have made more curious Discoveries in the way of Astronomy, and Natural History, than at any one Time have been presented to the learned World for these fifty years past'. After being formally presented to George III in early August, Banks and Solander subsequently met the king several times in Kew, where they brought plants from the voyage for the royal gardens.

Not until 16 August was there any mention of the humble lieutenant who had commanded the *Endeavour*. Then the *London Evening Post* reported that he had been presented to the king, and promoted to commander, but the wording – 'Lieutenant Cook, of the Royal Navy, who sailed round the Globe with Messrs. Solander, Banks etc.' – gave the impression that somehow they had been in charge and that he had been a passenger. For Cook, more important issues were his warm reception by the Admiralty, his promotion and his appointment to command a second expedition to the Pacific to search for the fabled southern continent. Even then, the popular press could not believe that the next voyage would be his. As the *Westminster Journal* reported at the end of August: 'The celebrated Mr. Banks will shortly make another voyage to St. George's Island, in the South Seas, and it is said, that the Government will allow him three ships, with men, arms and provisions, in order to plant and settle a colony there.'[59] Within days of his return Banks had become a celebrity, on terms with the king and several of the nobility, the recipient (with Solander) of an honorary degree from Oxford and the object of distant admiration by Linnaeus, who suggested that the southern continent should be named 'Banksia' in his honour.[60] From Uppsala the scholar wrote: 'If I were not bound fast here by 64 years of age, and a worn-out body, I would this very day set out for London.'[61] As the months went by, Linnaeus waited in vain for Banks and Solander to send him plant specimens from the *Endeavour* voyage, and in later years he fulminated against an 'ungrateful Solander' who neglected to provide him with 'one single herb or insect of all those he collected'.[62]

Meanwhile, both Solander and Banks were preparing to accompany Cook on his new Pacific expedition, a sign that British discovery voyages from this time on would be expected to have a scientific dimension. Banks took for granted that he would occupy an altogether more elevated position than he had held on the *Endeavour*. Without, it seems, consulting Cook or the Admiralty, he brought together scientists, artists, secretaries – even two musicians – amounting to seventeen in all. It was a sign of the competition between the various applicants to sail with him that one naval officer withdrew when he found that 'Explorers of an equal zeal can't bear one another anymore than rival Beauties'.[63] Among those accepted was Dr James Lind, an Edinburgh physician with exper-tise in astronomy, who was also recommended by the Royal Society for 'his great Knowledge in Mineralogy, Chemistry, Mechanics, and various branches of Natural Philosophy; and also from his having spent several years in different climates, in the Indies'.[64] Lind was to receive from the normally parsimonious government the substantial sum of £4,000 for his expenses on the voyage. The applicants seemed to have assumed that the forthcoming voyage was, to all intents and purposes, Banks's. But there was trouble ahead, for the ship Cook and the Admiralty had chosen for the voyage was another collier, the *Marquis of Granby*, renamed the *Resolution*. It was larger than the *Endeavour*, but not large enough for Banks and his retinue, and in any case much too lowly a vessel for his taste – 'not fit for a gentleman to embark in', as he said. He insisted on alterations being made including the addition of an extra upper deck and a round-house on top to accommodate Cook (Banks was to have the great cabin, the traditional preserve of the captain). A trial trip out of the Thames almost led to the top-heavy vessel capsizing, and one of her lieutenants told Banks that, although he would put to sea in a grog tub if necessary, he jibbed at the prospect of sailing in 'the most unsafe ship I ever saw or heard of'.[65] The response of the Admiralty was immediate and drastic. All the new superstructure was to be removed, and the ship restored as far as possible to its original state.

When Banks journeyed down to Sheerness and saw the alterations, he was apoplectic. A midshipman remembered: 'He swore and stamp'd upon the Warfe, like a Mad Man; and instantly order'd his servants, and all the things out of the Ship.'[66] After engaging in futile protests to the Admiralty, Banks chartered a brig and went to Iceland, accompanied by Solander. It was a mildly interesting excursion, but not to be compared with a voyage to the South Pacific, where Banks had hoped 'to set my heel upon ye Pole! and turn myself round 360 degrees in a second'.[67] There is an intriguing postscript to the story of Banks and the *Resolution*. When the vessel reached Madeira in July 1772, Cook was informed that a 'Mr Burnett' had arrived on the island three months earlier and had spent time there botanising; but on hearing that Banks would not be on the *Resolution*, he had left Madeira to return to England. Cook reported: 'Every part of Mr Burnetts behaviour and every action tended to prove that he was a Woman.'[68] One is reminded of Solander's comment that Banks intended 'to make life agreeable during the Voyage'.[69] If indeed he was responsible for this arrangement, it says little for Banks's respect for naval conventions. As a midshipman on the *Resolution* wrote in a later memoir, 'unless this Lady had been *very prudent indeed*, she might have been the cause of much mischief.'[70] The publication in 1773 of Hawkesworth's edition of Cook's first voyage, with its frank descriptions of Polynesian customs, gave the satirists of London's literary circles ample material with which to portray Banks as a sexual profligate. Even the sober Solander was dragged in, as one anonymous versifier wrote

> Ye who o'er Southern Oceans wander
>   With simpling B—ks or sly S—r:
>   Who so familiarly describe
>   The Frolicks of the wanton Tribe,
>   And think that simple Fornication
>   Requires no sort of Palliation.
>   Let Wanton Dames and Demireps,
>   To *Otaheite* guide their Steps.[71]

Meanwhile, there was serious work to be done; to sort out, classify and publish the natural-history specimens brought back in the *Endeavour*. Of greatest popular interest was the kangaroo skin, which formed the basis for George Stubbs's painting of a stationary animal. More lifelike than this representation were Parkinson's rough sketches of a bounding, leaping kangaroo (Pl. 9),[72] but these remained unpublished, and it was Stubbs's painting that was engraved for publication in Hawkesworth's *Voyages*. Oddly, although Solander had examined three specimens, he did not recognise them as marsupials. Some indication of the size of the natural-history haul brought back was given by Banks in a letter to a French scientific acquaintance: 'The Number of Natural productions discover'd in this Voyage is incredible: about 1000 Species of Plants that have not been at all describ'd by any Botanical author; 500 fish, as many Birds, and insects Sea and Land innumerable.'[73] From the beginning Banks had grand ideas about publication. There was to be no modest serial issue in instalments, but the publication of a colossal work running to eight hundred full-page illustrations accompanied by explanatory texts. However, there were problems. Parkinson's death deprived the project of an artist who had seen the living plants, and the team of artists and engravers subsequently hired by Banks could not compensate for his loss. Moreover, Banks became involved in a time-consuming dispute with Parkinson's unstable brother, Stanfield, over ownership of the artist's drawings and notes. Banks was busy – voyaging to Iceland, attending the Royal Society where he had been elected to the Council, managing his Lincolnshire estates and paying attention to his onerous new responsibilities of supervising the Royal Gardens at Kew – and he left much of the work on the publication to Solander. But he too had other commitments, since he had been promoted to keeper of the Natural History Department of the British Museum. In addition, he was responsible for organising Banks's magnificent herbarium at New Burlington Street, London, which in 1777 was moved to the more spacious and more famous house at 32 Soho Square, Banks's London home for the rest of his life.

We are fortunate in having a detailed description by the Rev. W. Sheffield, keeper of the Ashmolean Museum, Oxford, of the collection as it appeared in December 1772. Banks's house 'is a perfect museum; every room contains an inestimable treasure. I passed almost a whole day there in perfect astonishment, could scarce credit my senses. Had I not been an eye-witness of this immense magazine of curiosities, I could not have thought it possible for him to have made a twentieth part of the collection.' The South Sea collection was arranged in three large rooms. The first, the Armoury, contained weapons, tools and utensils, all made from wood or bone. The second displayed the different clothes of the island peoples together with – a very Banksian touch this – the raw materials from which they were made. It also included plants – 'about 3000, 100 of which are new genera, and 1300 new species which were never seen or heard of before in Europe'. The third room began with 'an almost numberless collection of animals; quadrupeds, birds, fish, amphibia, reptiles, insects and vermin, preserved in spirits'. Finally came a glimpse of what the scientific world had been waiting for:

> The choicest collection of drawings in Natural History that perhaps ever enriched any cabinet, public or private: – 987 plants drawn and coloured by Parkinson, and 1300 or 1400 more drawn with each of them a flower, a leaf, and a portion of the stalk, coloured by the same hand; besides a number of other drawings of animals, birds, fish, etc. and what is still more extraordinary still [*sic*], all the new genera and species contained in this vast collection are accurately described, the descriptions fairly transcribed and fit to be put to the press.[74]

For Linnaeus in Sweden, the withdrawal of Banks and Solander from Cook's second voyage had come as an immense relief. After their return in the *Endeavour* he had received the unwelcome news that the pair were to set sail again, abandoning Europe and their collections, perhaps for several years, perhaps for ever. 'This report,' Linnaeus told his informant, John Ellis, 'has affected me so much as almost entirely to deprive me of sleep ... all their matchless and truly astonishing collection, such as has never been seen before, nor may ever be seen again, is to be put aside untouched, to be thrust into some corner, to become perhaps the prey of insects and of destruction.'[75] At one level Linnaeus had little reason to worry. As Sheffield's account shows, Banks's herbarium was on display in London, where its specimens were mounted and labelled, and were open to inspection by all serious enquirers. But there was no rush to publication; the very size and novelty of the collections militated against it. Full-size coloured drawings had to be made from Parkinson's sketches by a team of artists working under Solander's supervision that included the Miller brothers and John Cleveley Jnr, and then copper engravings made. Accompanying each plate was a scientific description based on Solander's original notes.

Time passed without any sign of the promised publication. The year 1778 saw the death of Linnaeus and Banks's election as president of the Royal Society, but the next glimpse of the work-in-progress did not appear until early 1782, when Banks noted: 'I have been engag'd in a Botanical work which I hope soon to publish, as I have now near 700 folio plates prepar'd: it is to give an account of all the new plants discovered in my voyage round the world, somewhere above 800.'[76] Weeks later, in May 1782, Solander died from a stroke at the age of forty-nine. A sign of how the *Endeavour* voyage lingered in his memory came from a friend, who reported that just before he was struck down he was recalling 'the danger he had undergone in Tierra del Fuego ... and his efforts through the snow' all those years earlier.[77] It was not quite the end of the great project, for in 1785 Banks was writing as though publication were imminent. Solander's name was to appear next to his on the title-page since 'There is hardly a single clause written in it, while he lived, in which he did not have a part.' Banks concluded: 'All that remains to do is so little that it can be completed in two months if only the engraver can be brought to put the finishing touches to it.'[78] These were never applied. Banks's biographers have puzzled over his failure to take the final step and astonish the learned world with a publication unsurpassed in scholarly thoroughness and beauty. It is possible that there was more to do than Banks's letter of 1785 indicated, and that without Solander at his side he let matters drift. Or perhaps the formidable costs of publication deterred him. One is reminded of the words of Alwyn Wheeler, the celebrated ichthyologist: 'There appears to be an optimum period during which the observations from a biological expedition have to be completed after which coherence is lost, results are anticipated by others, and enthusiasm dwindles as other interests present themselves ... The stores of museums, laboratories and universities all over the world are cluttered with collections which are "going to be worked up".'[79]

Banks himself seemed unconcerned by the failure to publish. Writing in the same year that Solander died, he seemed to overlook the non-appearance of the promised work, and ignored his predecessors such as Commerson and Steller, when he told a correspondent: 'I may flatter myself that being the first man of scientific education who undertook a voyage of discovery and that voyage of discovery being the first which turned out satisfactorily in this enlightened age, I was in some measure the first who gave that turn to such voyages.'[80] His collection was used by scholars in a piecemeal way, but not until the late twentieth century was his florilegium of 743 plants published in its magnificent entirety – two hundred years too late. As Banks's biographer has written, the plates of the florilegium remained 'a sort of buried treasure whose site was known but whose wealth was tapped only by the learned few – those who made the pilgrimage to Soho Square and those very few to whom selected proofs were sent.'[81]

CHAPTER 5

'A kind of Linnaean being'
The Woes of Johann Reinhold Forster

THE PRECIPITATE WITHDRAWAL of Joseph Banks and his entourage from the *Resolution* in May 1772 left a huge gap in the scientific arrangements for Cook's second voyage. To find a naturalist who could substitute for Banks and Solander at short notice, and who would be willing to leave home and family for three years, was not a straightforward task. The replacement candidate – as far as we know the only one considered – was to prove controversial. Six years earlier Johann Reinhold Forster had arrived in England as, in his own words at the time, 'a Foreigner & but an obscure Person'.[1] Born in Danzig, West Prussia, in 1729, Forster was educated in Berlin and at the University of Halle before returning to the Danzig area as a curate in the Reformed Church, a lowly position but one that allowed him ample time for studying the impressive collection of scholarly books he was acquiring. In 1765 he made an abrupt change of career when, accompanied by his young son, Georg, he entered the service of Catherine II of Russia, charged with the task of surveying the new settlements of German families on the Volga. Forster interpreted his mission in terms that in some ways antici-pate his years on the *Resolution*, for he was determined to examine 'the ground, plants and animals of those regions, together with the climate and its influence upon people, animals, plants and productions'.[2] He published descriptions of more than two hundred plants, as well as of birds, mammals, reptiles and birds, although the initiative in collecting usually came from his son. His father later recollected that as a boy 'when he [Georg] saw the insects and new plants coming out in the garden with the first signs of spring, he wanted to know in detail the name of each insect, each flower, each bird'. To cope with his son's questions, Forster bought a copy of the German edition of Linnaeus's *Systema naturae*, and with this in hand 'dictated the names as well as the peculiarities, economy and characteristics of the plants and animals' to Georg.[3]

Forster's report, critical of administration on the Volga, was not well received by the Russian government, and he and his son (still only twelve years old) left Russia without preferment, determined to try their luck in what they hoped would be the more sympathetic environment of Georgian England. Forster quickly entered the London world of learned societies, attending meetings of the Royal Society, the Society of Antiquaries and the Society for the Encouragement of Arts, Manufactures and Commerce. He hoped to obtain a post at the new British Museum, but in 1767 financial problems forced him to accept a teaching job at the dissenting Warrington Academy in Lancashire. There he spent three increasingly unhappy and financially straitened years before returning to London, where he renewed his acquaintance with men of standing in the world of natural history such as the antiquarian Daines Barrington and the naturalist Thomas Pennant. It was to the latter that Forster wrote, in uncharacteristically modest terms, explaining that although he had not studied under Linnaeus he understood his methods, and so 'reckon myself to be a kind of Linnaean being'.[4]

In the winter of 1771–2 he was elected a Fellow of the Royal Society; among his sponsors were Banks and Solander, newly returned from the Pacific. Forster continued to publish learned articles on natural history, but much of his time was spent earning modest sums of money by translating into English (with the help of his talented son) the published narratives of botanical travellers sent to all corners of the world by Linnaeus from his centre in Uppsala. A project that had more of a connection with Forster's immediate future was his edition of Bougainville's *Voyage autour du monde*. Praise of this in the *Critical Review* was probably written by Forster himself as it argued that an editor so well versed in natural history should join Banks and Solander on their next voyage since he 'must be eminently qualified to form a triumvirate'.[5] In September 1771, Forster left Banks in no doubt about his eagerness to accompany him on his return to the Pacific: 'how ready I am in keenest strength of mind and body to sustain the labours of a new journey that you are contemplating ... to endure the same perils with you, to share with you the same pleasures the study of nature gives, under a new sky.'[6]

Banks and Solander showed no desire to accept as their travelling companion a man who had already earned a reputation for being difficult, but when they withdrew from the voyage the Forsters seemed ready-made replacements. In his journal Cook wrote that Johann Reinhold 'from the first was desirous of going the Voyage and therefore no sooner heard that Mr Banks had given it up then he applied to go'.[7] Forster's application was supported by the well-connected Daines Barrington, and more pertinently it found favour with the earl of Sandwich, the First Lord of the Admiralty. Sandwich had been at the heart of the dispute with Banks, and in a draft letter to him he regretted in particular that Dr James Lind had also withdrawn from the expedition, but explained that 'another person has been found who is as well qualified, and more so in one particular, as he carry's his son with him who is a very able designer [artist], and will of course be extremely usefull in that part of the business'.[8] Lind's grant from Parliament of £4,000, Barrington reminded the prime minister, Lord North, had not specifically been awarded to Lind but 'to any other person going on the same expedition'.[9] With this assurance, George III cut through the intricacies of parliamentary approval by authorising a payment of £1,795 from the Civil List to meet Forster's initial expenses. As soon as he heard the news, Forster wrote that 'from this day I could first consider myself as being appointed by his Majesty on the Expedition'.[10] In the months and years to come, his shipmates would tire of hearing this claim. Meanwhile, with the Treasury paying out the promised sum within days, Forster had more ready cash at his disposal than at any previous time in his life – he never received full payment from the Russian government, and his salary at the Warrington Academy had been only £60 a year. There also seems to have been a verbal understanding, negotiated by Barrington, that Forster would write the official history of the voyage, and keep the profits from the publication. In time this agreement, if such it ever was, would result in much acrimony on all sides. Banks, still smarting from his withdrawal from the expedition, gave little help to Forster, and rejected his request to examine the plant drawings brought back on the *Endeavour*. He was, Forster complained, a 'false friend'.[11] With only ten days to prepare for the voyage, Forster hurriedly spent £1,200 on 'Equipment', ranging from books and instruments to suitable clothing, most of it, his son later complained, charged at double the normal price.[12]

The Forsters joined the *Resolution* at Plymouth on 3 July 1772. Ten days later, accompanied by the *Adventure* commanded by Captain Furneaux, Cook set sail on one of the greatest of all seaborne exploration voyages. In his three years away he demolished the theory of a great southern continent, took the *Resolution* closer to the South Pole than any other vessel had been and touched on a multitude of lands – Tahiti and New Zealand again, and for the first time Easter Island, the Marquesas, Vanuatu and New Caledonia. For Forster, it should have been the opportunity of a lifetime. He sailed with the good wishes of Linnaeus, who told him that he showed 'a spirit of heroism which rivals that of the heroes of war' and that his participation on the voyage would 'turn the eyes and minds of all botanists in your direction'.[13] Forster, his son Georg, the astronomer William Wales and the artist William Hodges were to be the 'experimental gentlemen' on a voyage whose task was to record the potential of unknown lands.[14] At the age of forty-two, Forster was a man of prodigious learning. With his knowledge of philology, zoology, ornithology, botany, ethnology, mineralogy, geography and history, and his reputed familiarity with seventeen languages, he was, in Linnaeus's words, 'a natural-born scientist'. He took on the voyage a large collection of books on natural history and travel, and a determination to settle those questions about the Pacific and its peoples that were engaging the scholars of Europe. Unfortunately, he was prickly and awkward in his relationships, and his defects of personality were to be magnified in the close quarters of the *Resolution*, so much so that Cook's biographer, J.C. Beaglehole, described him as 'one of the Admiralty's vast mistakes. From first to last on the voyage, and afterwards, he was an incubus.'[15] Perhaps greater allowance should be made for the fact that, like Steller, Commerson and Banks before him, Forster was to spend years in the unfamiliar and harsh environment of a man-of-war, far from the normal surroundings of a study-bound scholar.

Forster's personal journal – for the most part a frank and uninhibited record – begins in a conventional way, with a description of the first port of call, Madeira, and its natural history. There a pattern of botanising was established in which Forster, accompanied by his servant, Ernst Scholient, searched and collected, bringing back plants for Georg to draw. It was already clear that Forster was willing to face discomfort, even danger, in his quest: 'I had been obliged to climb up the steepest precipices, often up to the Ancles in mud, and through briars & thorns. My little bottle was broke from two hard falls I had, though my wine had been spent before, my Forces began to flag & I eat a couple of sour Aples.'[16] He added that Scholient, who was carrying the box of plants, was 'quite spent'. Collecting was only the beginning of the naturalists' task, for after the younger Forster had drawn the plants they had to find a suitable space on shipboard to dry them, all the time protecting them from saltwater and the ravages of insects and weevils. Once this was done, they had to label the dried flowers and leaves before placing them between layers of absorbent paper and storing them in boxes.

As the ships continued their course southwards through the Atlantic, relations between Forster and Cook were good. Forster advised Cook on a number of matters, and offered his services as an interpreter in the Cape Verde Islands. Not all his shipmates met with Forster's approval, however. When a swallow that had made its home on the ship disappeared, Forster suspected it had been caught and given to a cat, 'for we had some very cruel & ill-natured people in the Ship; who made it their business to disturb other peoples happiness & enjoyment'. More seems to be at stake here than the fate of the swallow as Forster complained that his ears 'thundered with thousand vollies of execrations, & oaths & rudeness'.[17] Towards most birds flying about the ship he was less sensitive as he recorded his 'good luck' in shooting two albatrosses and several other birds.

During a three-week stay at the Cape of Good Hope, Forster persuaded a Swedish botanist (and physician), Anders Sparrman, who had studied under Linnaeus, to join the expedition. To gain the services of a specialist botanist was a stroke of good fortune for the Forsters, although the arrangement was made at some personal financial cost, for as Sparrman explained they 'offered me my voyage gratis, with part of such natural curiosities as they might choose to collect'.[18] In a prefiguration of worries to come, Forster – only a few months out from England – wrote to Pennant from the Cape asking him to make sure that his name was not forgotten, a concern given a sharper edge when he wrote that 'Mr Banks was here expected, but every soul was disappointed to find him not in the Ship'.[19] However, the recruiting of Sparrman proved a considerable bonus. Throughout the rest of the voyage he shared the duties of sketching and painting specimens with Georg, while his botanical expertise allowed the latter's father to spend more time on his zoological and ornithological investigations.

From the Cape the *Resolution* embarked on a gruelling four-month cruise in far southerly latitudes. In heavy seas Forster's cabin flooded, a misfortune that prompted the first of many grumbles about his inadequate accommodation. Three years after the *Resolution*'s return Georg Forster was still exercised enough about the matter to air his complaints in a published *Letter* to Lord Sandwich. He pointed out that Cook, the first lieutenant, Joseph Cooper, the master, Joseph Gilbert, and the astronomer, William Wales, all had roomy, light cabins, whereas as 'the last comers' the Forsters were allocated two dark cabins which were so small that their precious books had to be stowed in the damp recesses of the steerage. For most on board the order in which the cabins had been allocated would have seemed absolutely appropriate, although the problems faced by the Forsters are not to be underestimated. Their cabins were never watertight: 'Morning and evening every day, on washing the decks, our cabins were filled with water over the ankles, and . . . when it rained, or as often as a wave struck over the ship, our beds were thoroughly drenched.'[20] In his journal Cook conceded that as the ships sailed south they encountered gales and squalls, but Georg Forster, having barely found his sea legs, made rather more of the weather: 'The people, and especially persons not brought up to sea-affairs, were ignorant how to behave in this new situation; the prodigious rolling of the vessel therefore daily made great havock among cups, saucers, glasses, bottles, dishes, plates, and every thing that was moveable . . . the howl of the storm in the rigging, the roar of the waves, added to the violent agitation of the vessel, which precluded almost every occupation, were new and awful scenes.'[21] Not all was hardship. Afraid that the wine on board might not last the voyage, Johann Reinhold Forster noted that 'we determined to omit toasting & drink after dinner 3 glasses only . . . A Voyage of the Nature of ours requires some Economy', while Sparrman wrote that Christmas Day was celebrated 'with punch, stout, port and

Madeira, claret, Cape and other wines, to which the captain and Mr Forster treated senior officers until late into the night'.[22]

By mid-December 1772 large masses of ice had been sighted, and Forster wrote, 'It requires really a steady, settled mind to view with composure the Quantities of Ice surrounding a Ship ... & to be constantly exposed to Snow & rain for several weeks without intermission.'[23] Often it was not clear whether they were sighting giant icebergs or ice-covered land. Week after week the *Resolution* sailed in far southerly latitudes, becoming the first ship to cross the Antarctic Circle, and losing company with the *Adventure* as she did so. Even Cook's normally laconic journal indulged in a touch of melodrama as its author contemplated the scene: 'The whole exhibits a View which can only be described by the pencil of an able painter and at once fills the mind with admiration and horror, the first is occasioned by the beautifulness of the Picture, and the latter by the danger attending it, for was a ship to fall aboard one of these large pieces of ice she would be dashed to pieces in a moment.'[24] For Cook, to sail farther south than any man had been before brought its own satisfaction. Forster, on the other hand, found it increasingly hard to come to terms with the fact that the voyage's main objective – the confirming or otherwise of the existence of a great southern continent – involved spending months sailing in southerly latitudes on a course that was neither compatible with personal comfort nor conducive to scientific investigation. A long journal entry on 15 March 1773 expressed his irritation on both counts. First, there was a further complaint about his accommodation: 'We have been obliged to prepare a better & warmer birth for two Ewes and a Ram, which we wished to bring safe to New Zealand, no more convenient place could be devised than the space between my & the Masters Cabin. I was now beset with cattle & stench on both sides ... peaceably bleating creatures, who on a stage raised as high as my bed, shit & pissed on one side, whilst 5 Goats did the same afore on the other side.' Having got that off his chest, Forster turned to the more serious matter of his frustration as a scientist: 'Had it not been for the pleasing hopes of making great discoveries in Natural History in this Expedition, I would never had so great an inclination of going on it. But instead of meeting with any object worthy of our attention, after having circumnavigated very near half the globe, we saw nothing, but water, Ice & Sky.' He went on to express his fear that they would winter, not in the balmy climes of the Society Islands, but in New Zealand's Queen Charlotte Sound, where the bitter cold would kill all plant life.[25]

The *Resolution* reached Dusky Sound near the southern tip of New Zealand's South Island at the end of March 1773. They had been at sea without sighting land for more than four months, much of the time without the reassurance of the *Adventure*'s company. Georg Forster sympathised with the crew's hardships to a much greater extent than either Cook or his father. The former presumably took a seaman's lot for granted, while Johann Reinhold was too concerned with his own discomfort to pay much attention to that of others. Georg, by contrast, was at pains to describe how 'our seamen and officers were exposed to rain, sleet, hail, and snow; our rigging was constantly encrusted with ice, which cut the hands of those who were obliged to touch it; our provision of fresh water was to be collected in lumps of ice floating on the sea, where the cold and sharp saline element alternately numbed and scarified the sailors' limbs'.[26] The *Resolution* stayed at Dusky Sound until mid-May, but when he went ashore Forster senior was disappointed by the scene that greeted him: 'I found myself quite tantalized with the sight of innumerable plants and Trees, all new ones, none of which had flowers at this Season & the fruits were quite unripe or already gone.'[27] There was some consolation in the number of bird and fish specimens he was able to catch: thirty-eight new species of birds, and twenty-five fish new to science (the latter drawn by Georg).

For an ungainly if muscular man in his mid-forties, unused to physical exertion and suffering from rheumatism, Forster showed commendable dedication in his collecting activities. On 6 April his hands were so swollen from insect bites that he could not hold a pen, and he spent a sleepless, feverish night; but the next morning he was up and about botanising on shore. There, Sparrman's journal entries for Dusky Sound give some idea of the difficulties of entering 'a dense forest which had been left in peace since the Flood ... The botanist, the shooter, or any other wanderer wanting to move about here, sank deep, sometimes to his thighs and even to his waist, in this damp mire of mouldered trunks and already formed humus of rotting leaves and trees and loose tussocks of moss.'[28] For Forster, matters were not helped by further problems with his cabin. Anchored under the trees at Dusky Sound, the ship was so dark that a candle had to be lit in the cabin even during the day, and as 'there was always a world of lumber before its door: sometimes I was pent in for hours together, or excluded from it during the same time'. The incessant rain added to Forster's problems, for his cabin 'was a Magazine of all the various kinds of plants, fish, birds, Shells, Seeds etc. hitherto collected: which made it vastly damp, dirty, crammed, & caused very noxious vapours, & an offensive smell'.[29] Forster was not a man to make light of his troubles, but it is hard not to sympathise with him and his son as they tried to examine, describe and draw their specimens in these cramped conditions, especially when their experience is compared with that of Banks, Solander and Parkinson, who had been given the run of the captain's great cabin on the *Endeavour* voyage. Many years later Georg Forster remembered how the cabins allocated to him and his father measured only 'six feet cubed in which a bunk, a locker and a writing desk leave just enough room for a folding stool'.[30] Nevertheless, at this time Forster senior's relations with Cook remained reasonable. He accompanied him on several shore excursions, and worried incessantly about the captain's health. During the stay at Dusky Sound, Cook was ill with a fever, and had pains in the groin and a rheumatic swelling of the right foot, all caused, Forster thought, 'by wading too frequently in the water & sitting too cold & wet in the boat'.[31]

His concern for Cook may be explained, in part, by the prospect of his replacement, First Lieutenant Robert Cooper – that he 'should then command the Ship if the Capt should die, is enough to frighten every living Soul . . . as none is a stranger to his illtemper, capricious & whimsical way of thinking, without any principles'.[32] Another occasion found Forster and Charles Clerke, the *Resolution*'s second lieutenant, threatening to fire at each other,[33] while Forster and the ship's astronomer, William Wales, were on especially bad terms. According to Wales, hardly a week went by without Forster quarrelling with someone on board, often warning the culprit that he would be reported to the king ('ze Kinck') on the ship's return. This threat was heard so often that 'it became a bye word amongst the seamen, whom I have frequently heard threaten each other with the same dreadful denunciation'.[34] More surprisingly, the Forsters seem to have had little time for the talented surgeon's mate and amateur naturalist on the *Resolution*, William Anderson, whom perhaps they regarded as a rival.[35] With Joseph

Gilbert, master of the *Resolution*, Forster had niggling disputes over accommodation from the earliest days of the voyage, and could not be reconciled to him even when he found that the master possessed a copy of Pope's Homer.[36] To offset this roll-call of offenders, Forster had nothing but praise for the *Adventure*'s astronomer, William Bayly, 'a Man of an agreeable, friendly Character, good parts, & an excellent mechanical genius, & of great learning in his profession'.[37] William Hodges, 'the ingenious Artist' whose drawings 'does honour to his Ability & to the Discernment in making Choice of so able a Painter', also seems to have enjoyed good relations with the Forsters,[38] as did James Patten, the *Resolution*'s surgeon.

One consolation for the Forsters during their otherwise frustrating stay at Dusky Sound was the chance to observe the local Maori. Georg Forster feared that the discovery voyages led to a loss of innocent lives, and also had a corrupting effect on the morals of 'little uncivilized communities'. The cannibalism that so shocked his shipmates he put into a perspective, reminiscent of Montaigne in the sixteenth century, when he pointed out that 'though we are too much polished to be cannibals, we do not find it unnaturally and savagely cruel to take the field, and to cut one another's throats by thousands'.[39] Although both Forsters showed sympathy to the Pacific peoples they encountered, neither had much patience with the more sentimental aspects of the primitivist school of European thinking. One of the most revealing passages in the elder Forster's journal came in April 1773 when he contemplated the bustling scene at the ship's anchorage at Dusky Sound, and described buildings and improvements completed by the crew in a few days that five hundred of the local inhabitants could not have managed in three months. This was proof to him of 'the superiority & advantages, which the use of Sciences, arts & mechanical improved trades, & the use of convenient tools give to civilized Nations over those that live in a more pure state of nature'.[40] On the homeward voyage Georg Forster was even more outspoken than his father as he regarded the shivering inhabitants of Tierra del Fuego: 'Till it can be proved, that a man in continual pain, from the rigour of climate, is happy, I shall not give credit to the eloquence of philosophers.'[41]

After rejoining the *Adventure*, Cook sailed north to Queen Charlotte Sound. There he visited the gardens established by Furneaux and his officers, who had put in potatoes, carrots, parsnips and turnips as well as salad greens. They were, he thought, in 'flourishing condition and if improved or taken care of by the natives might prove of great use to them'. Furneaux added that at various places along the coast he had transplanted several hundred cabbages.[42] It is clear from their journals that both Cook and Furneaux took a personal interest in the establishment of gardens in selected spots, although perhaps not with the motives advanced by a modern critic of Cook's voyages who has insisted that 'wherever he [Cook] goes he plants English gardens. The act is primarily symbolic, supplanting the disorderly way of savage peoples with ordered landscapes on the English model.'[43] Cook's intentions were almost certainly more modest than this, and he was often disappointed with the results of his efforts. On his next visit to Queen Charlotte Sound in October 1774 he found the gardens 'allmost in a state of Nature and had been wholy neglected by the Inhabitants, nevertheless many Articles were in a flourishing state'.[44] Forster's journal mentions these horticultural activities, but gives no impression of personal involvement. So, on 1 June 1773 on Motuara Island in Queen Charlotte Sound, he writes about Cook digging up ground which he then sowed with wheat, beans, French beans and peas while 'I picked in the mean time some Coralines & shells'.[45] Collecting exotic items rather than attempting to grow plants familiar in Europe seems to have been Forster's main preoccupation.

During his botanising activities in New Zealand, Forster paid a more gracious tribute to his predecessor on Cook's first voyage than Banks's unhelpful behaviour to him before he left England perhaps deserved: 'We saw a new plant, very minute & prostrated, which proved to be a Species of a new Genus we had called *Banksia*, in honour of *Joseph Banks* Esqr, who on board the Endeavour was the first Naturalist, that ever searched the South-Seas & especially *New-Zealand*, & enriched Natural History with more than 800 new plants and 200 or 300 new Animals: an Addition which never one single Man made to this branch of Learning.'[46] Later, during the ship's stay at Tahiti, he became even more deferential towards the distant Banks when he downplayed the significance of his own observations, 'as I suppose that by this time the Description of the manners, customs, religion, cultures & manufactures of the Inhabitants of Otaheite are in the hands of the public, written by a far abler hand than mine'. It is true that much of Banks's journal had just appeared in the Hawkesworth volumes, but Forster was being altogether too flattering when he described his predecessor as 'a perfect master' of the Tahitians' language.[47]

Tahiti had a breathtaking impact on some of the newcomers. Georg Forster wrote of his first sighting, on 15 August 1773, how 'In the evening, about sun-set, we plainly saw the mountains of that desirable island before us, half emerging from the gilded clouds on the horizon', while the next morning a faint breeze 'wafted a delicious perfume from the land'.[48] His father was less impressed. He was injured helping with the capstan – not a normal task for a supernumerary – as the *Resolution* bumped against a reef, and once ashore grumbled that 'we are come here as it seems in the wrong Season, or in the wrong place'. Nor was it easy to preserve those few new plants he was able to collect, for so many islanders crowded on board that he could not dry his plants on deck, and he had to place them in an oven.[49] One or other of the Forsters usually accompanied Cook on his shore excursions, but at the neighbouring island of Raiatea, Johann Reinhold became embroiled in an angry dispute with Cook after he had shot and wounded an islander who had attempted to seize his son's musket. In Forster's words, 'hot & unguarded Expressions came out on both sides & he sent me by Force out of his cabin.'[50] Three days afterwards the two men shook hands, but later in the voyage a generalised complaint by Forster may have been aimed, in part at least, at Cook when he wrote: 'People who know nothing of Sciences & hate them, never care whether they are enlarged & knowledge increases or not.'[51] At Huahine, Sparrman was attacked and badly beaten by two islanders while botanising, and his usually placid journal recorded his resentment that Cook blamed him for his 'solitary' excursion: 'On New Zealand and elsewhere we frequently risked our lives while gathering our plants among hundreds of savages, like pulling burning embers out of a fire.'[52]

At the beginning of October 1773 the ships reached the Tongan Islands, where Forster noted that there were few new plants for him. As in New Zealand and the Society Islands his journal was increasingly taken up with observations on the peoples and their lifestyle. In Tonga he praised the sophistication of the islanders' music and then wrote: 'let us examine their Ingenuity, their contrivances, the Simplicity of their tools, their Arts, their manufactures, their cultivation, their fishing, their Navigation, their Shipbuilding, their knowledge of the Stars, & all this must convince us, that they have more civilization than we at first outset think.'[53] At Eua, Cook gave some garden seeds to the chief, but his main hopes in this respect were centred on the more temperate climatic conditions of New Zealand. Back in Queen Charlotte Sound on the expedition's return to southerly latitudes, Forster was again disappointed in his haul of plants, especially since the ship arrived at the more promising time for the botanists of early spring. As the *Resolution* neared the antipodes of London (she had once more lost company with the *Adventure*, which made her own way back to England), Forster seemed rather more excited about what lay ahead as they approached 'a place where no other Ship has ever been since the Creation'.[54] But as they skirted the Antarctic ice shelf in fog and freezing weather his mood changed to one of total dejection. On 21 December, after the *Resolution* crossed the Antarctic Circle for the second time on the voyage, he opened his heart in a journal entry several pages long. First, there was the standard complaint about his cabin, 'cold & open to the piercing winds, full of unwholesome effluvia & vapours, every thing I touch is moist & mouldy & looks more like a subterraneous mansion for the dead than a habitation for the living'. Nor was Cook much better off, for 'in the Captain's Cabin there are broken panes, the apartment full of currents & smoke, a parcel of damp Sails spread, & a couple of Sailmakers at work, now and then discharging the mephitic Air from the pease & Sower-krout they have eaten, & besides 5 or 6 other people constantly in it'.

More galling than these physical discomforts was Forster's frustration at what he feared would be seen as the failure of his mission as a scientist who hoped that his investigations would 'be useful to mankind in general & to the Dominions of Great-Britain in particular ... but having toiled for more than 18 Months, we have seen nothing that has not been seen before; for I believe all the few plants & Animals we could come at during our short Stay ashore, were probably all observed by Mr *Banks* & Dr *Solander*'. Increasingly, if unnecessarily, Forster was obsessed by the fear that his researches would appear second best to those of his predecessors who had sailed on the *Endeavour*. Unaware of the delays in London that were hindering completion of their natural-history publication, Forster reflected gloomily that Banks and Solander would long since have published their discoveries, 'assisted by the opulence of the one, & the great skill in Natural History of the other'.[55]

As Cook doggedly continued on his course, Forster pointed out that the tracks they had followed for two seasons had eliminated all possibility of a great southern continent: 'What helps it therefore to harass the Ship, the rigging & the crew in these turbulent Seas beating to windward. If to satisfy Government & the public that no land is left behind: it will not suffice the incredulous part of the public if the whole Ocean were ploughed up.'[56] Suffering agonies from rheumatism in his legs, Forster blamed an unnamed person for making 'this cruising wilfully longer, in order to satisfy interest & vanity'. For Forster, it was the lowest point of the voyage, and his journal almost reverted to shorthand as he noted: 'I do not live, not even vegetate, I wither, I dwindle away.'[57] On 30 January 1774 the *Resolution* encountered an 'immense field' of ice that Cook suspected extended to the Pole. They were in latitude 71°10´S., the farthest south any ship had reached to that time. In words that unwittingly echoed Forster's diatribe of some days earlier, Cook wrote: 'I who had Ambition not only to go farther than any one had done before, but as far as it was possible for man to go . . . could not proceed one Inch farther to the South.' A relieved Forster simply wrote: 'We put about & went again to the North.'[58]

One of Forster's characteristics that did little to endear him to his shipmates was his readiness to pronounce on matters that they regarded as falling outside his province. For example, as the *Resolution* crossed the central Pacific in March and April 1774, Forster made some justified criticisms of the haphazard way in which previous European explorers had named islands of which they assumed they were first discoverers. He pointed out that, as a result of the same island sometimes being given several different names, there had been much confusion as to the identity of islands in the Marquesas and Tuamotus: 'Had all the former Navigators taken the prudent Step to inquire the Natives, for the Names of the Islands they saw, we might be able to ascertain with certainty, what are new discoveries & what not.'[59] This commonsense recommendation would later be the standard approach of hydrographers on the Admiralty's survey voyages.

In this final stage of the voyage, Forster often took risks as he continued to collect specimens. On a return visit to Tahiti in April 1774 he decided that in an effort to find new plants (by which he usually meant ones that had not been seen by Banks and Solander) he would climb to the summit of the hills that lay inland from Matavai Bay. It was an arduous business, 'for the continual clouds that are lying on these mountains make the paths extremely slippery, & the path going always on the ridge of hills which have a very steep descent on both Sides it is very dangerous to make a slip'. His servant, Scholient, had been ill for some months, so Forster was accompanied by a Tahitian youth who carried the plant box. After a night's rest he continued to climb, collecting a few plants while 'standing on the sloping sides of the hills which are so steep, as to come next to a precipice'. In heavy rain 'I fell down & hurt my thigh in such a manner that I was near fainting for pain', and only then did his Tahitian helper agree to follow him.[60] He had suffered a rupture that would bother him for years to come, but he was delighted that he had found eight plants that he thought had probably not been collected by Banks and Solander. At Niue (named Savage Island by Cook) the Forsters and Sparrman had just landed on the rocky shore when they came under attack, and had to retreat to the ship carrying only one plant. Tana presented other dangers. There a volcano belched out black ash in the form of tiny needle-like splinters that 'proved remarkably dangerous to our eyes in botanising, as every leaf on the island was entirely covered with them'.[61]

Time and again Forster grumbled that he was denied the opportunity of collecting as they sailed past a succession of islands in the Tongan group. That these were islands unvisited by Banks on the *Endeavour* voyage only added to his frustration. Off the island of Vatoa in the Fiji group he set out his objections in general terms:

> All the money of the public is as it were thrown away & my mission absolutely made useless for want of opportunities to collect new plants, which grow plentifully in the new Isles ... either we stay but a day or 2 in one place, go late ashore, come early off again ... What good can arise from seeing 2 or 3 Isles more in the South-Sea? without knowing its products & and the Nature of its Soil & the Disposition of its Inhabitants; all which cannot be learned by staring afar off at an Isle.[62]

By this time Forster was directing his anger at Cook in person. Denied a boat to go ashore at Vatoa, he expressed his frustration at seeing 'an Isle full of plants as it were within reach', and blamed those 'who have nothing in view, but the aggrandisement of their rank & fortune, without any tincture of Science'.[63] On one frustrating day off Eromanga he expressed the perennial complaint of naturalists on the discovery voyages: 'We are to float on the water for ever, to have very few relaxations a shore & it is as if envy would have it so, that lands should be discovered, but none of its productions.'[64] For landlubbers such as the Forsters readily admitted to being, their wooden world was a place of trials and tribulations, and there is much to admire in their unrelenting determination to carry out the tasks they had set themselves. Even when allowed on shore, they did not always have good fortune. One morning at Tana could stand for many. In swamps near their landing place Johann Reinhold Forster wrote that he and his son 'saw several black Parrokeets with red heads & breasts, but they are amazingly shy & we could therefore shoot none, though we stood 2 hours on the same spot & were most miserably bitten by the *Mosquitos*. I found a new plant, & saw three Ducks, but my piece unhappily snapping twice, they all flew away.'[65]

Forster the scientific collector faced other annoyances, for many on board were acquiring 'Curiosities' to sell when they got home. His irritation increased when a sailor demanded from him the payment of a half-gallon of brandy (equivalent to half a guinea in money terms) for six imperfect shells. Among the crew the gunner and carpenter had made 'vast Collections' of several thousand shells. These references are reminders of the vogue for collecting shells that had reached its height in the demand for the beautiful specimens brought back on the ships of the Dutch East India Company, and then gained a new lease of life thanks to the Pacific voyages of the later eighteenth century. Unlike most natural-history specimens, shells were easy to collect, preserve and display, and the best had a ready saleable value. Rare specimens reached fabulous prices and were secured in the locked glass cabinets of wealthy collectors, where they were arranged more to show off their appearance than to accord with any scientific principles. When one of the *Resolution*'s sailors arrived back in England in 1775 he wrote to Joseph Banks, 'begging pardon for my boldness', and sending him 'a few curiosities as good as could be expected from a person of my capacity. Together with a small assortment of shells. Such as was esteem'd by pretended Judges of Shells.'[66]

On his collecting excursions ashore, the elder Forster more than once challenged the authority of Cook and his officers. Even Charles Clerke, second lieutenant on the *Resolution*, and usually regarded as the most congenial of men, became embroiled in a dispute with Forster on Tana over the latter's treatment of an islander. In his journal Forster complained that Clerke threatened that he would order a sentry to shoot him if he did not obey orders. This seems improbable, but even more so is Georg Forster's version of the incident in which he claimed that his father drew a pistol to force Clerke to desist. Both men took their quarrel to Cook, who 'seemed not to believe neither the one nor the other' – perhaps out of sheer incredulity.[67] As the voyage progressed the grant of £4,000 made to Forster became more widely known among the crew, and may have been responsible for some of the ill feeling. 'I am the object of their Envy,' Forster wrote, '& they hinder me in the pursuit of Natural History, where they can, from base & mean, dirty principles.'[68]

Forster's habitual inclination to criticise and complain can obscure the more positive features of the naturalists' work. During the expedition's stay at Balade in New Caledonia, Cook wrote, rather dryly: 'Our Botanists did not complain for want of employment while we lay here, every day brought in some thing new, either in Botany or Natural History.'[69] Viewed from the Forsters' standpoint, however, Balade provided yet another example of unfair treatment. Georg Forster described how at a time when he and his father were ill one of the surgeon's mates (almost certainly William Anderson) collected many new plants and shells, 'but the meanest and most unreasonable envy taught him to conceal these discoveries from us'. This incident provoked a diatribe in Georg's published account, four years later, that appears out of all proportion to the offence: 'It may seem extraordinary, that men of science, sent out in a ship belonging to the most enlightened nation in the world, should be cramped and deprived of the means of pursuing knowledge, in a manner that would only become a set of barbarians.'[70] On the other hand, relations with Cook were not always as fraught as some passages in the elder Forster's journal would suggest. On the homeward voyage, as the *Resolution* sailed in far southerly latitudes in the Atlantic, Cook seems to have accepted Forster's suggestion that the large island discovered between latitudes 54°S. and 55°S. should be named South Georgia.[71] Two weeks later Cook named a deep bay in the South Sandwich Islands 'Forsters Bay' (today Forsters Passage). It had taken him almost three years to make this gesture, but it is worth noting that neither William Wales nor William Hodges, the other 'experimental gentlemen' on the ship, was so honoured.

At the Cape, Sparrman, who had been an invaluable help to the Forsters, left the expedition. His first book dwelt more on his South African experiences than on his years on the voyage, and the fact that he had little to say about the hardships that bulked so large in Forster's journal made his grumbles about the state of the *Resolution*'s provisions as she reached the Cape in March 1775 the more telling:

> Our bread was, and had been for a long time, both musty and mouldy; and at the same time swarming with two different sorts of little brown grubs ... their larvas, or maggots were found in such quantities in the peas-soup, as if they had been strewed over our plates on purpose ... our salt meat, now almost three years old, having been kept on board during the whole of this period, was the more dried and shrunk up, as the salt had so much the longer time to absorb to itself and dry up all the moisture and juices.[72]

As the *Resolution* neared home waters, Forster's apprehensions increased: 'I am no more young, if I should be unfortunate enough to have lost in my Absence my best Friends & Patrons, I must begin life as it were again.'[73] Events were to show that there were some grounds for Forster's fears, although perhaps not in the way he envisaged. The *Resolution* anchored at Spithead on 30 July 1775 and, together with Cook, the Forsters immediately left for London. There was much to do: family and friends to be seen; specimens to be unpacked and sorted; the world of learning to be apprised of the scientific import of the voyage; and, of course, the account of the voyage to be written. At first, all seemed well. Naturalists in England and abroad showed their interest in the Forsters' researches and their impressive haul of specimens. In August the elder Forster was received by the king, and a few days later he presented to the queen live animals he had bought at the Cape. In November, Oxford University conferred the honorary degree of Doctor of Laws on him, recognition, the naturalist Gilbert White wrote, that he had 'hazarded his life in a circumnavigation in the pursuit of Natural knowledge'.[74] He anticipated that the publication of his account of the voyage would set the seal on his fame, and secure his financial situation, but within months Forster's optimistic plans fell to pieces. From the early stages of the voyage, there had been a misunderstanding between Cook and Forster about the writing of the official account of the voyage. The latter was convinced that the Admiralty had agreed that the task, and the rewards, were to be his, and with this is mind he kept a daily journal. Cook, on the other hand, bruised by the offhand treatment of his *Endeavour* journal by Hawkesworth, who in his published account of Cook's first voyage had often preferred to quote Banks's journal, was determined to give his own account of his second voyage. For most of the voyage Cook kept three different versions of his journal, in addition to a number of manuscript fragments of varying length. The several revisions, additions and subtractions are evidence that he was writing, not simply for his superiors at the Admiralty, but for the public at large. As he said, although not an author as such, he was 'a man Zealously employed in the Service of his Country and obliged to give the best account he is able of his proceedings'.[75] The conflict between Forster's assumptions and those of Cook was concealed for a while, and during the winter after the *Resolution*'s return both men were busy working on their respective narratives. In April 1776, Sandwich tried to sort matters out. At a meeting attended by Cook and Forster a compromise agreement was drawn up that envisaged a two-volume work. Cook would write the first volume, 'containing a Narrative of the late Voyage, with his nautical Observations, and also his Remarks upon the Customs and Manners of the natives of the Islands he touched at'. Forster would be responsible for the second volume, containing his 'Observations upon Natural History, and upon the Manners, Customs, Genius, and Language of the natives of the several Islands, with his philosophical remarks in the course of the voyage'. The profits were to be shared equally between the two men.[76]

Even the few words of the agreement covering the division of subject matter between the two volumes promised trouble, for it seemed that both men would be describing 'the Customs and Manners' of the Pacific Islanders. And this section, for a reading public that had been fascinated by Banks's amorous exploits in Tahiti, and intrigued more recently by the arrival on the *Adventure* of a Society Islander named 'Mai, would be bound to form a popular part of the promised narrative; 'nautical Observations' might not have the same appeal. Matters never reached this stage, however, for Forster – against Sandwich's expectations – wrote a specimen chapter in the form of a narrative rather than a scientific discourse, thus setting himself in direct competition with Cook's efforts. Worse was to follow, for Sandwich found the chapter's literary style unsatisfactory and handed the manuscript to Daines Barrington to revise. His suggested changes Forster found totally unacceptable, and he reacted furiously, accusing his former friend and patron of 'mutilating' his manuscript. From this point on there was no question of joint authorship. Indeed, by the early summer of 1776 Cook must have finished his narrative and handed it to his editor, Dr John Douglas, for in July he left Plymouth on his third, and last, Pacific voyage.

Denied any part in the official account of the voyage, the Forsters adopted a new strategy. Georg, who was unhindered by any agreements with the Admiralty, would write a narrative loosely based on his father's daily journal. This, it was hoped, would not only command good sales, but would leave Johann Reinhold free to pen a more philosophical disquisition on the voyage. By early 1777, Georg Forster had completed his task, and his *Voyage round the World* was published in March in two volumes, six weeks before Cook's own account, *A Voyage towards the South Pole*, also in two volumes. Any expectations the Forsters had of large profits from the book soon evaporated. In his *Letter to the Right Honourable the Earl of Sandwich*, which published the whole sorry story from the Forsters' point of view, Georg accused the First Lord of the Admiralty of deliberately setting the price of the official account, which included sixty-three plates based on Hodges's drawings, at too low a level at two guineas – the same price as his own book, which had no plates. In addition, the Admiralty covered the paper and printing costs of Cook's volumes, so that, the younger Forster concluded, 'By thus lavishing the public money, you were enabled to retail the books below prime cost . . . and I am the loser by two-thirds of what I must otherwise have earned.'[77] His father went further when he complained that the book 'has not brought us in one single Shilling', and that there remained 570 unsold copies.[78]

A Voyage round the World was favourably reviewed by most London periodicals of the day. Georg Forster's determination to avoid nautical details – 'how often we reefed, or split a sail in a storm, how many times we tacked to weather a point' – probably helped, although it did not prevent Dr Johnson, no great lover of exploration accounts from Hawkesworth onwards, telling Boswell that Forster 'makes me turn over many leaves at a time'.[79] The book was written in a hurry, for the younger Forster had only nine months to produce a text that ran to twelve hundred pages. The task was only possible because Georg had his father's daily journal in front of him, although his presence hovering at the author's shoulder may have been as much a hindrance as a help. The book was written in an altogether more measured and conciliatory style than the elder Forster's journal, but past disputes were revived after the book's appearance when, in 1778, William Wales published a long pamphlet of critical *Remarks on Mr. Forster's Account*, which was answered by Georg Forster in his *Reply to Mr Wales's Remarks*.

More importantly, that year also saw the publication of Johann Reinhold Forster's *Observations Made during a Voyage round the World*, a survey of the philosophical implications of the voyage. Almost four hundred pages were devoted to 'Remarks on the Human Species in the South-Sea Isles', which began with a resounding quotation from Alexander Pope – 'The Proper Study of Mankind is MAN'. This was followed by an assertion by Forster that helps to show why he was not the best loved of scholars: 'Though we have many accounts of distant regions, it has been a general misfortune, that their authors were either too ignorant to collect any valuable or useful observations, or desirous of making a shew with a superficial knowledge.'[80] There is much that is perceptive in the book: a realisation that the Pacific Islands had been peopled from Asia, and a recognition of the importance of population growth in explaining the development of societies. Forster's thesis anticipated Humboldt with its recognition of the influence of diet and custom as well as climate in shaping societies, and sixty years before Darwin he argued that volcanoes 'cause great changes on the surface of our globe', and speculated that coral formations were 'the most probable cause of THE ORIGIN of all THE TROPICAL LOW ISLES over the whole South-Sea'.[81]

Diversity, comparative customs and the climatic environment became the new catchphrases, bandied about as much in literary circles as in the writings of the philosopher-travellers.[82] In his *Outlines of a Philosophy of the History of Man*, the renowned German philosopher Johann Gottfried Herder praised Forster for presenting 'such a learned and intelligent account of the species and varieties of the human race in them, that we cannot but wish we had similar materials for a *philosophic-physical geography* of other parts of the world, as foundations for a history of man'.[83] A more recent assessment describes the *Observations* as 'an astonishingly rich and wide-ranging meditation on the voyage's scientific and above all proto-anthropological findings'.[84] For a book written by the official naturalist on the voyage, it contains remarkably little natural history, if that is defined simply in terms of describing the animal and vegetable kingdoms. Rather, in drawing up his account, Forster might have had in mind the advice of Linnaeus who had written in 1760: 'We must pursue the great chain of nature till we arrive at its origin; we should begin to contemplate her operations in the human frame, and from thence continue our researches through the various tribes of quadrupeds, birds, reptiles, fishes, insects and worms, until we arrive at the vegetable creation.'[85] Georg Forster put the matter more bluntly when he remarked that Parliament had not sent out 'my father as a naturalist, who was merely to bring home a collection of butterflies and dried plants ... from him they expected a philosophical history of the voyage'.[86] There is no evidence that Parliament, government ministers or the Admiralty expected any such thing; but that is what they got.

The story of the reception and disposal of the Forsters' collections from Cook's second voyage is a melancholy one. A month before the *Resolution*'s return to England, Solander, relying on letters from the Cape, informed Banks that the Forsters had gathered many new plants and animals.[87] This haul was listed in detail in Forster's *Observations*. They had collected 330 plants, if which 220 were new to science, together with 104 birds, half of them aquatic, and 74 fishes unknown to science. Despite Forster's grumbles, the long periods spent at sea or in the bleaker regions of the southern ocean were not altogether wasted, and his published monographs on albatrosses and penguins have been widely recognised as pioneering studies. In a work he wrote many years later dedicated to his son, Forster outlined the division of labour between himself, Georg and Sparrman: 'In sketching plants in particular we used as an assistant our good friend Sparrman: it was your [i.e. Georg's] task to put his work in order, and at the same time to describe the plants.

It was my particular province to examine more closely these efforts here and there, and to correct them in a very few places, [and] to describe all the animals.' In all, Forster estimated, about five hundred plants and three hundred animals had been sketched: 'Any understanding person will be amazed that so much work could have been completed by one man and a youth not yet 20, with only one single assistant.'[88] The Forsters had collected more than one specimen of many of the plants they brought back, and they were generous in giving duplicates to individuals and institutions. They had also obtained more than five hundred ethno-graphic artefacts, or 'artificial curiosities'. These they had collected with care and often at some expense, but all thoughts of method seem to have disappeared when it came to preserving their haul. Some items were sold, others were given away, often without record, and today at least fourteen different museums hold artefacts collected by the Forsters.[89]

Outstanding among the museums in Britain that exhibited both natural-history and ethnographic items from the discovery voyages was Sir Ashton Lever's Holophusicon which opened in Leicester Square, London, in February 1775, a few months before the return of the *Resolution*.[90] It was a serious rival to the British Museum, although it received no official funding and was regarded with disfavour by Joseph Banks. Lever, whose first museum had been in Manchester, acquired large collections from Cook's voyages (some, possibly, as a gift from Cook himself) and made them accessible to the fee-paying public. Unlike in the British Museum, the items were carefully labelled. A foreign visitor wrote: 'What I like about it is, that one can walk around for hours without a guide, because every case contains the name of the article in a small painted label stuck to the glass.' Among the celebrities who visited the museum was John Wesley, who thought that 'for natural curiosities, it is not to be excelled by any museum in Europe, and all the beasts, birds, reptiles, and insects are admirably well ranged and preserved'. Wesley's comments indicate that dried plants and seeds were not much in evidence among the museum's displays – they were more likely to be found in the herbaria of Banks and other botanists. The items, both natural history and ethnographic, were drawn and painted by artists, notably Sarah Stone, who made several hundred sketches and watercolours that form an invaluable record of the museum's collections. Despite the museum's popularity, it ran into financial trouble, and shortly before Lever's death in 1788 moved to the Rotunda across the Thames at Blackfriars Bridge under the new ownership of James Parkinson. Hindered by its new and less fashionable location, the Leverian Museum, as it was now called, survived only until 1806. Its collection was then dispersed in a sale that lasted more than two months. The largest collection among the seven thousand lots was bought on behalf of Emperor Franz I of Austria for his Imperial Cabinet. As well as 230 ethnographic items, it included large numbers of natural-history specimens, among them 276 birds. Adrienne Kaeppler's detective work has traced the fortunes and established the present locations of an extraordinary number of other items sold in 1806, although some may be lost for ever.

For the naturalists on Cook's voyages, the task of collecting, describing and drawing hundreds of specimens was already demanding enough; to publish them in acceptable scholarly form was to prove even more arduous and costly. At first, Johann Reinhold Forster was full of confidence. Within three weeks of his return he was announcing plans to publish two works, *Genera plantarum novarum* and *Descriptionibus zoologicus* and *botanicus*,[91] but these never appeared in their proposed form. Instead, a volume containing Sparrman's descriptions and Georg Forster's drawings of many, but by no means all, of the botanical specimens collected on the voyage, *Characteres generum plantarum*, was issued in quarto format in 1776. It was a hurried and incomplete work that the elder Forster later regretted publishing. He intended it as a preliminary to fuller and more systematic volumes, a kind of advance notice, but those further volumes never materialised. The disputes over the official narrative of the voyage, and the precarious financial circumstances that afflicted the now jobless Forster and his family, put their publication out of sight. It was a sign of Forster's desperation that in August 1776 he sold several volumes of Georg's botanical and zoological drawings to Banks for 400 guineas, while the most important of his manuscripts, the zoological investigations contained in his 'Descriptiones animalium', fell by the wayside. Many years later Georg Forster arranged for the latter to be published in Göttingen, but his father's financial and other demands led the publisher to abandon the project, and the work was not published until 1844. In a passage that could be applied to many other naturalists, the manuscript's nineteenth-century editor remarked that, if this pioneering work had been published during Forster's lifetime, 'nothing would be detracted from the glory of a great man, glory which many decades later English and French zoologists have won with easy labour'.[92]

In 1778, Georg left for Germany in an effort to restore his family's finances, but his post as professor of natural history at Kassel provided only a minimal income. He had taken with him a considerable collection of specimens and artefacts, but many were lost when the ship on which he was travelling was wrecked. Johann Reinhold, encumbered by debt brought about (his son conceded) by his 'imprudence & want of Oeconomy', finally accepted a university post in Halle, although his salary was hardly commensurate with his grand title of professor of philosophy, natural history and mineralogy. He managed to take with him most of his collection of books (some he had been forced to sell) and his natural-history specimens. The herbarium he established at Halle was estimated to contain three thousand rare plants. During his impoverished years in London, Forster had relied heavily on Banks for loans, the largest of which was never repaid. As the elder Forster was about to leave for Germany, Georg wrote to Banks thanking him for his generosity to his father 'at a time when he was forsaken by almost every body', and apologised for 'that Singular fieriness and impatience of temper in which my father is so particularly unhappy'.[93] Even during his deepest financial troubles, when he spent much time evading bailiffs and sending out begging letters, Forster never stopped writing. As Georg realised, he seemed 'only spurred on by difficulties'. He continued to submit learned articles, translated Swedish and German scientific books, and engaged with Buffon's theories of the earth's evolution.[94] Forster's review of Buffon's *Epoques de la nature*, published in 1779, indicated his own views on the future of natural history. He and his fellows, he admitted, spent too much time 'obscurely rummaging around', counting hairs, feathers and fins. They were treating 'natural history too microscopically, and so it was high time for Buffon to teach us to exchange the naturalist's microscopic view for a telescopic one'.[95]

Forster was to survive his son by more than four years, but it was Georg who struck out in new directions. At first, as he moved from one academic post to another, he remained committed to the Pacific. The German edition of his narrative of Cook's second voyage was well

received, as was his translation into German of the official account of
Cook's third and last voyage. This included as an introduction 'Cook der
Entdecker' ('Cook the Discoverer'), a substantial piece of more than
twenty thousand words. A striking and unexpected feature, given the
disputes on the *Resolution*, was Georg Forster's unqualified admiration
for Cook. His handling of officers and men, his attention to shipboard
health, and his sense of when discipline must be rigidly enforced and
when it might be relaxed, were all praised by Forster. Above all, he was
impressed by Cook's 'iron perseverance' on his voyages which helped
him to transform Europe's knowledge of the Pacific and its peoples: 'he
remains, as a mariner and discoverer, incomparable and unique, the
pride of his century.'[96]

Writing about the Pacific was second best for the younger Forster.
As he told a friend, for years he thought about little else than making
another voyage. He hoped to sail on a planned Austrian expedition
organised by a Dutch merchant, William Bolts; when that fell through,
he prepared to join a party of savants on an ambitious Russian expedition commanded by an officer in the Russian navy, Grigory Ivanovich
Mulovsky. The expedition was to visit both the South and North Pacific,
and 'everywhere', he was assured, 'our zeal for Science will be unhindered',[97] but the voyage was cancelled when war broke out between
Russia and Turkey in 1787. For Forster, it was the end of his dreams,
and health worries and family concerns persuaded him to take up the
position of university librarian in Mainz. A final glimpse of his time on
the *Resolution* comes from Alexander von Humboldt, who as a young
man accompanied Georg down the Rhine and across to England in
1790. Many years later Humboldt remembered how Forster's words on
that trip about his voyage with Cook 'helped determine the travel plans
I had been hatching since I was eighteen years old'.[98] In the last years of
his life Georg Forster became embroiled in the revolutionary politics of
the war-torn Rhineland, and in 1794 he died in Paris, aged only forty.

Johann Reinhold spent the rest of his life in Halle, where despite
disputes and continuing financial problems he became a respected
member of the university community. In all, he was a member or fellow
of twenty learned societies. His work rate remained impressive, but the
books on the natural history of Cook's second voyage that he and Georg
had planned never reached fruition. He continued to produce a steady

stream of articles and essays, mostly in the fields of geology and mineralogy, but his best-known work in his early years in Halle was his *History of the Voyages and Discoveries Made in the North* (the title of the English edition, published in 1786). He then turned, for reasons both scholarly and financial, to the task of editing, translating and otherwise preparing for publication a comprehensive corpus of travel literature. He was general editor of a series that ran to no fewer than sixteen volumes in the 1790s, and provided German readers with their first coherent collection of travel accounts. Fittingly, at the time of his death in 1798, he was translating the account by Milet-Mureau of La Pérouse's voyage to the Pacific, and waiting impatiently to receive a copy of the journal kept by George Vancouver, a young shipmate on the *Resolution* a quarter of a century earlier, describing his great surveying voyage to the North Pacific. In his own mind he would always remain, as Herder once described him, the 'Ulysses of the South Seas'.[99]

CHAPTER 6

'Curse scientists, and all science into the bargain'

Cook, Vancouver and 'experimental gentlemen'

FROM THE MOMENT that the news of Captain Cook's death reached England in January 1780 interest in his last Pacific voyage was dominated by the fatal encounter at Kealakekua Bay, Hawaii, the previous February. Up until the present day, scholars have argued interminably about the circumstances of his death, while other, equally puzzling aspects of the voyage have received less attention. Even before Cook left Plymouth in July 1776 in search of the Northwest Passage, he seemed to be behaving out of character. On his first two Pacific voyages he had destroyed the illusion of speculative geographers that a large temperate continent existed in the southern hemisphere, treating their entrancing hypotheses with a professional scepticism. As he wrote on one occasion, 'hanging clowds and a thick horizon are certainly no known Signs of a Continent.'[1] Yet on his third voyage he revealed a credulous reliance on speculative cartography that took a whole season of frustrating exploration along the northwest coast of America to correct. By this time Cook himself was responsible for drawing up the main part of his Admiralty instructions, and for the North Pacific these included reliance on a fictitious map of Russian trading voyages that showed Alaska as an island, with a wide strait between its eastern coast and the American mainland seeming to offer an easy passage into the Arctic Ocean.[2] Once through this strait, Cook intended to turn east along the continent's northern shores, into the Atlantic Ocean – and home. This was not the only improbable part of his projected route.

Despite his experiences in Baltic and North American waters, and more recently in the Antarctic, he accepted the theory of a handful of self-proclaimed experts that the north polar sea was ice-free in summer. Consequently, neither of his two ships, the *Resolution* (once more) and the *Discovery*, was strengthened to meet ice. Almost as puzzling, Cook took no Russian-speakers with him, although he was heading for a region where he would expect to meet Russian traders and seamen; nor is there any evidence that he or the Admiralty had consulted scholars in Britain such as William Robertson and William Coxe who were collecting up-to-date information about the Russian voyages.[3]

It is in this context that the scientific part of the voyage needs to be considered. We are reliant, perhaps unduly so, on a remarkable outburst on the subject attributed to Cook in the weeks before he sailed. One of the most interesting of the appointments to the voyage was James King, second lieutenant of the *Resolution*. Still in his mid-twenties, while on half-pay he had spent time at Oxford with Thomas Hornsby, professor of astronomy, who recommended him for the voyage.[4] With his training in astronomy added to Cook's expertise, there was no need for a civilian astronomer on the *Resolution*, although William Bayly was on the voyage again, sailing on the consort vessel, the *Discovery*. In addition to his scientific training, King was a perceptive observer who kept an excellent journal, and the official account of the voyage edited by Dr John Douglas relied heavily on this for events after Cook's death. It must have jolted King when, on reporting to Cook, he expressed his regret at the lack of scientists on the forthcoming voyage, only to be told, 'Curse the scientists, and all science into the bargain.' The evidence for this statement is not quite straightforward. It comes five years later from Johann Reinhold Forster, hardly a dispassionate witness on the subject of scientists and Cook. In his preface to the German edition of an anonymous account of Cook's third voyage, Forster recalled being told by King of 'this discourteous reply' only a day later, adding that King's 'respect for the man under whose command he was to sail was considerably diminished' by it. By his own account, Forster played – unusually for him – a conciliatory role, although his words were hardly as reassuring as he imagined: 'I took the opportunity of setting things right by describing Cook's character and pointing out that it was in reality not as bad as it appeared, but that he was a cross-grained fellow who sometimes showed a mean disposition and was carried away by a hasty temper.'[5] One could wish for a supporting witness to the conversation between Cook and King; in its absence the plain facts of the situation on the third voyage can be set out. There was no team of civilian scientists, no equivalent of Banks and Solander on the first voyage, or of the Forsters and Sparrman on the second. It may be that the arrogant behaviour of Banks after his return from the first voyage, and the rancorous presence of Johann Reinhold Forster on the second, had turned Cook against taking on board any more 'experimental gentlemen', but of this we have no further proof.

Astronomy was left in the capable hands of Cook and Bayly, with King in support. Natural history was entrusted to William Anderson, surgeon on the *Resolution* and, in King's words, 'by far the most accurate & inquisitive Person on board'.[6] He had sailed on the second voyage at the age of twenty-two as surgeon's mate, and there showed his interest both in natural history and in ethnography. He compiled vocabularies in most of the Pacific islands visited by the expedition, and his long Tahitian vocabulary was printed in the official account of the voyage. There is a glimpse of his collecting activities in Georg Forster's narrative of Cook's second voyage when he complained that one of the surgeon's mates (clearly Anderson) had collected and hidden away 'a prodigious variety of new and curious shells' in New Caledonia as well as 'many new species of plants' while on an excursion that included neither of the Forsters, who were ill at the time.[7] When the *Resolution* returned to England in July 1775 the vessel was visited by Solander, who wrote to Banks that he was 'told that Mr Anderson, one of the Surgeons Mates, has made a good Botanical Collection', although he had not seen it.[8] Before sailing on the third voyage, Anderson sent Sir John Pringle, president of the Royal Society, 'An Account of Some Poisonous Fish in the South Seas', and this had the distinction of being printed in the Society's *Philosophical Transactions* for 1776.

Looked at from Cook's point of view, then, Anderson's appointment to sail on the third voyage as surgeon on the *Resolution* had advantages. Cook knew, and clearly respected, his abilities, and although his main task would be as surgeon he could be relied upon to follow his natural-history interests in his spare time. And, unlike Banks, Solander and the Forsters, he would – thankfully – be subject to naval discipline. To assist him, David Nelson, a gardener from Kew, was to sail on the *Discovery* as a civilian supernumerary, his wages paid by Banks. His function was a narrow but useful one. James Lee, a well-known nurseryman, told Banks that Nelson was 'a proper person for the purpos you told me of, he knows the general run of our collection of plants about London understands something of botany but doe's not pretend to much knowledge in it'.[9] It is perhaps a sign of Nelson's limitations that, after the voyage, Banks complained that he had failed to bring back 'good specimens of the Bread fruit & Flax'.[10]

The supernumerary who achieved the highest reputation for his work on the voyage was John Webber, employed by the Admiralty as a landscape painter. Although he drew birds and fishes, natural history was not his main responsibility. Rather, as Cook explained, 'Mr. Webber was pitched upon ... for the express purpose of supplying the unavoidable imperfections of written accounts, by enabling us to preserve, and to bring home, such drawings of the most memorable scenes of our transactions, as could be executed by a professed and skilful artist.'[11] In this, Webber was hugely successful. The official account of the voyage included sixty-one engravings taken from his drawings, while his later painting of the death of Cook became an iconic representation of the fracas at Kealakekua Bay. Less noticed were his careful natural-history paintings; the British Museum holds fifty-six of these, mostly of birds. Also busy sketching and painting was a young amateur artist, William Ellis, surgeon's mate on the *Discovery*, who brought back 114 watercolours, again mostly of birds. There was no equivalent of the hundreds of plant drawings provided by Parkinson on Cook's first voyage, and by Georg Forster on the second, but natural history was not entirely neglected.

The questions remain: had Cook taken against science and men of science by the time of his third voyage, and is this reflected in the personnel of his final expedition? Although Cook's instructions contained the standard recommendations on astronomical, ethnographic and natural-history observations, there is no doubt that, of all his three Pacific voyages, Cook considered this to be purely one of geographical discovery. For him, the search for the Northwest Passage was the overriding objective. To command the first ships to sail through the passage would bring not only fame and honour but a substantial part of the £20,000 reward offered by Parliament for the discovery, a motive of some importance for one who was moving in circles of society well beyond the means of a naval officer of humble origins and limited resources. Shortly after his appointment Cook wrote to his old Whitby master, John Walker, telling him about the new voyage, and adding, 'If I am fortunate to get safe home, theres no doubt but it will be greatly to my advantage.'[12] These last words suggest a series of quiet commitments and pledges by those in authority. The provision for scientific work on the voyage was adequate, no more. In their manuscript form Cook's instructions were spread over a dozen pages, but those dealing with natural history took up only a few lines.[13] The lack of any full-time observers apart from Bayly stands in sharp contrast with the large contingents of scientists taken on the voyages of La Pérouse, d'Entrecasteaux and Malaspina that followed Cook's. An intriguing snatch of conversation with James Boswell possibly hints better at Cook's feelings on the scientific aspect of the discovery expeditions than his reported outburst to King. Dr Johnson's biographer recalled that they discussed a plan for the government to send 'men of enquiry' to live for several years in the Pacific, where they would learn the local languages and investigate the lifestyle of the people.[14] If accurate, this conversation suggests that Cook realised the limitations of ethnographic investigations carried out on hurried visits by naval vessels. It brings to mind his grumble at Tonga on the third voyage when he thought that his questioning of the islanders resulted in 'a hundred mistakes' and, fatally to the cause of scientific enquiry, the arrival of the ships brought the normal processes of living to a halt. 'It was always holyday,' Cook reflected gloomily.[15]

On the *Resolution*, Anderson kept an excellent journal – official in the sense that it was handed to the Admiralty at the end of the voyage as part of the ship's records, unofficial in the sense that it was a personal record, not a surgeon's log. The two volumes that survive run to 2 September 1777, when the ships were at Vaitepiha Bay, Tahiti. A third volume continued until 3 June 1778, when the expedition was exploring Cook Inlet, Alaska, but that has disappeared. As J.C. Beaglehole has noted, in places 'we learn far more from Anderson than we do from Cook'.[16] Anderson's death on 3 August 1778, after a lingering illness during which he predicted his end with sad accuracy, was widely mourned on the ships – only Cook's own death more so. For Anderson, Cook had 'a very great regard', a rare compliment from one normally reserved in his personal remarks. 'He was a Sensible Young Man, an agreeable companion, well skilld in his profession, and had acquired much knowledge in other Sciences.' In the introduction he wrote for inclusion in the published account of his voyage, Cook acknowledged that Anderson's contributions had enabled him 'to enrich my relation of that voyage with various useful remarks on men and things'.[17] For Captain Charles Clerke on the *Discovery*, his death left 'a Void in the Voyage much to be regretted'. James King felt Anderson's death most of all. Apart from his sense of personal grief, he lamented the loss to the world of scholarship as he wrote how Anderson's knowledge

> took in all natural objects; his application was constant & regular, & far too great for his health; & the Spirit & temper with which he apply'd & studied the different Sciences of Natural History & of the human Species ... was such as must have given a real pleasure to good men & men of Science ... he was the freest from that narrow confind spirit which is fond of hiding its light under a bush [*sic*] of any man I ever knew.[18]

Early entries in Anderson's journal give us some indication of his qualities as an independent obsever. At Tenerife he thought it 'rather presumptuous' to describe the place from a stay of only three days. When the *Resolution* crossed the Equator on 1 September 1776 he wrote of 'the old ridiculous ceremony' of ducking those who had not already crossed the Line that it was 'one of those absurd customs which craft and inconsiderate levity has impos'd on mankind and which every sensible person who has it in his power ought to suppress instead of encouraging'. And when, at the end of 1776, Cook, following Admiralty instructions, took possession of desolate, uninhabited Kerguelen Island, Anderson wrote of the ceremony that it was 'truely ridiculous, and perhaps fitter to excite laughter than indignation'.[19] This remark may have been born out of frustration, for he told Cook that 'perhaps no place hitherto discovered in either Hemisphere under the same parallel of latitude affords so scanty a field for the naturalist as this barren spot'.[20] It was not quite without interest, for in one boggy area Anderson discovered and named (after the president of the Royal Society) a new

genus, the Kerguelen cabbage, *Pringlea antiscorbutica*, while he passed
on for inclusion in Cook's journal perceptive descriptions of the island's
penguins.[21]

Like Banks and the Forsters, as the voyage progressed Anderson
spent more time in describing the peoples of the lands visited than in
listing natural products. When the *Discovery* joined the *Resolution* at the
Cape of Good Hope, Anderson had written to Banks that he was 'happy
to find there is a person in her who understands botany, as he will be
able to procure you every new article in that branch, a task which I have
not vanity enough to suppose myself equal to'. The reference was
presumably to David Nelson, and although Anderson continued to
collect, the letter to Banks gives some indication of his sense of priorities. During the ships' brief stay at Adventure Bay, Tasmania, Anderson's
description of the inhabitants is more detailed than Cook's, with some
deft touches. All the journal-keepers commented on the nakedness of
the Tasmanians, but it was Anderson who gleefully compared their
unabashed behaviour with that of the natives of New Caledonia, whom
he had seen on Cook's second voyage: 'Their men absolutely play with
their Penis as a child would with any Bauble or a Man twirl about the
key of his Watch while conversing with you.' Anderson also supplied an
account of the natural history of the area, with particular attention to
the ubiquitous eucalyptus, specimens of which were brought back and
are now in the Natural History Museum, London. The same formula
was followed during the expedition's longer stay at Queen Charlotte
Sound, where Anderson's descriptions of the trees and plants, especially
the 'useful' ones, were followed by an account of the inhabitants more
detailed than that given by Cook, who, it must be remembered, had
written much about the area on his previous visits to New Zealand. At
Atiu the islanders seized from Anderson plants that he had just collected,
but their unruly behaviour presented him with an even more vexing
problem, for he could get no further than a hundred yards from the
beach: 'It was an oppurtunity I had long wish'd for, to see a people
following the dictates of nature without being bias'd by education or
corrupted by an intercourse with more polish'd nations, and to observe
them at leisure, but was here disappointed.'[22]

In the Cook Islands both Cook and Anderson wrote about the coral
formations. Cook speculated in dispassionate, workaday terms about

how coral might be formed, and how it was subject to a natural expansion through 'the spreading of the Coral bank or reef into the Sea, which in my opinion, is continually though imperceptibly effected, the waves receeding [sic] with it leaves a dry rock behind for the reception of broken coral, sand &c.' Anderson was less concerned with the coral's origins than with its beguiling appearance, and his observations anticipated many later descriptions:

> There was a large bed of Coral almost even with the surface, which afforded perhaps one of the most enchanting prospects that nature has any where produc'd . . . it seem'd suspended in the water which deepend so suddenly that at the distance of a few yards there might be seven or eight fathom. It was at this time quite unruffled, and the sun shining bright expos'd the various sorts of coral in the most beautiful order, some branching into the water with great luxuriance, others lying collected in round balls & of various other figures, all of which were greatly heightened by spangles of the richest colours that glowd from a number of large clams which were every where interspers'd and now half open . . . But the appearance of these was still inferior to the number of fishes that glided gently along . . . The colours of the different sorts were the most beautiful that can be imagin'd, the yellow, blue, red, black &c far exceeding anything art can produce.[23]

In places, Anderson seemed to distance himself from Cook's actions. At Nomuka he was unhappy that a man flogged for stealing an iron bolt remained tied up until a ransom was paid: 'that he should be confin'd in a painfull posture for some hours after, or a ransom demanded after proper punishment for the crime had been inflicted I believe will scarcely be found consonant with the principles of justice or humanity.'[24] In his account of the severe punishments meted out by Cook in Tonga, Anderson noted dryly that the islanders 'were so audacious as to throw cocoa nuts at some of our people', while he thought the slashing of one offender's arm with a knife was 'the least excusable part of the punishment'. About Cook's controversial behaviour at the sacred *inasi* ceremony, where he stripped to the waist, Anderson obviously had reservations: 'it appear'd the natives would rather have dispens'd with such a proof of our curiosity had they not been averse to giving us offence.'[25] In his description of Tongan canoes, Anderson seemed to be aiming a satirical arrow, possibly at Cook, when, after criticising the poor construction of Tongan huts, he wrote that the islanders compensated for this 'by their attention to and great dexterity in Naval Architecture, if I may be allow'd to give it that name'.[26]

From Tonga the ships sailed to Tahiti, and soon after their arrival there Anderson began the third volume of his journal, which has disappeared, although his most important natural-history observations and vocabularies from this manuscript were included in the official account of the voyage, again edited by Dr John Douglas. Some passages were directly acknowledged as being by Anderson: descriptions of Kauai and Nihau in the newly discovered Sandwich Islands; observations on the inhabitants of Nootka Sound – 'I owe everything to him that relates to their language,' Cook wrote; and then, only two months before his death, a short vocabulary and 'many remarks' by Anderson collected at Prince William Sound in Alaska.[27] That Anderson continued to observe and collect almost until the end is a tribute to his courage and sense of duty. On 10 January 1778, as the ships reached the Hawaiian Islands, Bayly had noted that all of Cook's crew were 'in good health except Mr Anderson the Surgeon who is very ill being in a Consumptive State'.[28] In several places in the published account, passages *sound* like Anderson, although they are unattributed. At Tahiti in October 1777 the ships were inundated with cockroaches: 'If food of any kind was exposed, only for a few minutes, it was covered with them; and they soon pierced it full of holes, resembling a honeycomb. They were particularly destructive to birds, which had been stuffed and preserved as curiosities; and, what was worse, were uncommonly fond of ink: so that writing on the labels, fastened to different articles, was quite eaten out.' Some confirmation that this aggrieved cry was Anderson's came in the supplementary note that two different sorts of cockroaches were responsible for the damage: *blatta orientalis*, still on board from the previous voyage, and *blatta germanica*, new arrivals from New Zealand.[29] Still, it is a matter of regret that we lack Anderson's direct observations on the Hawaiian Islands and the northwest coast of America, areas never before visited by British vessels.

There is a sad postscript to the story of Anderson on the third voyage, paralleled by that of Captain Charles Clerke, as told by Lieutenant James Burney of the *Discovery*. Both men were suffering from tuberculosis, and when the ships reached Tahiti in August 1777 before sailing to the North Pacific, they realised that a spell in Arctic waters would almost certainly be their death warrant. They therefore agreed to ask Cook whether they could resign their positions and stay at Tahiti, whose benign climate might restore their health. However, they delayed approaching him because Clerke had not got his 'papers and accounts' in order, and then decided to wait until the ships reached Huahine. There the same problem occurred, and again at Raiatea, though the ships remained at the island for a month. Finally, they agreed to put the matter to Cook and leave the expedition at Borabora, their last port of call in the Society Islands, only for the expedition's stay there to last only a matter of hours; although Cook went ashore neither Clerke nor Anderson accompanied him to put their case. We can guess that Clerke was the more reluctant of the two sick men to leave the expedition. As J.C. Beaglehole puts it, 'what he failed to master was not his accounts but his sense of duty and his loyalty to his own commander.'[30]

Anderson left his collection of natural-history specimens and ethnographic items to Joseph Banks, together with a handwritten list, 'Genera Nova Plantarum ... ad hoc incognitarum ... 1776, 1777', that gave Linnaean specifications in Latin (with an English translation) of a dozen plants collected in the early part of the voyage at Kerguelen Island, Van Diemen's Land and New Zealand.[31] Whether Anderson continued these field notes is not known. In a period well before the establishment of university or museum departments of botany and anthropology, Anderson's donation of his specimens to Banks and his herbarium at 32 Soho Square was entirely predictable. Since his bad-tempered refusal to sail on Cook's second voyage on any terms except his own, Banks had repaired relations with the explorer, and had become the custodian of his reputation as well as the protector of the interests of those who had sailed with him. The third voyage showed this quite clearly. It was Banks to whom Lieutenant Gore entrusted the care of his child if he did not return; Banks to whom Captain Clerke wrote his last, sad letter before his death in Kamchatka; Banks who saw all the official despatches from the voyage; and, most decisively, Banks who became 'the principal adviser and co-ordinating authority' for the editing and publishing of the official account of the voyage.[32] It is little wonder that, in late 1780, James King, who had brought the *Discovery* home after

Clerke's death, wrote to Banks: 'I look up to you as the common center of we discoverers.'[33]

During the 1780s, Banks – by now baronet, president of the Royal Society, confidant of George III and Queen Charlotte, adviser of cabinet ministers, patron of the sciences on an international scale – became the promoter of enterprises associated with Cook's discoveries. Breadfruit from Tahiti, settlement at Botany Bay, the fur trade of the northwest coast of America, the southern whale fishery: all attracted Banks's attention. Away from Cook's old stamping grounds, Banks in his garb as botanist preached the gospel of plant interchange to help develop imperial self-sufficiency. Naturalists on board discovery vessels formed a well-publicised but relatively minor part of this global enterprise; over the years at least 126 overseas collectors were commissioned to send plant specimens to Banks for his herbarium at 32 Soho Square or for the more spacious grounds of the Royal Botanic Gardens at Kew which flourished under his guidance.[34] 'At the peripheries of British power, the age of Banks generated professional opportunities for men of science which did not exist at home. The medical service and surveys of the East India Company, the hydrography of the Admiralty, collecting, and botanic gardens, offered unmoneyed young men the chance to make natural history their trade.'[35] Banks was involved in attempts, sometimes haphazard, occasionally bizarre, to transfer cotton seeds from India to the Caribbean, tea plants from China to Bengal, cochineal insects from South America to Madras and, of course, breadfruit from Tahiti to the Caribbean – to list only a few of his projects.[36] In the period of national reorganisation after the War of American Independence, Banks was a one-man correspondence centre, academic institution, finance house. His only rival in Europe was the Comte de Buffon at the Jardin du Roi, who with royal support employed a series of distinguished savants in Paris, and encouraged the collection of plants and other specimens by French naval officers and colonial officials overseas.

In many ways Cook's third voyage had been a disappointment. When Captain Clerke's letter reporting Cook's death reached London, the king is said to have wept at the news, while at the Admiralty the earl of Sandwich wrote to Banks: 'What is uppermost in our minds always must come first, poor captaine [*sic*] Cooke is no more.'[37] Attempts in 1778 and 1779 to push through the Bering Strait had both failed, with the ships narrowly escaping from the great masses of ice that bore down upon them. The only compensation for a voyage in which both captains had lost their lives, and which had failed in its main objective, was that it drew attention to the ease with which sea-otter pelts could be obtained along North America's Pacific coast from Nootka Sound to Alaska. In the official account King noted that prime sea-otter skins bartered for a few beads or other trinkets on the coast fetched ninety or a hundred dollars apiece on the China market. The account included a drawing by John Webber of a young sea otter killed at Nootka, with a description of how its fur was 'softer and finer' than that of any other known creature (Pl. 16). News of the sea-otter pelts brought back by the crews who had sailed with Bering and Chirikov had lured *promyshlenniki*, or Russian fur traders, into Alaskan waters, where they caused havoc among both the human and animal populations, but in Western Europe little was known about their activities. King's report drew aside the veil of secrecy, and the traders from Europe and the United States were quick to respond to the opportunities offered in the North Pacific. By the mid-1780s, British merchants in India and China were fitting out vessels for the northwest coast, and others from Europe and Boston soon followed.

Banks was interested from the beginning in the possibility of trading voyages to the northwest coast, primarily through his support of a new enterprise, the King George's Sound Company, headed by Richard Cadman Etches. As Richard's brother, John, put it, Banks was a leading patron of schemes 'for prosecuting and converting to national utility the discoveries of the late Captain Cook'.[38] In all, the Etches brothers sent four ships to the northwest coast: the *King George* and *Queen Charlotte*, commanded by Nathaniel Portlock and George Dixon, in 1785, and the *Prince of Wales* and *Princess Royal*, commanded by James Colnett and Charles Duncan, in 1786. All the captains except Duncan had sailed with Cook. A sign of Banks's continuing interest in botanical investigations on these expeditions was shown by the appointment as surgeon and botanist on the *Prince of Wales* of Archibald Menzies, recommended to Banks by Dr John Hope, regius professor of medicine and botany at the University of Edinburgh, who had been one of the first British naturalists to teach the Linnaean system of classification. He told Banks that Menzies was 'early acquainted with the culture of plants and acquired the principles of Botany by attending my Lectures' before serving for several years as surgeon's mate on a naval vessel on the Halifax station where he 'paid unremitting attention to his favourite Study of Botany'.[39]

Menzies had already made himself known to Banks through correspondence from the other side of the Atlantic, and in June 1786 he sent him a package of fifty-six different seeds from the Bahamas, excusing any deficiencies in their classification by explaining that 'they were examined and described by candlelight in the center of a noisy cockpit'.[40] In August he wrote again to Banks, telling him that if he were appointed to the *Prince of Wales* it would 'gratify one of my greatest earthly ambitions & afford one of the best opportunities of collecting Seeds and other objects of Natural History'.[41] Before sailing, Menzies visited Banks's herbarium where he familiarised himself with specimens collected on the northwest coast on Cook's third voyage.

If Menzies kept a journal on the *Prince of Wales* it has not survived, and we are left with only fragmentary information about his activities on the voyage. During calls on the vessel's passage south in the Atlantic towards Cape Horn, he sent two letters to Banks. In November 1786 he wrote from the Cape Verde Islands, assuring Banks that 'the west coast of N. America presents to me a new & an extensive field for Botanical researches as well as other branches of Natural History'. Then, in February 1787, he wrote from Staten Island (Isla de los Estados) off the coast of Tierra del Fuego that 'notwithstanding the severity of the weather in this wild & inhospitable clime I have had several pleasing excursions which enabled me to examine & collect many rare & curious plants some of which are not described even in the 14th Edition of the Linnean System'.[42] The long voyage must have taxed Menzies's medical skills to the full. When the *Prince of Wales* reached the northwest coast in July 1787, Colnett noted in his journal that the crews were suffering badly from scurvy, and that land was 'a pleasing sight to us who had not seen a bush for one & twenty weeks[;] the very thought revivd the drooping spirits of those that were not able to crawl on deck to see it for no one remain'd below that was able to get up on their hands & knees'.[43]

The two seasons spent trading on the northwest coast, interspersed with a stay in the Hawaiian Islands during the winter of 1787–8, brought in a good collection of pelts, though at prices much lower than those obtained by Cook's crews ten years earlier. There are occasional references to Menzies accompanying shore parties, but it is not clear how much time he was able to spend collecting. Although Richard Etches had told Banks that Menzies's activities were to be unrestricted 'so far as can have any tendency to be beneficial to science', the voyage was a commercial undertaking, and the encounters between the crews and the coastal peoples were dominated on both sides by the imperatives of trade. Relations were anxious, sometimes fraught, and misunderstandings were frequent. In his determination to obtain pelts, provisions and wood, Colnett often violated rules of native etiquette and ownership, while the persistent stealing of items from the ships, ranging from needles to a longboat, angered the crews. As the encounters turned violent, it was the native peoples who suffered most, but there was also strain on Colnett's men. When the *Prince of Wales* seemed irretrievably wedged off Banks Island, her boatswain, 'every inch a Sailor, and one from his Countenance and manner one would suppose never entertained fear in his Life ... Cryed like a Child'.[44]

As the ship reached home waters in the summer of 1789, Menzies wrote to Banks, telling him that he was forwarding 'a small box containing dryed Specimens of all the plants I collected in our late voyage round the world ... What seeds I collected are not yet packed up, but you may expect them in the course of a few days.'[45] The reference to 'a small box' suggests that Menzies was not bringing back a great haul either of ethnographic items or plant specimens, and it may well be that difficulties in preserving live plants persuaded Banks that he had to provide better accommodation for such specimens on Menzies's next, and more significant, voyage to the Pacific. Meanwhile, Menzies studied in Banks's herbarium at 32 Soho Square, where he met James Edward Smith, first president of the Linnean Society, who had bought and transferred to London the whole of Linnaeus's library and herbarium.

During Menzies's absence in the Pacific, Banks had been a prime mover in two of the country's most important Pacific projects of the late eighteenth century: the voyage of the *Bounty* to Tahiti to collect breadfruit for the West Indies, and the sailing of the First Fleet to New South Wales to establish a convict settlement. As though to demonstrate that his outlook was global rather than regional, he still retained his interest in the North Pacific, where by the late 1780s the northwest coast was attracting the attention of governments as well as of merchants and geographers. Despite the insistence to the contrary of Cook in 1778, the surveys of the maritime fur traders seemed to show that a strait leading deep into the interior might yet be found. The coast's furs could provide the missing link needed to make viable a northern network of trade encompassing China and Japan to the west and the territories of the Hudson's Bay Company to the east. In London speculation and plans on the subject were brought to an abrupt halt when, at the beginning of February 1790, news arrived from Madrid of the Spanish seizure of British vessels and property at Nootka Sound the previous summer. Among the initial responses of the British government was a decision to send a naval expedition to the northwest coast, commanded by Henry Roberts, who had sailed on Cook's second and third voyages. Two ships, the *Discovery* and *Chatham*, were to go by way of New South Wales, where they would take on board thirty men to establish a settlement on the northwest coast. Menzies, back from his Pacific voyage in the *Prince of Wales*, was appointed as botanist and naturalist to the planned expedition, but it was abandoned as the crisis with Spain heightened and a general mobilisation of the fleet was ordered.

Following the Nootka Sound Convention of October 1790, the naval expedition was reinstated, but with different instructions, and a different commander, George Vancouver, who had also sailed on Cook's last two voyages. His task was partly exploratory – to carry out a detailed survey of the northwest coast – and partly diplomatic – to receive restitution of the land seized by Spain at Nootka in 1789. Vancouver's expectation was that he was to locate a suitable place for a settlement on the coast, but the Spanish reluctance to cede more than a tiny strip of land at Nootka, and a change of policy by the British government about the desirability of settlement on the northwest coast, brought him frustration and a sense of failure.[46] Menzies again asked to serve on the expedition, but this time as surgeon, a post in which 'my vacant hours from my professional charge' would be spent as a naturalist. It is not clear why Menzies requested this dual role, nor why Vancouver objected to it, insisting that Alexander Cranstoun should be appointed surgeon of the *Discovery*. It was an awkward beginning to a relationship that would endure many twists and turns before the end of the voyage. One particular source of contention between Vancouver and Menzies was the wood and glass plant frame, in effect a small greenhouse, that was placed on the quarterdeck of the *Discovery*, larger and more cumbersome than the plant cases supplied to the La Pérouse expedition. The plan of the structure was Banks's, who had already ordered one for the *Guardian* before it sailed with plants for the infant colony of New South Wales. It seemed to meet Menzies's request to Banks that 'a suitable place may be constructed for bringing home the plants in safety', undoubtedly a plea stemming from problems on his voyage on the *Prince of Wales*.[47] Naturalists on the discovery voyages trying to bring home their material faced persistent problems in fending off the ravages of insects and rodents, and in protecting them from the disruptive effects of climate changes and saltwater. Pressed specimens could be kept in sheets of paper but seeds were more difficult to preserve, and live plants, which needed both protection and watering, presented the greatest challenge of all. The scale of the problem was demonstrated in the publications of the 1770s by the botanist and merchant John Ellis, who was concerned with the casualty rate of plants and seeds brought back from the Eastern Seas. Of seeds obtained in China, he wrote, 'scarce one in fifty came to any thing'.[48] His solution was a series of boxes with wire fronts and moveable shutters, but for long sea voyages they did not provide sufficient protection if exposed to the elements on deck (Pl. 17). A ship's great cabin was the best location for plants, but its availability depended on the goodwill of the captain and his officers. An alternative was a plant frame on deck, given over exclusively to the protection and care of the plants, and such was supplied to Vancouver's *Discovery* in 1791, and to Matthew Flinders's *Investigator* some years later.

However, for Vancouver, and indeed for any naval captain sensitive about his authority, the location of a large greenhouse on his quarterdeck, filled with tubs and pots, and needing constant attention and watering, was an affront. So might have been the 'Heads of instructions for Mr Vancouvers Conduct towards Mr Menzies' drawn up by Banks and sent to Evan Nepean, undersecretary of state at the Home Office, if Vancouver ever saw them. He was to supply men to help Menzies carry luggage and water; no lumber should be put into 'the Plant hutch', any broken glass panels should be 'immediately' repaired, and so on. Even more irritating to Vancouver, a vastly experienced hydrographer, would have been the instructions on surveying drawn up by Banks, though again it is not certain whether Vancouver saw them.[49] Other documents among the Banks papers indicate frayed relations between the two men. In December 1790, Captain Portlock wrote to Banks that he had 'deliverd your Message to Captain Vancouver respecting his waiting on you but think you may not Expect a call from him'.[50] Most revealing of all was a note Banks wrote to Menzies after the expedition had sailed: 'How Capt Van will behave towards you is more than I can guess unless I was to judge by his Conduct towards me which was not such as I am usd to receive from Persons in his situation.' A continuing irritant was Banks's inability to persuade Nepean to let him have a copy of the instructions Vancouver had been given respecting Menzies. As he told Nepean, these would be essential 'if the difference between these gentlemen when they went to Sea Should Continue during their Voyage'.[51]

Banks drew up detailed instructions for Menzies. He was the only naturalist on the expedition, but his remit was no less than 'an investigation of the whole of the Natural History of the Countries you are to visit'. Apart from standard instructions about describing and collecting plants, he was to report on which 'Grains, Pulse and Fruits cultivated in Europe are likely to thrive' in case it should 'hereafter be deemed expedient to send out Settlers from England' – an indication that Banks had not abandoned his hopes of establishing some sort of settlement on the coast. Finally, he was to keep a journal, and this, together with the ethnographic items and natural-history specimens he had collected, was to be delivered to the home secretary on his return. As far as Menzies's journal was concerned, this seemingly innocuous order was to cause considerable trouble towards the end of the voyage.[52] For botanists, the journal itself, although a lively record of the voyage, and complete up until March 1795 when the ships were off the coast of Chile on their homeward voyage, is disappointing in two ways. Recent research has shown that it is a clean copy written after the return of the expedition to England, when Banks's request for 'a full and continued narrative of the voyage' undoubtedly led to considerable editorial work by Menzies on his original manuscript, no trace of which remains.[53] Secondly, and probably because of Banks's desire for a publishable narrative of the voyage that would pre-empt Vancouver's official account, the journal in places contains surprisingly little detail on Menzies's botanical activities (especially for the 1793 and 1794 surveying seasons). The assumption is that these were recorded by Menzies in a separate journal or notes, which have disappeared. Some confirmation of the existence of such a record comes from Vancouver's protest to the Admiralty after his return at the suggestion that 'the observations made by Mr. Menzies, on the various subjects of the Natural History of the different Country we visited, should in some way be connected with my account of the Voyage'.[54] Painstaking reconstruction of his botanical work has been carried out using the labels attached to the dried plants that he brought back with him, but it is far from being a complete survey.[55]

The *Discovery* and *Chatham* sailed from England on 1 April 1791, a date that later prompted a derisory comment from Vancouver, always sceptical about the chances of finding a northwest passage in temperate latitudes. For Menzies, the voyage out was marked by a significant addition to his duties. After the ships left the Cape of Good Hope, the *Discovery*'s surgeon, Alexander Cranstoun, was struck down with a paralytic stroke and Vancouver asked Menzies to take over as surgeon, an arrangement that was formalised when the ships reached Nootka Sound. Menzies was already acting as personal physician to Vancouver, who was not in good health, but he only agreed to Vancouver's 'earnest solicitation' to take up official duties with some reluctance. From this time on, as a regular member of the ship's company rather than as a civilian supernumerary, he would come under Vancouver's direct orders. As compensation he gained an extra cabin, which he intended to use to store his collections.[56] These soon accumulated as the expedition called at King George Sound in the Great Australian Bight and at Dusky Sound in New Zealand. In Tahiti, Menzies was very aware of his predecessors: at Matavai Bay he saw shaddock trees 'in a very flourishing state' planted by Banks, and orange trees two feet high planted by Bligh.

In April 1792 the ships reached the northwest coast of America and began the main part of their mission by surveying Puget Sound and the Strait of Juan de Fuca. On a boat excursion to Discovery Bay at the southeast extremity of the Strait, Menzies first saw the striking madrona or arbutus tree – *Arbutus menziesii* – originally named by him the oriental strawberry tree. It was, he wrote of this distinctive tree of the region, 'a peculiar ornament to the Forest by its large clusters of whitish flowers & ever green leaves, but its peculiar smooth bark of a reddish brown colour will at all times attract the Notice of the most superficial observer'. The region was a botanist's paradise, with its wide stretches of lawn bursting with flowers and marked by tall pines, 'the whole seeming as if it had been laid out from the premeditated plan of a judicious designer'.[57] A few days later Menzies saw 'vast abundance in full blossom' of *Rhododendron macrophyllum* or Pacific rhododendron – later to become the official flower of the state of Washington. Confirmation that his journal was being kept with an eye on the general reader rather than the botanical specialist came in an entry that described and named two striking plants – the *Cypripedium bulbosum* (*Calypso bulbosa* or false lady-slipper) and the *Rhodendron ponticum* (*Rhododendron californicum* or large-flowered rhododendron) – but was followed by a statement by Menzies that he saw 'a variety of other plants which would be too tedious here to enumerate'.[58] Accompanying Lieutenant Peter Puget on an expedition along the intricate coastline of Puget Sound, Menzies grumbled that 'these surveying Cruises were not very favorable for my pursuits as it afforded me so little time on shore at the different places we landed',[59] but on that one-week trip he went ashore fourteen times, collected a dozen plants unknown to science and, before long, realised that such excursions took him to places otherwise out of reach. Whether accompanied by his servant or on his own, Menzies had to keep within hailing distance or musket shot of the boat. It is difficult to judge Vancouver's reaction to these botanising trips, but it is doubtful whether he gave them much priority. Describing this Puget Sound excursion, Vancouver added a sardonic note to the effect that 'Mr. Menzies found constant amusement; and, I believe, was enabled to make some addition to the catalogue of plants'.[60]

From the idyllic surroundings of Puget Sound the ships headed up the Strait of Georgia. Already it was clear that the *Discovery* was too large, and the *Chatham* too clumsy, for close inshore work. The survey would have to be carried out from the ships' boats, and during its three-year extent there were forty-six boat excursions, some of them lasting weeks at a time. This change of method condemned officers and men alike to gruelling, tiring work, especially as they moved farther north. Menzies was full of admiration for the men – 'in open Boats exposd to the cold rigorous blasts of a high northern situation with high dreary snowy mountains on every side, performing tiresome labor on their Oars in the day, & alternatively watching for their own safety at night, with no other Couch to repose upon than the Cold Stony Beach or the wet mossy Turf . . . & enduring at times the tormenting pangs of both hunger and thirst'.[61] On these excursions, it always seemed to be raining, food often ran short, the canoes appearing out of the mist might or might not be friendly. The crews encountered the coastal peoples in all their variety, Salish, Kwakiutl, Haida and Tlingit among them. For Vancouver's men, they were an ever-present but unpredictable accompaniment to the business of surveying, often a distraction and an irritant, sometimes a threat. When the ships headed up the Strait of Georgia the mountains began to press in so close to the shore that the boat parties often found it difficult to find any level ground where they could spend the night. Puget's description of one day's experience can stand for many. It was eleven o'clock at night before they landed 'after a most disagreeable & laborious Row, the Boats and their Furniture were all wet nor was there a Spot to shelter us from the Inclemency of the Weather . . . as it was equally uncomfortable either remaining in the Water afloat or on Shore'.[62] Although he had the help of two surgeon's mates, Menzies's medical duties prevented him from accompanying as many of the boat excursions as he would have wished, but he was able to continue his botanising work nearer the ship. It was a tribute to his medical skills that, of the six deaths on the long voyage, only one was from illness and none from scurvy.

During this first season of the survey in the Strait of Juan de Fuca, Vancouver was dismayed to find two Spanish vessels engaged in the same task. They had been sent by Alejandro Malaspina, in command of a Spanish surveying expedition, to complete explorations begun in the Strait two years earlier. To Vancouver, convinced that he was the first European to sail in these waterways, the sight was a mortifying one, but relations between the two groups were good as they exchanged charts and information. In August, Vancouver sailed around the northern tip of the large island later to be named after him and reached Nootka, where he engaged in amicable but inconclusive negotiations with the Spanish commandant, Bodega y Quadra, over the restitution of British property. Menzies was impressed by the botanical work being done there by José Moziño and Atanasio Echeverría y Godoy, 'who as a Natural History Painter had great merit'. As Menzies pointed out, their presence showed that the Spaniards 'mean to shake off entirely that odium of indolence & secrecy with which they have long been accused',[63] and it may well have struck him as ironic that the Vancouver expedition had no professional artist to help him record his discoveries. In all, Moziño listed 352 plants as well as mammals, birds, fish, shellfish and insects, many of which were drawn by Echeverría. He used his observations at Nootka to compile 'Noticias de Nutka', described by its modern editor as 'a unique ethnographic and historical study', although it remained unpublished until 1970.[64]

From Nootka, Vancouver's ships sailed south to California before spending the winter in Hawaii. At Monterey, Menzies was able to send a report to Banks, advising him against the idea of a settlement in the Strait of Juan de Fuca, where few sea otters had been seen. The Queen Charlotte Islands and the ocean coast of Vancouver Island offered better trading prospects, and the sooner a post was established there, the greater would be the advantages. Although Menzies had treated Vancouver for his bouts of ill health, and the two men had often accompanied each other on shore excursions, he clearly had reservations about his captain. There were many 'strange things' happening on the voyage, he told Banks, and he criticised Vancouver for having 'already perpetuated his name on the Coast' by naming the great offshore island after himself and Bodega y Quadra. The plant frame had been a disappointment, 'for if it is uncovered in raining weather to admit air, the dripping from the rigging impregnated with Tar and Turpentine hurts their foliage & soil – and if the Side Lights are opened Goats – Cats – Pigeons – Poultry &c &c are ever creeping in & destroying the Plants'.[65]

For Vancouver and his crews, life followed an annual cycle: winter in the Hawaiian Islands, summer spent in arduous small-boat work on the northwest coast, followed by a call at Nootka in the hope that there would be further instructions from London. The winter stays in the islands provided much needed rest for the crews, but for Menzies the Hawaiian visits were at the wrong time of year. Although he found many new trees and plants, 'few of them being at this time in bloom, my researches were in a great measure fruitless … my situation [is] the most vexatious & tantalizing that a scrutinizing Botanist could ever be

A View of the Watering Place at TENIAN.

1 Lieutenant Brett's drawing of the watering place on the island of Tinian in the North Pacific where Commodore Anson's *Centurion* stopped during the circumnavigation of 1740–4. Palms and citrus trees are shown, but dominating the foreground is a breadfruit tree, whose fruit 'was constantly eaten by us during our stay upon the Island instead of bread, and so universally preferred to it, that no ship's bread was expended.'

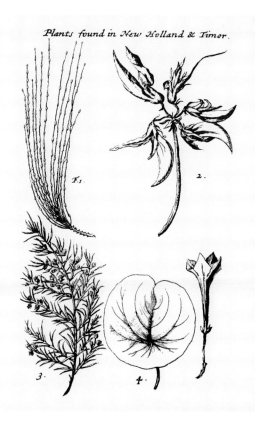

Plants found in New Holland & Timor.

2 An unknown artist's drawings in William Dampier's *Voyage to New Holland* of 'Plants found in New Holland and Timor'. Fig. 2 is of the plant later known as Sturt's desert pea. The accompanying notes by the botanist John Ray describe the plant as having no leaves, but scarlet flowers with a deep purple spot.

3 Top right is of the dried specimen of Sturt's desert pea brought back by William Dampier. It shows how close to the original was the drawing of Dampier's unknown artist (Pl. 2), but also how cumbersome was Ray's classification system. He named the plant *Colutea Novae Hollandiae floribus amplis coccineis, umbellatim dispositis macula purpurea notates.*

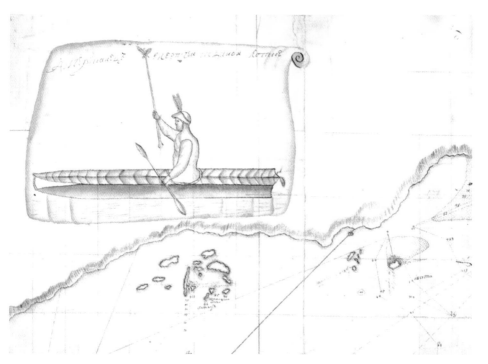

4 Sven Waxell's drawing of an Aleut in his sealskin craft near Bird Island, Alaska. He is gesturing to Bering's ship with a spruce stick; both the stick and the paddler's cap have falcon feathers attached.

5 Drawings from Sven Waxell's chart of Bering's second Kamchatka expedition. The sketch of the long-extinct sea-cow, *Hydrodamalis gigas* (A), is the only one known to have been drawn by (or under the supervision of) someone who had actually seen the mammal, and settled the question of whether the creature had a forked tail.

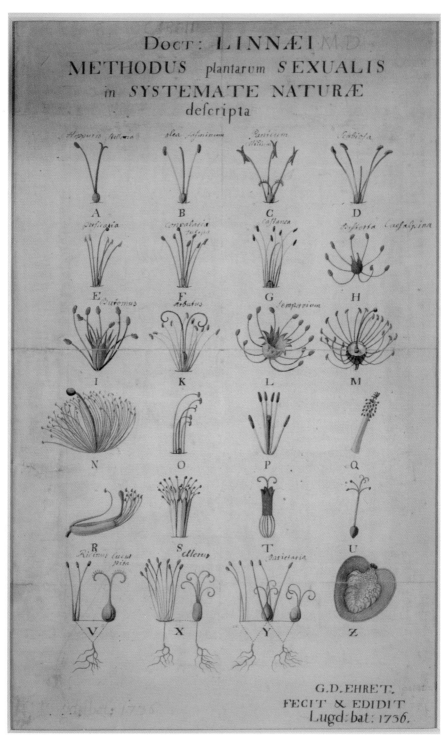

6 A drawing by Georg Dionysus Ehret to illustrate Linnaeus's system of classification, 1736. Linnaeus's twenty-four classes are represented by letters of the alphabet.

Calyxis ternaria.

Sydney Parkinson pinx. 1768.

Brasil

7 Sydney Parkinson's painting of a bougainvillea plant, *Bougainvillea spectabilis*, first described by Commerson in Rio de Janeiro in 1767; this specimen was collected by Joseph Banks during the *Endeavour*'s visit the following year.

Stodium altile.

Otaheite

Sydney Parkinson pinxt 1769.

8 Sydney Parkinson's painting of a Tahitian breadfruit, *Artocarpus atilis*, 1769. On his first visit to Tahiti Cook praised breadfruit as being among those 'articles the Earth almost spontaneously produces or at least they are rais'd with very little labour, in the article of food these people may almost be said to be exempt from the curse of our fore fathers.'

9 A pencil drawing by Sydney Parkinson, entitled 'Kanguru, Endeavour's River'. It fits with Banks's description of the creature: 'Its fore legs are extremely short and of no use to it in walking, its hind again as disproportionately long; with these it hops 7 or 8 feet at each hop.'

VII

10 This Australian parrot was brought to England by Joseph Banks after Cook's first voyage. It was said to have belonged to Tupaia, the Polynesian priest-navigator who died at Batavia, before it came into Banks's possession. After the parrot's death it was presented by Banks to Marmaduke Tunstall's private museum where it was painted by Peter Brown, a leading zoological artist.

11 This caricature of Joseph Banks as 'The Fly-Catching Macaroni' shows him as a foolish dandy, adorned with an ass's ears, trying to catch a butterfly while precariously balanced on twin globes of the world.

12 This caricature of Daniel Solander as 'The Simpling Macaroni' shows him holding a botanist's curved knife in one hand, and a dug-up plant in the other.

13 Drawing by George Forster on Cook's second voyage of a chinstrap penguin on an ice floe, December 1772. This seems to have been the only shot on this particular day, for in his journal Forster wrote that 'The chace of penguins proved very unsuccessful...the birds dived so frequently, continued so long under water, and at times skipped continually into and out of the water, making way with such astonishing velocity in a strait line, that we were obliged to give over the pursuit.'

14 Painting by John Francis Rigaud showing George Forster in Tahiti drawing a bird held by his father, Johann Reinhold Forster.

15 A devilfish or giant ray, caught by Johann Reinhold Forster on 10 May 1774 in Matavai Bay, Tahiti, and drawn by his son, George, the next day.

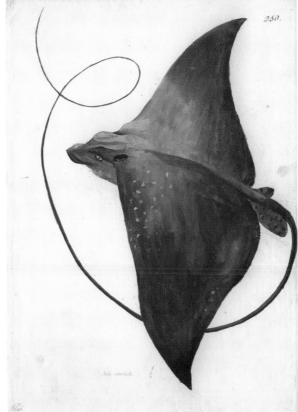

16 The official account of Cook's third voyage described how at Nootka Sound in 1778 the expedition's artist, John Webber, drew a dead sea otter purchased from local hunters. The specimen was 'rather young, weighing only twenty-five pounds, of a shining or glossy black colour, but many of the hairs were tipt with white, gave it a greyish cast at first sight.'

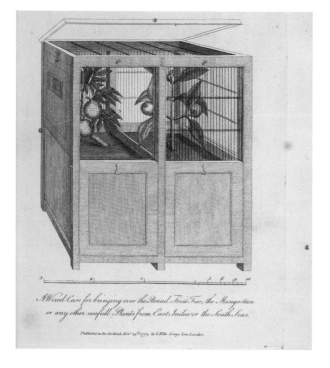

A Wired Case for bringing over the Bread Fruit Tree; the Mangostan or any other usefull Plants from East India or the South Seas.

Published in the Act directs Nov. 19th 1774 by I. Ellis Grays Inn London.

17 The problem of transporting live plants at sea was addressed by the merchant and collector John Ellis, who had noticed that of the many seeds sent to England from China 'scarce one in fifty ever comes to any thing'. His 1775 tract showed a wire cage for holding breadfruit and other plants, and included 'Observations and Instructions for Captains of Ships, Surgeons, and others, who are unacquainted with Botany.'

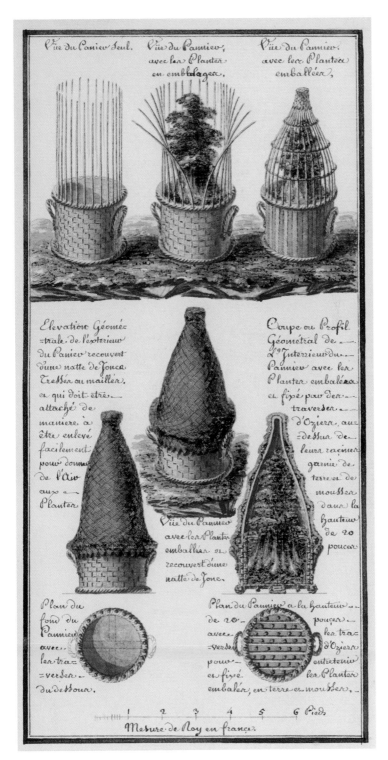

18 A variety of receptacles, designed by André Thouin, head gardener at the Jardin du Roi, Paris, for transporting plants on shipboard during the La Pérouse voyage.

Massacre of De Langles.

19 A melodramatic representation by Nicolas Ozanne of the 'Massacre of De Langle, Lamanon and Ten Others of the Two Crews' from the La Pérouse expedition at Tutuila, Samoa, 11 December 1787.

20 Part of Joseph Banks's herbarium and library in the spacious quarters of his London house at 32 Soho Square which held the bulk of his collections from the 1770s to his death in 1820.

EUCALYPTUS GLOBULUS.

21 The blue gum collected by the botanist Labillardière in May 1792 during the visit of the d'Entrecasteaux expedition to Van Diemen's Land. Timber from the tree was used by the ship's carpenter to repair the *Espérance*. It is now the floral emblem of Tasmania.

SAUVAGES DU CAP DE DIEMEN PRÉPARANT LEUR REPAS.

22 Jean Piron's sketch of 'Sauvages du cap de Diemen préparant leur repas', showing friendly relations between d'Entrecasteaux's crew and Tasmanian Aborigines as they prepare a meal. It has been suggested that the figure third from the left, wearing a cap and pantaloons, is Piron himself.

23 José Cardero's painting is a rare depiction of naturalists at work, in this case from Malaspina's expedition, as they search for specimens on the beach of Isla Naos off the city of Panama in late November or early December 1790.

24 José Cardero's painting of the imposing wooden edifices at a Tlingit burial ground, Port Mulgrave, Alaska, during the visit of the Malaspina expedition, June 1791.

25 Juan Ravenet's representation of the death of Antonio Pineda, chief naturalist of the Malaspina expedition, on 23 June 1792 at Badoc on the island of Luzon.

26 Juan Ravenet's painting of Malaspina (the figure on the far right) and his assistants carrying out gravity experiments at Port Egmont, West Falkland, June 1794, with their portable observatory shown on the left.

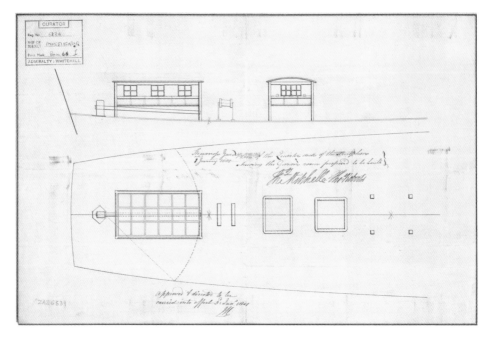

27 Plan of the quarter-deck of Matthew Flinders's *Investigator* showing the prefabricated plant frame that was erected at Port Jackson in 1802 to accommodate live plants. It was probably similar in size and location to that on Vancouver's *Discovery* which caused such friction between Archibald Menzies and the captain.

28 Engraving from Charles-Alexandre Lesueur's drawing of the zoologist, François Péron, wearing the broad-brimmed hat favoured by most naturalists of the period (see also Pl. 23), as he excavates an Aboriginal cremation site at Cape des Tombeaux, Maria Island, Tasmania.

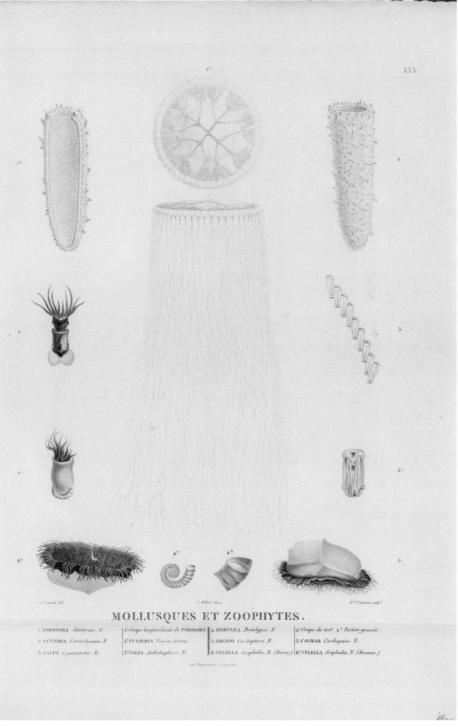

29 Engraving from Charles-Alexandre Lesueur's detailed drawing of molluscs and zoophytes from the Baudin expedition.

30 Engraving from Charles-Alexandre Lesueur's portrait of François Péron, completed a few days before his death, showing him reading the proofs of the second volume of his *Voyage*.

31 The grounds of Empress Josephine's chateau at Malmaison, with kangaroos, emus, black swans, and a variety of Australian flora brought back by the Baudin expedition.

32 This palm is a fine example of one of Ferdinand Bauer's botanical watercolours from the *Investigator* voyage. It probably began as a pencil sketch made on the spot with references to his colour coding system, and was then completed on his return to England.

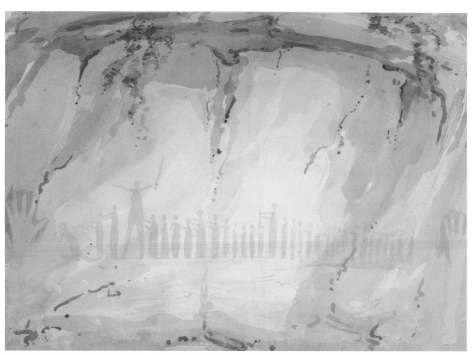

33 One of the earliest copies of an Aboriginal rock painting, made by William Westall on the *Investigator* in 1803, showing a party of Aborigines in single file following a kangaroo.

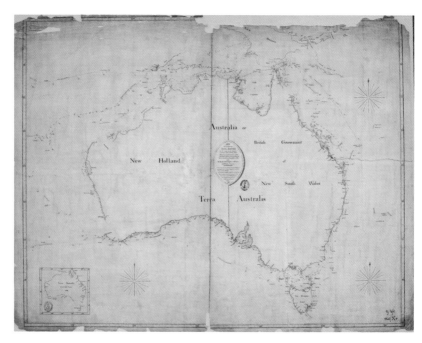

34 The general chart of 'Australia or Terra Australis', completed by Matthew Flinders in 1804 while he was detained on Isle de France, but not published until 1814.

35 A Fuegian of the Tekeenica tribe with his canoe, dog and family shelter. In his *Narrative* Robert Fitzroy described the natives of the southeast part of Tierra del Fuego as 'low in stature, ill-looking and badly proportioned...Their rough, coarse, and extremely dirty black hair half hides yet heightens a villainous expression of the worst description of savage features. 'They were', he concluded, 'satires upon mankind'.

36 Ruins of the cathedral at Concepción after the great earthquake of 20 February 1835 as drawn by Lieutenant Wickham of HMS *Beagle* two weeks later. In his *Narrative* Robert Fitzroy noted that the cathedral's walls were 'four feet in thickness, supported by great buttresses.'

differing from the Progne purpurea of both Americas, only in being rather duller coloured, smaller, and slenderer, is considered by Mr. Gould as specifically distinct. Fifthly, there are three species of mocking-thrush—a form highly characteristic of America. The remaining land-birds form a most singular group of finches, related to each other in the structure of their beaks, short tails, form of body, and plumage: there are thirteen species, which Mr. Gould has divided into four sub-groups. All these species are peculiar to this archipelago; and so is the whole group, with the exception of one species of the sub-group Cactornis, lately brought from Bow island, in the Low Archipelago. Of Cactornis, the two species may be often seen climbing about the flowers of the great cactus-trees; but all the other species of this group of finches, mingled together in flocks, feed on the dry and sterile ground of the lower districts. The males of all, or certainly of the greater number, are jet black; and the females (with perhaps one or two exceptions) are brown. The most curious fact is the perfect gradation in the size of the beaks in the different species of Geospiza, from one as

1. Geospiza magnirostris. 2. Geospiza fortis.
3. Geospiza parvula.   4. Certhidea olivaea.

large as that of a hawfinch to that of a chaffinch, and (if Mr. Gould is right in including his sub-group, Certhidea, in the main

37 Four Galapagos finches with different beaks, specimens from the voyage of HMS *Beagle*, as illustrated in the 1870 edition of Charles Darwin's *Journal of Researches* in which he wrote that the 'preservation of favourable individual differences and variations, and the destruction of those which are injurious, I have called Natural Selection, or the Survival of the Fittest.'

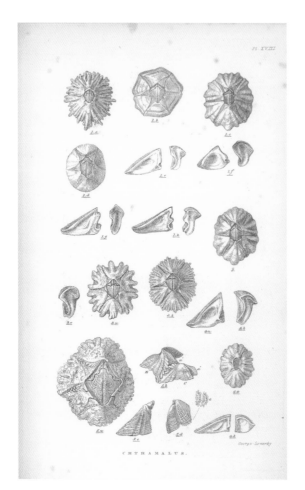

CHTHAMALUS.

38 A selection of the barnacles studied by Charles Darwin in the 1850s, drawn by the natural history artist, George Sowerby.

39 A Bancks dissecting microscope, believed to be the one taken by Charles Darwin on the voyage of HMS *Beagle*, and today on display at his home at Down House, Kent.

placed in, surrounded on all sides by new & rare objects & yet destitute of the means of obtaining a knowledge of them'. At Maui in March 1793 he had rather better luck. Again, many plants were not in seed or flower but those that were 'were quite new & undescribed by any Botanist whatever'.[66]

In May the ships returned to the northernmost point of their explorations the previous year, with the boats much improved. Menzies, who went on the first of that season's boat expeditions, described how the stern of each boat was protected by a curtained canopy and the rest of the boat by a large awning. A tent with a thick floor was large enough to shelter the crew at night, while their spare clothes and provisions were packed in watertight canvas bags. Wheat and soup gave the crew two hot meals a day, while the officer in charge had an allowance of spirits to be distributed among the men as he thought fit. Even so, the work was hard. Much of the survey was carried out in drenching rain, and Vancouver himself was ill for long periods at a time. Along one especially fractured stretch of coast, the survey boats were away from the ship for twenty-three days, and charted seven hundred miles of shoreline – all to advance knowledge of the mainland coast by only sixty miles. The several hundred place names given by Vancouver mostly followed conventional forms – royalty, the peerage, ministers and patrons, fellow officers – but a few hint at something more heartfelt. There is Desolation Sound, Foulweather Bluff, Traitors Cove, Escape Island, Destruction Island, Poison Cove, all of them reminders of dangers and hardships; and an unofficial name given by the crew of the *Chatham* that never found its way onto the charts – Starve-Gut Cove. Vancouver's account of the voyage recorded the survey in meticulous detail, but Menzies's journal has fuller descriptions of the wildlife of the area. At Observatory Inlet near the estuary of the Nass River in latitude 55°N., Vancouver simply noted a 'great abundance' of salmon. In his entry Menzies wrote:

This being about the beginning of the Spawning season for Salmon on the Coast, a stream of fresh water that entered into the bay swarm'd with them, though there was scarcely water to cover them, so that they could be caught with the hand in any quantity we chose; in shooting the Seine across the entrance of the Stream it surrounded such a prodigious quantity of these Salmon that it was impossible to haul them all on shore without endangering the Seine.[67]

Farther north, at Port Stewart, Menzies found 'a plant with blue berries', *Clintonia uniflora*. His treatment of the plant followed best botanical practice: he sketched it, prepared a dried specimen for his collection and dug up several live plants for the frame on the quarterdeck.[68]

At the end of the 1793 season the ships again headed south along the Californian coast before spending the winter in the Hawaiian Islands. At Santa Barbara, Menzies was once more frustrated in his botanising, for although he found new plants at 'almost every step', few were in flower or seed. His main problems, however, were on board rather than on shore. While on the Californian coast he wrote to Banks, complaining of Vancouver's 'passionate & illiberal' abuse whenever the subject of the plant frame arose, and he enclosed a copy of a letter he had sent to the captain. It began in unpromising fashion: 'It is really become so unpleasant to me to represent to You verbally any thing relative to the Plant-frame on the Quarter-deck that I have now adopted this method.'[69] He asked that the frame be covered with strong netting to protect the plants against the depredations of poultry and pigeons, and for the services of a crew member to look after the plants while he and his servant were off the ship. The first request, Menzies reported to Banks, was met – two months after the sending of the letter – but not the second, and he often had to ask an officer to look after the frame while he was away. For good measure, Menzies added that he had the services of a boat for his botanical excursions only four or five times during the whole voyage, and then only for a few hours at a time.

At Hawaii, Menzies and three crew members climbed the island's highest mountain, the formidable Mauna Loa, the first ascent by Europeans. 'After the most persevering and hazardous struggle that can possibly be conceived', Menzies used a portable barometer to estimate the height of the mountain as 13,634 feet, within a hundred feet of its actual size. For Vancouver, the stay was marked by an agreement in February 1794 with the island's most powerful chief, Kamehameha, to transfer the sovereignty of Hawaii to George III, although it was never acknowledged by the home government, deeply engaged in war with revolutionary France by the time Vancouver returned.

The final season of the survey took Vancouver to the Alaskan coast, where he discovered that 'Cook's River' ended only a little distance beyond the farthest point reached by Cook eighteen years before. For Menzies, the time spent in the freezing temperatures was disastrous, for most of the live plants from California and Hawaii that he kept in the plant frame were killed by the cold. From the renamed Cook Inlet the ships headed east along the coast until they reached the northernmost limit of their survey of the previous year. On the *Discovery*, frostbite, outbreaks of food poisoning and early symptoms of scurvy kept Menzies busy with his surgeon's duties. He had no time to go on the first boat excursions of the season, although while in Tlingit territory west of Chatham Strait he was fascinated to see from the ship patches of ground where, he thought, tobacco was being grown. In rather self-righteous vein he commented: 'Here then we see the first dawns of Agriculture excited among these savages, not in rearing any article of real utility either to their comfort or support, as might naturally be expected, but in cultivating a mere drug to satisfy the cravings of a fanciful appetite.'[70] During his spells ashore he continued to collect local plants to fill the spaces in the plant frame emptied by the frosts of Cook Inlet, and he also picked any greens and berries he could find to add to the men's diet. On 19 August 1794 the boats returned from their final survey, and an extra allowance of grog was served. There was no way through the towering range of mountains, and among the celebrations Vancouver wrote that 'no small portion of facetious mirth passed amongst the seamen, in consequence of our having sailed from old England on the *first of April*, for the purpose of discovering a north-west passage'.[71]

The first leg of the homeward voyage followed a familiar route: Nootka, then the Californian coast. At Monterey, Menzies found that the region 'abounded with a variety of new and curious plants', and collected forty-two species, including the giant Californian coast redwood and the commercially important Monterey pine.[72] With them he was able to fill spaces in the plant frame where plants brought on board at Nootka had been washed away by high waves. A call at the Tres Marias Islands was less successful. Menzies collected a number of plants new to him, but they were all lost in high surf that swamped the boats. At Albemarle Island, one of the Galapagos group, the shore parties saw none of the giant tortoises from which the islands took their name, but there was much else to fascinate them. Menzies wrote that 'to our great astonishment [we] found these rugged shores on the equator, inhabited with Seals & Penguins in vast abundance, while the surface of the sea that laved them with its surf, swarmed with large lizards swimming around in different directions'.[73] On the ships' visit to Santiago, Menzies obtained specimens of the monkey puzzle tree, *Araucaria araucana*, in an unusual way: he pocketed some kernels that were served as dessert at a dinner given by the governor; on board ship they sprouted and arrived as seedlings in England where several were planted at Kew Gardens. One, known to King William IV as 'Sir Joseph Banks' tree', lived until 1892.[74]

Not all of Menzies's plants survived these final stages of the voyage, for, when the *Discovery* left St Helena, Vancouver had to transfer so many men to a Dutch ship taken as a prize that he assigned Menzies's servant to watch-keeping duties. For Menzies, the man's priority was to look after the plant frame; given the incessant wrangling over the structure during the voyage, there was a sad inevitability about what followed. As bad weather loomed the servant was engaged elsewhere in the vessel and neglected to put the panels in place 'till a very heavy and sudden deluge of rain crushed down the tender Shoots of many of the plants that never afterwards recovered it'. A distraught Menzies wrote to Banks: 'I can now only show the dead stumps! of many that were Alive and in a flourishing state when we crossed the Equator for the last time.'[75] Unfortunate though this particular episode was, the plant frame had not been a great success. Sudden climatic changes, saltwater washing over the frame in heavy seas, and the depredations of animals and birds, all made the preservation of the young plants a chancy business.

Vancouver refused to discipline the errant servant, who would have been acting under his orders. When Menzies protested, '[he] immediately flew in a rage, and his passionate behaviour and abusive language on the occasion, prevented any further explanation'. Refusing to apologise, Menzies was placed under arrest. Vancouver's version of events was given in a letter to Evan Nepean, by now secretary to the Admiralty, in which he demanded that Menzies should be court-martialled. He insisted that Menzies had 'behaved to me on the quarter-deck, with great contempt and disrespect', and had refused to withdraw his 'harsh and improper expressions'.[76] Their relations worsened when Menzies declined to give up his journal, as the rest of the crew had done, insisting that he must wait for instructions from Joseph Banks or the home secretary. There are several possible explanations for Vancouver's behaviour towards Menzies. It is clear that even before the voyage began he resented Banks's interference, as he regarded it, on behalf of Menzies. On Cook's second voyage, Vancouver, then a midshipman, would have witnessed some of the difficulties associated with Johann Reinhold Forster, and this may have influenced his attitude towards Menzies. Added to this, his health was poor, the survey work was gruelling and he was left without any up-to-date instructions from home. He suffered, as one medical historian has put it, from an interplay of physical and psychological pressures.[77] Menzies was not the only one to suffer the captain's wrath. At various times on the voyage, officers, midshipmen and crew alike were punished or tongue-lashed for their perceived shortcomings. In one incident Thomas Manby, a competent officer who commanded many boat expeditions, was greeted with such abuse by Vancouver when he arrived back at the ship before his captain that 'his salutation I can never forget, and his language I will never forgive'; although it is worth noting that at the end of the voyage Vancouver recommended Manby to the Admiralty for promotion as a 'very active, diligent, and deserving' officer.[78]

Vancouver's homecoming was not a happy one. There were delays in settling his back pay, problems in securing prize money for the capture of the Dutch ship, issues about the financing of the publication of his account of the voyage, and violent public confrontations with the Hon. William Pitt (by this time Lord Camelford), an unstable midshipman whose insubordinate behaviour on the voyage had led to him being flogged, and then sent home. The quarrel with Menzies had been settled with an apology, and the botanist was allowed to leave the *Discovery* with his journal and plants. For Vancouver, whose health was deteriorating fast, that was presumably the end of the matter, but Banks, not a man to forget his brusque treatment by Vancouver five years earlier, had other ideas. When Vancouver objected to the proposal that Menzies's journal should in some way be amalgamated with his own,[79] Banks seems to have decided that the botanist's journal should be published as soon as possible, so as to forestall any account by Vancouver. The difficulty was that the journal as it existed at the end of the voyage was not in publishable form and, as Menzies admitted, he was 'but a slow hand at the pen'. In January 1798, more than two years after the return of the expedition, he was still struggling to complete 'a full and continued narrative' of the voyage. In a plaintive letter he told Banks that he began work on it at 5 a.m. and continued until 6 or 7 p.m. To complete his journal before Vancouver's 'is what I most ardently wish, for more reasons than one' (an intriguing aside), but it was not to be.[80] Despite his wretched health, Vancouver as an author was well in advance of Menzies, and was helped by his brother, John, to write the final stages of his account. When he died in May 1798, most of his narrative was already set in type, and the three-volume account was published later that summer. At this moment, Menzies may simply have given up work on his journal, of which the last entry in any surviving copy is for 18 March 1795, six months before the end of the voyage.

In addition to writing up his journal, Menzies was also busy listing the plants and artefacts he had brought home on the *Discovery*. Once his catalogue was completed, the duke of Portland, home secretary at the time, decided that the 'curiosities and natural productions' should go to the British Museum, except for the plants and seeds, which should join the live plants surviving from the voyage which had already been planted at Kew Gardens.[81] Although a member of the Linnean Society, and often consulted on botanical matters, Menzies published little himself – just two papers, on mosses, in 1798. As a fellow botanist noted in 1806, 'I am sorry that Mr. Menzies apparently neglects his discoveries . . . How immense unknown riches may still remain concealed in [his] collections.'[82] Despite this complaint, in time most of Menzies's botanical discoveries were published, although by other naturalists, especially his correspondent and friend William Hooker, who cited more than 190 species attributed to Menzies. On the west coast of America the scientific names of several notable species of trees, including the Douglas-fir and the madrone tree, still bear his name, as well as many plants, although because of the delays in publishing his discoveries not all of his given names survive. For example, while botanising in Birch Bay at the southern end of the Strait of Georgia in the 1792 season, Menzies wrote, 'here I found in full bloom diffusing its sweetness that beautiful shrub the Philadelphus coronarius'; but in Friedrich Pursh's *Flora Americae* of 1814 the plant is described as *Philadelphus lewisii* because it was noted in the published account of the Lewis and Clark expedition of 1804–6.[83] It has not helped a proper assessment of Menzies's work as a naturalist that no field book kept by him has been found, and that the labels on many of his specimens are without a precise provenance.

In the North Pacific botanists such as David Douglas of the Hudson's Bay Company and John Scouler, collecting for the Royal Horticultural Society, visited in the 1820s and 1830s many of the places familiar to Menzies on his two voyages, and sent home seeds and live plants he had first seen or had brought back as herbarium specimens. They looked to him for advice, and in 1828 there was a notable meeting of the three men, as related by Douglas: 'I was all day with Mr. Menzies . . . and just on the evening who came in but Scouler . . . three North Americans under the one roof.'[84] Six years later Douglas was in Hawaii and wrote that some of the islanders still remembered Menzies – if in rather unflattering terms – as 'the red-faced man who cut off the limbs of men and gathered grass'.[85]

After further spells as a ship's surgeon, Menzies left the navy in 1802 and set up a medical practice in London. He retired from this in 1826, and died in 1842, aged eighty-eight. Two years before his death he looked back at the voyage with Vancouver. Time had softened memories. It was Vancouver's strengths rather than his weaknesses that he remembered as he recalled his captain and his other shipmates: 'What days they were – a fine group of officers – all gone now – a credit to the Captain – he chose them all, except me . . . He was a great Captain.'[86]

CHAPTER 7

'Devilish fellows who test patience to the very limit'
Naturalists with La Pérouse and d'Entrecasteaux

ORE THAN ALMOST any other exploring voyage of the eighteenth century, the French expedition commanded by the Comte de la Pérouse reflected both the spirit of scientific enquiry of the Enlightenment and the great-power rivalries of the period. Much of Cook's third voyage had taken place in wartime under the protection of a safe conduct issued by the French government in recognition of the scientific importance of his expedition, and the explorer's death at Hawaii had been mourned by many outside his own country. Even so, once the War of American Independence had come to an end, preparations began in France for a voyage that was designed to rival if not eclipse those of Cook. It was to be entrusted to one of the most respected of French naval officers, Jean-François de Galaup de la Pérouse, who had served in the Seven Years' War, and in peacetime in the Indian Ocean, and was steadily promoted until in 1781 he became *capitaine de vaisseau* (post captain). The next year, as the War of American Independence drew to a close, he commanded a squadron of three ships sent to raid British fur-trading posts in Hudson Bay. Ice and shoals were more of a threat to the French ships than the unmanned cannon of the 'forts' of the Hudson's Bay Company; but La Pérouse showed both decisiveness and humanity in destroying the posts while leaving enough provisions and ammunition to keep any survivors and their native suppliers alive during the winter.

On his three voyages Cook had established the main outlines of the Pacific from New Zealand to Alaska, but there was still much in the way of detail to be determined, and the French intended to complete the survey of the great ocean that Cook had begun. They aimed to carry out scientific investigations into the lands, peoples and products of the Pacific that would dwarf those of Cook's voyages. The voyage of La Pérouse was to be the 'réplique française' to Cook's.[1] More than that, it was the extension to the farthest reaches of the globe of the collecting activities centred on the Jardin du Roi under the patronage of the Comte de Buffon, director from 1740 to 1788, that in the second half of the century led to a dramatic increase in the exchange of spices and other valuable plants between France's overseas territories. Planning the expedition involved influential men inside and outside the government: the Maréchal de Castries, minister of the Marine; Claret de Fleurieu, director of ports and arsenals who was also a distinguished geographer;[2] Jean-Nicolas Buache de la Neuville, the royal geographer; and the king himself. Louis XVI had long been interested in exploration, and he took an active part in finalising the instructions for the expedition, making numerous marginal notes on the draft plan submitted by Fleurieu. La Pérouse was instructed to sail by way of Cape Horn into the South Pacific, where he would spend twenty months in survey work along the coast of New Holland and in the ocean's island groups. He was then to head to the North Pacific, completing Cook's survey of the northwest coast of America, including the stretch between latitudes 50°N. and 55°N., where Buache de la Neuville was among those geographers who thought that the entrance to a Northwest Passage might be found. The king added a marginal exhortation at this point: 'This is the area that must be more carefully explored.'[3] After surveying the coasts of Japan, the Kuriles, Korea and Formosa, the ships would sail to the Moluccas and Isle de France (Mauritius). Even once he reached the Atlantic on the homeward voyage, La Pérouse was given surveying work to do. 'It was a voyage of exploration, cast in the traditional mould, a grandiose version of all that had gone before.'[4] It was also so far-reaching that the suspicion has arisen that the daunting range of objectives was intended to mislead other nations, especially Britain, as to what might have been the real purpose of the expedition – establishing a French presence on the northwest coast of America to exploit the burgeoning maritime fur trade.

Once the general plan had been approved, consultations began to determine the expedition's scientific programme. The Académie des Sciences provided a long list of recommendations for the savants taken on the expedition,[5] Buffon met La Pérouse to advise on natural-history investigations, while the senior gardener at the Jardin du Roi, André Thouin, drew up detailed instructions on collecting and preserving plants for the young gardener sailing on the expedition, Jean-Nicolas Collignon.[6] He was not only to collect in remote regions; he took great quantities of seeds, bulbs and live plants with him to distribute in suitable locations. To help preserve specimens on both the outgoing and homeward voyages, special plant cases were designed with pitched roofs (Pl. 18). They had holes in the side for watering, and their fragile contents were protected by separate layers of wire mesh, glass panels and wooden shutters.[7] In all, the Académie's recommendations ran to almost two hundred pages. The naturalists were advised to make a special search for the paper mulberry, *Morus papyrifera*, 'which is used in China to make paper and on the island of Tahiti to make cloth'; the Chilean strawberry, *Fragaria chiloensis*, first brought back by Frézier in 1714; and, above all, New Zealand flax, *Phormium tenax*: 'They use its fibers to make sail cloth, cordage, and various fabrics. Captain James Cook brought back a large supply of the plant's seeds to England, none of which came up. Transporting the roots of that plant could provide one of the handsomest presents that voyagers could make for our climate.'[8] The more general instructions to La Pérouse on the role of the naturalists again stressed the utility of their work: 'He will ensure that natural, terrestrial and marine curiosities are collected ... he will get them classified according to category, and will have drawn up, for each species, a descriptive catalogue indicating where they have been found, and the use to which the local natives put them.' Finally, La Pérouse was to ensure that in addition to their landscape sketches the artists were to draw 'portraits of natives in the various countries, their dress, ceremonies, games, buildings, seagoing crafts and all the products of the land and sea'.[9] When the expedition arrived back in France, the natural-history collections, the drawings by the artists and the scientific observations were to be sent by La Pérouse to the king. There were to be none of the disputes that had soured the return of earlier discovery expeditions.

On a voyage as long and arduous as this one was expected to be, the choice of both senior officers and scientists was important. La Pérouse was captain of the *Boussole*, a capacious storeship of 450 tons, while Paul-Antoine Fleuriot de Langle, who had accompanied him on the Hudson Bay expedition, commanded the *Astrolabe*, another storeship. (After they had been converted for the long voyage ahead, both vessels were given the courtesy title of frigate.) Most of the officers had served in the recent war and were known personally to the two captains. The same could not be said of the fifteen scientists who sailed with La Pérouse or de Langle, several of whom proved to be awkward and fractious members of the expedition. Not all were civilian supernumeraries. On the *Boussole* the senior surgeon, Claude-Nicolas Rollin, also served as a naturalist, as did Simon-Pierre Lavaux, surgeon on the *Astrolabe*, while the chaplains, Father Receveur and Father Mongez, doubled as naturalists. Some of the savants worked in subject areas that would later be regarded as separate disciplines. For example, the Chevalier de Lamanon joined the *Boussole* as physicist, meteorologist, mineralogist and botanist. His commitment to Rousseauist ideas on the nobility of primitive peoples led to differences between him and La Pérouse, and eventually, it can be argued, to his death. This was a risk all naturalists on the discovery voyages took, for in their collecting zeal they were the most likely to wander away from the ships and encounter native peoples in areas far from the protection offered by armed crew members. The senior botanist was Joseph-Hughes de Lamartinière, a qualified doctor who had turned to botany when he moved to Paris, and who began his commission on the *Astrolabe* by scouring the neighbourhood of Brest for unusual plants before the ships left harbour. The task of drawing the botanical and other specimens was left to the Prévosts, uncle and nephew; the former was to prove one of the more unfortunate appointments to the expedition. With the addition of astronomers, meteorologists and other specialists, the contingent of scientists was the largest yet taken on a discovery voyage to the Pacific. By contrast, as we have seen, on his final voyage Cook took only two civilian supernumeraries, the artist John Webber and the astronomer William Bayly, while natural-history observations and collections were left as a part-time activity for the *Resolution*'s surgeon, William Anderson. Unlike Cook on his voyage to the North Pacific, La Pérouse took the precaution of including a Russian-speaking crew member, the young and promising Barthélémy de Lesseps, whose father had been French consul-general at St Petersburg.

The *Boussole* and the *Astrolabe* set sail from Brest on 1 August 1785. Their passage into the South Atlantic was uneventful, although a tiff between La Pérouse and Lamanon at Tenerife hinted at future trouble. The scientist led a party that set out to climb the island's famous volcanic peak, Pico de Teide, in an effort to obtain an accurate measurement of its height. It was a legitimate scientific objective, and Lamanon assumed that the cost of mules and porters would be met from official funds. The group failed to reach the summit, and La Pérouse informed Lamanon that he would have to cover the considerable cost himself. The episode was recorded by La Pérouse in his journal with little additional comment, but a private letter to Fleurieu revealed his lack of respect for the scientist. He was a hothead, with a mean character, and although he claimed to know more about the formation of the world than Buffon, away from his own subject of physics he was 'as ignorant as a monk'. This La Pérouse illustrated with an example presumably invented to tickle Fleurieu's sense of humour, for he reported that Lamanon 'has been aiming his spyglass at the tropic ever since the apprentice pilots told him that it could be seen from a hundred leagues off'.[10] A letter to the minister of the marine from Lamanon at Tenerife gives a rather different impression of the naturalist. Lamanon enclosed seeds of a white bean that he suggested might replace the haricot bean, which traditionally formed the basic diet of peasants in his native Provence, but which had recently been afflicted by blight. He too was not without a sense of humour, writing: 'we are now back, my Lord, from the peak of Teneriffe . . . We were there on the eve of the feast of St Louis and we drank the King's health there; none of his subjects has ever celebrated his feast in such a high place.'[11] By the time the ships reached their next port of call at Santa Catarina off the coast of Brazil, good relations seem to have been restored, and La Pérouse reported himself 'very happy' with his scientists, although he could not resist a final dig: 'There is some rivalry between them today, each one favouring his own special subject, but I expected this.'[12]

The passage into the Pacific around Cape Horn was unusually calm, and on 24 February 1786 the ships reached the Chilean port of

Concepción where they stayed for three weeks. It was a pleasant stay, with much goodwill shown by the Spanish authorities, and La Pérouse's only anxiety was whether he had been foolish in promising the crews time ashore once repairs had been completed and food, water and wood taken on board: 'because wine is easily available in Chile, every house is a tavern and the common women of Chile are almost as free with their favours as those of Tahiti. There was however no unruliness.'[13] The naturalists were presumably busy ashore, but there is no mention of their activities by La Pérouse because, as he explained in the preface to his manuscript journal, all such matters would be described in the separate works being written by Lamanon, Lamartinière and their colleagues. Unfortunately, unlike his journal, their observations would never reach France. Ever practical, La Pérouse explained that his journal would contain details only of plants useful as antiscorbutics. From the *Astrolabe*, de Langle wrote to the minister of the Marine praising both his naturalists: Lamartinière was 'an untiring and passionate botanist', and Father Receveur was assiduous in his spiritual and scientific duties, and 'a gracious and intelligent person'. The only problem was the elder Prévost, disinclined to carry out his duties as an artist unless given direct orders.[14]

During his stay La Pérouse decided to reverse the route laid down in his instructions, and instead to sail first to the northwest coast of America. Explaining his decision to Fleurieu, he reminded him that before they left Brest news had reached France that Captain John Gore, who had sailed on Cook's third voyage, was preparing an expedition to the northwest coast. This was intended, La Pérouse wrote, 'to set up some establishment behind Hudson's Bay and I think it good to get there before him'. The rumoured expedition never took place, but in 1786 six British-owned trading vessels reached the coast to exploit the bountiful sea-otter resources reported by Cook's men. On his route north, La Pérouse intended to visit Easter Island and the Hawaiian Islands, both places of unusual interest. The expedition stayed only a day at Easter Island, comparing their reactions to the people and their mysterious statues to those of Cook's officers on their visit to the treeless island twelve years earlier. Like Cook, La Pérouse thought that the giant statues had been moved with the help of wooden levers a few feet long, although each would take a hundred men to shift. The gardener,

Collignon, accompanied de Langle inland until he found a fertile patch of ground, where he sowed vegetable and fruit-tree seeds. Reports from the voyages of Bougainville and Cook had led the crews to expect a certain amount of stealing once they reached Polynesia, but they were surprised and baffled to find that at Easter Island their hats and hand-kerchiefs were the main targets of the light-fingered thieves. When La Pérouse returned to the ships' anchorage, 'I found almost everyone hatless and with no handkerchiefs ... an islander who had helped me down from a platform took away my hat as soon as he had rendered me that service, and he fled as fast as he could.'[15]

The expedition's stay in the Hawaiian Islands was also brief. La Pérouse avoided Hawaii altogether because of its association with the death of Cook seven years earlier. Instead he landed at Maui, where he took on board pigs and other provisions before setting sail on 1 June 1786 for the northwest coast of America. Just over three weeks later, as the fog cleared, the immense peak of Bering's Mount St Elias was visible above the clouds. The expedition had arrived in the northern hemisphere summer on a coast that Louis XVI had ordered must be 'more carefully explored', and where the accounts of Spanish voyages, fictitious and real, suggested that the entrance to a Northwest Passage lay.[16] La Pérouse's journal entries seemed deliberately to dampen expec-tations as the ships approached the coast, 'which once upon a time was the setting for geographical tales that are too readily accepted by modern geographers'. Even the sight of land failed to raise his spirits as his 'eye rested painfully upon all this snow covering a sterile and treeless land'. After nine days of sailing south, La Pérouse took his ships through a hair-raising series of tidal rips into an inlet in latitude 58°52′N. that had not been sighted by the Spaniards or Cook. He named the inlet Port de Français (today's Lituya Bay), and suggested that, if the French decided to establish a trading post on the coast, this might be the spot. So plen-tiful were the sea otters that La Pérouse thought that ten thousand could be caught in a single year. One was taken and stuffed by the *Boussole's* surgeon, but it was a young one, weighing only a few pounds.

For the botanists, frustrated by the lack of collecting opportunities in the long months at sea, the inlet was a disappointment. Its shores were covered with trees, shrubs and plants, but almost all were familiar. La Pérouse wrote that Lamartinière found 'only three plants which he believes to be new, and one knows that a botanist can make a similar find in the environs of Paris'. For his part, Lamanon climbed the cliffs above the inlet and was intrigued to find scallop shells four hundred metres above sea level. Accompanied by Lamartinière, Mongez, Receveur and Collignon, he also struggled up the difficult higher slopes around the inlet in search of mineral specimens. La Pérouse was full of praise for their efforts: 'the most tireless and zealous of naturalists, [they] were unable to reach the summit, but they went up to a reasonable height, at the cost of incredible exertions – no stone, no pebble escaped their vigilance.' The Tlingit inhabitants of the inlet were 'as rough and barbarous as their soil is stony and untilled'. La Pérouse followed this dismissal with a passage that hinted that he had already formed his views on Rousseauite ideas of natural man, although they may have been sharpened in debates and disputes with the naturalists on board: 'Philosophers may well protest against this description; they write their books by their fireside: I have been sailing for thirty years. I am a witness of the meanness and deceit of these people who are presented to us as being so good because they live so close to Nature.'

Boat parties found that Port de Français culminated in a dead-end, and to disappointment was added tragedy when two of the ships' boats capsized near the entrance with the total loss of their crews – six officers and fifteen men. Not a single body was recovered. Contrary winds and a lingering hope that some survivors might yet be found kept the expedition in the bay until the end of July. As the ships finally sailed south, La Pérouse realised that the three months set aside in his instructions for the exploration of the northwest coast were totally inadequate. Given the intricate nature of the shoreline, the frequent fogs and the hazards of tide and current, a proper survey would take several seasons to complete. As the ships bore away for California, La Pérouse scribbled journal entries that reflected his irritation at this part of his mission, for the idea of a navigable Northwest Passage, he decided, was as 'absurd' as those 'pious frauds' of a more credulous age.[17]

On the Californian coast the expedition spent a relaxing week at the Spanish port of Monterey, founded only fifteen years earlier. It was a fertile area, and Collignon left a variety of seeds with the mission station there. 'For their part, our botanists did not lose a minute before increasing their collections of plants, but the season was not favourable – the summer heat had dried them out and their seeds were spread over the ground.'[18] On the long ocean crossing from Monterey to Macau which took more than three months, the naturalists would have had little to do. Because none of their journals has survived we can only guess at the frustrations and irritations that built up on the voyage, but some indication of them can be gained from a quarrel at Macau between La Pérouse and several of the scientists led by Lamanon when they left the ship and took up quarters ashore without informing him. He in turn decided not to forward any social invitations from the Portuguese authorities to them, and when they objected put them under arrest for twenty-four hours. The relationship between seamen and scientists was, in Catherine Gaziello's words, 'une difficile cohabitation'.[19] A flurry of letters to France outlined the grievances of both sides. In a despatch to Fleurieu, La Pérouse complained of the 'pretentious' behaviour of Lamanon and his colleagues, 'devilish fellows who test my patience to the very limit', while the elder Prévost was 'useless' and 'unmanageable' and had not drawn more than fifteen plants on the whole voyage. In turn, the astronomer Dagelet grumbled: 'We have just completed a voyage lasting a hundred days. It has seemed like a thousand.'[20] In a letter to Castries, Lamanon apologised for not sending the minister a proper account of his findings, pointing out that 'I would have to send you whole volumes'.[21] Instead, he had to be content with sending a few extracts from his observations to be forwarded to the Académie des Sciences.

From Macau the ships headed for Cavite (the port of Manila) in the Philippines, and after a stay of six weeks sailed north to spend the summer exploring the little-known coasts of northeast Asia. From now on La Pérouse hoped to spend more time ashore, for of the first eighteen months of the voyage fifteen had been spent at sea, 'and this long period at sea does not appeal to our botanists and our mineralogists who can only exercise on land their talents'.[22] On the ships' passage north through the Strait of Tartary between Sakhalin and the mainland, the naturalists and other scientists were put ashore on several occasions, but La Pérouse's notes on their activities are brief.[23] There were times when their casual behaviour reduced La Pérouse to despair. At Kastri Bay, near the expedition's farthest northern point in the Strait of Tartary, Collignon was busy sowing European seeds when a sudden shower soaked him. He decided to light a fire to dry his clothes, but 'unwisely used gunpowder to start it; the fire spread to the powder-flask he was holding, the explosion broke the bone in his thumb and he was so seriously wounded that his arm was saved only by the skill of our senior surgeon, Mr Rollin'.[24] Valuable survey work had been done along fogbound coasts that were almost entirely unknown to Europeans, and the strait between Sakhalin and Hokkaido which the ships sailed through on their way northeast to Kamchatka still bears the name of La Pérouse. To Fleurieu an exhausted La Pérouse wrote, 'I would find it difficult to describe the hardships of this part of the campaign, when I did not undress once.'[25]

On 6 September 1787 the ships reached Kamchatka and anchored in the harbour of St Peter and St Paul (Petropavlosk), where there were signs of earlier voyagers. The Frenchmen were shown the grave of Louis de la Croyère, who had died on Bering's Great Northern Expedition, and the more recent grave of Captain Charles Clerke, Cook's second-in-command on his last voyage. Among the reports sent by La Pérouse to the ministry in Paris was one criticising the scientists of the expedition: 'in general they belong to a class of people who are so full of self-esteem and vanity that they are very difficult to lead in long campaigns of this kind.'[26] During their stay Fathers Mongez and Receveur, together with the surveyor Gérault-Sébastien Bernizet, were able to expend some of their pent-up energy on a gruelling climb to the crater of the towering volcano, Mount Avachinskaya. Accompanied by eight Cossacks, they reached the edge of the crater; as they looked back down to the bay, the French ships appeared smaller than canoes. Even La Pérouse was impressed: 'There may never have been such a strenuous [journey] undertaken for science ... The appearance of the mountain made me think it was impossible – one could see no greenery, but only a snow-covered rocky surface with a slope of not less than 40 degrees.'[27]

In general, the botanists remained disgruntled: most of the plants were familiar, and had been described by Steller on Bering's voyage or by Stepan Krasheninnikov in his history of Kamchatka. Among the scientists Father Receveur was perhaps the most congenial. At about this time he wrote: 'I have never been bored. I have found the days very short with the varied and successive occupations. M. Vicomte de Langle, our captain, has given me much friendship and has contributed not inconsiderably to making me pass the days agreeably.'[28] The high point of the stay was the arrival of mail from Okhotsk, the first news from home for two years. Among the personal and family letters were official communications from the minister of the marine. There were several promotions: La Pérouse was made commodore. He was also informed that the British were planning a settlement at Botany Bay in New South Wales, which he was ordered to investigate. His original instructions to sail through Polynesia to New Zealand were set aside. Instead he was to take a direct route to the coast of eastern Australia, and within days the ships had sailed. Before they left, Barthélémy de Lesseps set out on an overland journey to France carrying official letters and journals. In La Pérouse's words, it would be a journey 'as strenuous as it would be long', and so it proved. After travelling across Siberia by horse, sleigh and boat for a year, the young Russian-speaking Frenchman reached St Petersburg in September 1788, and Paris on 17 October. He was still only twenty-two.

The long voyage south across the Pacific, sailing between extremes of cold in Kamchatka and of stifling heat in the tropics, was tedious for all concerned, and a threat to the health of many. 'Nothing broke the monotony of the long crossing,' La Pérouse wrote. In early December the ships reached the Samoan Islands, named the Navigators Archipelago by Bougainville twenty years earlier (because 'all their travel is made by canoe and they never walk from one village to another'). The islanders were prepared to trade; La Pérouse's main problem was finding a safe anchorage on the north coast of the island of Tutuila. Worried about the exposed roadstead where the ships were anchored, rolling at night as though they were in the open ocean, La Pérouse decided to spend only the next morning ashore, trading as much in the way of pigs, hens and fruit as possible. Occasionally, scuffles broke out, some stones were thrown, and La Pérouse felt he had to demonstrate the power of European firearms by shooting some pigeons and splintering wooden planks with pistol shots. When Lamartinière and Collignon ventured too far from the beach in their search for plants the islanders demanded a glass bead for every plant collected, and pelted Lamartinière with stones as he swam back to the boats with a sack of plants on his back. It was time to leave, but de Langle had found 'a small harbour ideal for boats, at the foot of a charming village and by a cascade of the clearest water', and he demanded to be allowed to take a boat party ashore the next morning to fill the water casks, insisting that fresh water was the best antiscorbutic. We have only La Pérouse's account of his argument with his closest friend on the expedition, but it was clearly a heated one:

> I told him that I found these islanders too turbulent to send ashore boats and longboats which could not be assisted by our ships' guns, that our moderation had inspired little respect for us on the part of these Indians who were colossi and looked only at our physical strength which was inferior to theirs; nothing could shake Mr de Langle's obstinacy and he added that my own would make me responsible for the progress of the scurvy that was beginning to manifest itself with some severity.[29]

Later, La Pérouse wrote that de Langle 'tore' rather than 'obtained' permission from him to go ashore.

Promising to be back within three hours, de Langle set off the next morning, 11 December, with two longboats armed with swivel guns and two smaller boats. Altogether they contained sixty-one men. Once they reached the cove they were out of sight of the ships and out of range of their guns. On de Langle's previous approach to the cove it had been high tide; on this second visit at low tide there was less than three feet of water over the coral reefs, and both longboats ran aground. As they waited for high tide more warriors gathered on the beach until by mid-afternoon there were a thousand or more; some waded into the shallow water to surround the boats and others began to hurl stones at the crews with stunning force. Still de Langle refused to give the order to open fire with the swivel guns, for 'fear of starting the hostilities and possibly being accused in Europe of barbarous behaviour', La Pérouse presumed, and although he belatedly fired his musket at the onrushing warriors he was knocked down and clubbed to death. This was the signal for a general assault (Pl. 19). The crews fired their muskets, killing up to thirty of their attackers, but had no time to reload. 'In less than five minutes there was not a single Frenchman left in either of the grounded boats, all those who escaped by swimming to the [other] two boats had received over ten wounds, almost all head wounds; those who had the misfortune of being knocked over the side of [sic] the Indians were finished off immediately with clubs.' In a frenzy of violence the warriors struck at the bodies of the dead time and time again, while others smashed the longboats to pieces. In the two remaining boats the crews threw the water casks overboard to make room for their wounded comrades, and managed to escape through the cove's narrow entrance. Ironically, as they edged towards safety with the bloodied bodies of the wounded sprawled across the thwarts, they found dozens of canoes quietly trading their provisions with the ships. It took all of La Pérouse's authority to stop his crews taking vengeance on 'the brothers, the children and the compatriots of these barbarous murderers'. In de Langle he had lost his 'friend of 30 years, a man full of wit, wisdom and knowledge [whose] humanity had brought about his death'. Also dead was Lamanon, who had accompanied the shore party to look for botanical specimens, and ten others, while twenty men were wounded. Father Receveur was among those in the boats, and received a wound near his eye that may have been more serious than was first realised. Collignon, who had intended to go ashore to dig up roots of the breadfruit plant, was wounded in several places, and wrote that 'we have been dupes of our own good nature'.[30] Writing to Fleurieu two months later, in a statement that was printed in Milet-Mureau's published account of the voyage and became a classic refutation of the arguments of the followers of Rousseau, La Pérouse declared: 'I am a hundred times more angry against the philosophers who so praise them as against the savages themselves. Lamanon, whom they murdered, was telling me the day before he died, that these men were worth more than us.'[31]

With the crews shaken and depressed, and the ships short-handed, La Pérouse steered as direct a course to Botany Bay as possible. Off Norfolk Island he hoped to allow the naturalists ashore because they had had so little to do since leaving Kamchatka, but the sea was too rough for any sort of landing. On 24 January 1788 they reached Botany Bay, to find a fleet of British ships already at anchor. Commanded by Governor Arthur Phillip, the First Fleet, carrying 750 convicts, had arrived only a few days earlier. Disappointed by Botany Bay, exposed as it was to easterly winds and with poor, sandy soil, the British vessels were about to move farther north along the coast to the fine harbour of Port Jackson (later Sydney). The British were more taken aback by this coincidental encounter than the French, but relations were amicable. As La Pérouse remarked, 'Europeans are all compatriots at such a great distance.'[32] The French stayed at Botany Bay for six weeks, building two longboats to replace those lost in Samoa. There was one more death, an especially sad one, for, on 17 February, Father Receveur died, possibly of the wound he had received at Tutuila two months earlier. In a touching tribute to the Frenchman, Governor Phillip later set up an inscribed copper plate to replace the wooden board on his grave that had fallen down after the departure of the French ships.[33]

Little is known about the final weeks of the French stay. On 1 February, British officers visited the French ships, where they were shown Lamartinière's 'very compleat' natural-history collection, and were told by La Pérouse that since he had stores for three years and expected to be back in France in fifteen months, 'he should be happy to oblige Mr Phillip with any that he might want'.[34] A week later a French officer arrived at Port Jackson and handed over La Pérouse's journal and despatches, together with other letters, with a request that they might be forwarded to the French ambassador in London by the first ship that sailed for England.[35] Among the letters written by La Pérouse was one to a friend that revealed the effects of the long voyage and the toll taken by the disasters of Lituya Bay and Tutuila: 'Whatever professional advantages this expedition may have brought me, you can be certain that few would want them at such cost ... When I return you will take me for a centenarian, I have no teeth and no hair left ... Tell your wife she will mistake me for my own grandfather.'[36] Work on the longboats was carried out behind a stockade guarded by guns; although the Aboriginal inhabitants of Botany Bay were poorly armed, there would be no repetition of the overconfident approach that had led to the Tutuila massacre. According to David Collins, judge-advocate of the new colony, the French held 'the most unfavourable idea of the country and its native inhabitants; the officers having been heard to declare, that in their whole voyage they no where found so poor a country, nor such wretched miserable people'.[37]

Despite his weariness and sadness, La Pérouse was determined to complete his explorations. From Botany Bay he intended to sail to Tonga, New Caledonia, the Solomon Islands and New Guinea, before surveying the northern and western coasts of New Holland. He hoped to reach Isle de France in December 1788 and France the following June. It was a schedule that would have taxed an expedition that had arrived fresh from Europe, let alone one that had been at sea for two and a half years. The *Boussole* and *Astrolabe* left Botany Bay on 10 March 1788 and disappeared into what one of La Pérouse's biographers has called 'Forty Years of Oblivion'.[38]

If La Pérouse had arrived at Isle de France by the end of 1788, as planned, then the news would have reached France three or four months later, and his ships should have been back at Brest by midsummer 1789. It was a time of political turmoil in France, marked by the fall of the Bastille on 14 July; even so, anxiety grew as the months passed with no news of the expedition. Britain and Spain were notified about the missing ships and asked to report any news of them, while in France there were efforts to mount private search expeditions. Helped by the appointment of the Comte de Fleurieu as minister of the marine in October 1790, and by pressure from relatives and colleagues of the lost men, an official search expedition was approved by the National Assembly in February 1791. The preamble to the Assembly's decree was a high-sounding declaration that concluded: 'No longer does the European penetrate the most distant latitudes to invade or to devastate, but rather to bring benefits and dividends; no longer the grasp for corrupting metals, but rather for the conquest of those useful plants that can render man's life easier and happier.'[39]

The expedition was placed under the command of Joseph-Antoine Bruny d'Entrecasteaux, an experienced naval officer who had worked closely with Fleurieu at the time he was planning the La Pérouse voyage. He was allocated two converted storeships, similar in type to those that had sailed with La Pérouse: the *Recherche*, captain Alexandre d'Hesmivy d'Auribeau, and the *Espérance*, captain Jean-Michel Huon de Kermadec. His instructions covered more than the search for La Pérouse, for they included a fully fledged programme of scientific research, some of it to be carried out in areas that were unlikely to have been visited by the missing ships. In this way, the Assembly's decree stated, the expedition would be 'useful and advantageous to navigation, to geography, to commerce, to the arts, and to the sciences'.[40] The scientific contingent included three naturalists, two hydrographers, two astronomers, a mineralogist, two artists and a gardener (two of this party abandoned the expedition at the Cape). Mineralogy, which as a discipline had lagged far behind botany and zoology, was represented for the first time in its own right on a discovery voyage in the person of Jean Blavier, chaplain on the *Recherche*. The naturalists were Jacques-Julien Houton de Labillardière, Louis-Auguste Deschamps and Claude Riche; like many other botanists of the period, they had trained as doctors. They were helped by a gardener-botanist, Félix Lahaye. All the naturalists were given lengthy instructions: fourteen foolscap pages on geology, eighteen on zoology, seventeen on botany and no fewer than twenty-nine to the gardener.[41]

Of the naturalists, Labillardière had botanised to good effect in the Levant, and in 1799 would publish the first narrative of the d'Entrecasteaux expedition, followed by a two-volume work on the plants of New Holland. He was, in the words of his biographer, 'one of the founders of botany, zoology and ethnography in Australia'.[42] Before sailing on the *Recherche*, he wrote to Sir Joseph Banks, whom he had met during a visit to England, asking his advice and stating: 'Botany will be my principal concern, however I shall try not to neglect completely other areas of natural history.' In reply, Banks pointed out: 'it is unusual for someone so experienced in botany to visit the countries where you are going, so it is essential to use the unique opportunity that you have.' Recalling his own practical problems on the *Endeavour* voyage a quarter of a century earlier, he also recommended that Labillardière take an 'enormous' supply of coarse paper for drying plants, and linen bags for holding them, with the contents identified by the number of knots on the string tying up the bag.[43] Labillardière had great difficulty in finding the amount of paper he wanted, for most had been issued to the artillery at Brest, where the ships were being fitted out.

To all outward appearances, the search expedition was well manned and sensibly equipped; but all three senior officers had health problems, and none survived the voyage. Furthermore, it became painfully clear as the voyage progressed that political events in France before the expedition sailed were having a disruptive impact on discipline. In origin the officers came mostly from the aristocratic ranks of the *Grand Corps* (a privileged group abolished in October 1789 in the early months of the French Revolution), while the scientists with the exception of Deschamps tended to have radical, republican sympathies. Many of the crews were from Saint-Malo or from Brest, whose municipal government was Jacobin, although care was taken to exclude any who had been caught up in a naval mutiny in the port in 1790.[44]

The ships sailed from Brest on 29 September 1791 and called at Tenerife, where a party including Labillardière and Deschamps struggled up the Pico de Teide – by now almost a standard initiation rite for naturalists on the discovery voyages. The slow-sailing ships arrived at the Cape of Good Hope on 18 January 1792 where d'Entrecasteaux received a report of a sighting made by a British naval officer, Captain John Hunter, on the Admiralty Islands (now part of the Bismarck Archipelago, northeast of New Guinea). Sailing on a Dutch ship after his own had been wrecked, Hunter was said to have seen natives 'covered with European cloth and, significantly, clothes which he identified as French uniforms ... Commodore Hunter was certain that these were the remains of the wreckage of M. de la Pérouse.'[45] Further investigation raised doubts as to whether Hunter had ever made such a report, and he had sailed from the Cape two hours after the arrival of the *Recherche* and *Espérance* – perhaps a sign that he had nothing of interest to tell the French expedition. In the account published two years later of his visit to the Admiralty Islands, Hunter made no mention of European clothing, and his only hint about the lost expedition was a rather inconsequential observation that one of the islanders 'made various motions for shaving, by holding up something in his hands, with which he frequently scraped his cheek and chin; this led me to conjecture that some European ship had been lately among them ... it might have been Mons. de la Perouse'.[46] Whatever his doubts, d'Entrecasteaux felt that he had no alternative but to investigate Hunter's supposed report. Before doing so, however, he had to deal with a situation in which six of the naturalists and artists demanded to be allowed to leave the expedition. They complained about arrangements to meet their expenses, and at the way in which the ships' surgeons were allowed to compete with them in collecting specimens. D'Entrecasteaux's private opinion was that science was like the air that one breathed and belonged to everyone, but he managed to deal with a threatened exodus that would have irreparably damaged the scientific objectives of the expedition; the two savants who left the expedition at the Cape did so only for reasons of ill health. Labillardière added to his Tenerife collection with a rich haul of specimens from the Cape, but was never to see them again. The French agent at the Cape entrusted with arrangements to send them to France died before doing so, and when five years later they were finally put on a French vessel it was captured by a British man-of-war and Labillardière's specimens were auctioned off.[47]

Influenced by Hunter's reported sighting, d'Entrecasteaux hoped to sail directly across the Indian Ocean to the Admiralty Islands, but after battling the easterly monsoon for three weeks he realised that this route was impossible. Instead, he intended to take the more circuitous, longer southern route around Australia, stopping at Van Diemen's Land (Tasmania) to make repairs. The expedition stayed at Recherche Bay on the southeast coast of Tasmania for five weeks, where the naturalists had ample time to investigate the immediate neighbourhood, although Labillardière grumbled at the meagre provisions they were given and the lack of a boat in which they could have ventured farther afield. In all, they collected five thousand botanical specimens, including about a hundred new species. It is not clear what proportion of these specimens was live as opposed to dried plants, or how many were simply seeds. Their haul was particularly rich in eucalyps, of which Labillardière published descriptions of seven new species, marvelling at the size of some and the way in which the Aborigines used the bark to roof their shelters. The trunks of one species, the Tasmanian bluegum, he thought might make ships' masts, and the carpenters used its wood to repair the *Espérance* (Pl. 21). In his journal d'Entrecasteaux welcomed the fact that the naturalists had gathered 'precious harvests among all species', and he seemed generally supportive of their efforts. When d'Auribeau ordered Labillardière to remove his plant presses from the *Recherche*'s great cabin, the latter successfully appealed to d'Entrecasteaux to countermand the instruction.[48] One of the youngest members of the expedition, Jurien de la Gravière, only nineteen years old at the beginning of the voyage, remembered many years later the tensions on board the ships: 'Conflict arose over geography and natural history. The admiral was accused of directing his favours towards hydrography. Where there had been only two deeply-divided camps there were now three, then four factions. A brawl of a size to try the patience of an angel broke out.'[49]

The naturalists were more interested in noting what they saw on their excursions than they were in their own appearance, but a description by one of the ship's officers offers a glimpse of Claude Riche's clothes and equipment as he went ashore:

> A large canvas vest, furnished with pockets front and back, serves him as clothing; an immense portfolio in the guise of a shooting pouch rests over his loins; a mineralogist's hammer hangs below, another string over the other shoulder suspends forceps padded with linen, which are destined to catch insects and butterflies; a pin-cushion covered with pins, long and fine, is attached to his buttonhole; a sabre or cutlass lies to his side. An umbrella-like broad-brimmed leather hat shades his head and protects his shoulders; leather gaiters ward his legs from all which can wound; and a gun usually serves to complete the equipment.[50]

The stay was notable for the detailed charting by the expedition's hydrographer, Charles-François Beautemps-Beaupré, of the D'Entrecasteaux Channel between Bruny Island and Van Diemen's Land, already sighted but not explored by British navigators. D'Entrecasteaux set sail from Recherche Bay on 28 May 1792, heading northeast to New Caledonia. With strong onshore winds blowing, it was impossible to land on the island's reef-strewn southern coast, and d'Entrecasteaux decided to set course for New Ireland. On the way he hoped to land on a small island sighted but not investigated in 1791 by Captain Edwards in the *Pandora*, and named by him Pitt's Island (Vanikoro). 'I was determined to investigate, reconnoitre, and compile a map,' d'Entrecasteaux wrote, but not sighting the island during the night of 7 July, 'I believed that it was not wise to lose a night of favourable wind for a minor reconnoitring, and I ordered that we continue on.'[51] It was the most poignant moment of the whole voyage, for Pitt's Island was where La Pérouse's ships had been wrecked four years earlier, and there were probably survivors still alive on the island. Instead, d'Entrecasteaux followed the false lead of Hunter's supposed report at the Cape, and from New Ireland sailed to the Admiralty Islands. There was momentary excitement off one of the Admiralties when 'a large tree lying across the reefs, from which rose a branch at one extremity and roots at the other, was mistaken by some to be the remains of a ship, with only the keel, the stem and part of the stern-post remaining', but it was an illusion, seen so often that d'Entrecasteaux wrote that it was almost as if a whole fleet had been wrecked.[52] As always, Labillardière was not afraid to speak his mind, so on a small island off the coast of New Ireland he stopped sailors from chopping down coconut trees as an easy way of getting at the fruit: 'had we suffered them to have their own way, there would not have remained a single cocoa-nut tree on the island.'[53]

September and the first half of October were spent at the small Dutch port of Amboina in the Moluccas. Time on land was much needed, for many among the crew were suffering from scurvy. Labillardière was allowed to make limited botanical excursions from the town, and later published some of his observations from this visit. He was especially interested in the production of cloves and nutmeg, and noted that the Dutch East India Company gave the growers only a small fraction of the price the spices commanded in Europe. He was struck by the beautiful collections of butterflies and shells in the possession of the governor and the secretary of his council, and described how they presented specimens to Company officers in Batavia and Europe as a way of gaining favour.[54] The region was hardly untrodden ground for the naturalists, for the mid-eighteenth century had seen the publication of the six-volume *Herbarium amboinense* by Georg Eberhard Rumphius, a German-born physician who had been in the service of the Dutch East India Company the previous century. Labillardière had a copy of this magnificent work, which described and illustrated nearly a thousand tropical plants, and he took care to visit Rumphius's tomb.[55] There was no recent news of events in France at the port, one of the most distant from Europe of all the Dutch possessions, but d'Entrecasteaux took the opportunity to draw up a report for Fleurieu of events on the voyage since the ships' stay at the Cape. (Together with other letters, it was never forwarded to France by the distrustful Dutch authorities.) Continuing complaints from Claude Riche about the lack of help he was receiving on his boat excursions perhaps accounted for one brief and unenthusiastic sentence in d'Entrecasteaux's despatch: 'I can give no account of the naturalists; I can appreciate neither their talents nor their work.' Only Deschamps was mentioned favourably, because of his 'avoidance of all factions' and the fact that he carried out his work 'peacefully'.[56]

From Amboina the ships sailed southwest, skirting Australia's west coast by a wide margin before turning east to the continent's south coast. There a short stay at Esperance Bay was marked by a near-escape from death by Riche. Wandering off in search of specimens, he mistook a large salt lake for the sea, became disorientated and spent more than fifty hours in scorching daytime temperatures with little food or water. As his strength failed, he threw away the plants and stones he had collected, although when he eventually sighted the ships in the distance, 'I started my collection once again; but my extreme weakness prevented me from burdening myself'.[57] Joy at Riche's safe return was short-lived, for the incident led to recriminations between officers and naturalists. The latter claimed that their interests were being overlooked in favour of the astronomers and hydrographers on board, that landings were few and far between, and when they were made they were denied crew members as escorts. In response, d'Entrecasteaux pointed out that Riche had stopped his servant accompanying him, and stated that normally he refused to allow the naturalists leave to join boat parties engaged on coastal survey work because they would be ashore for only short periods, and in addition overloaded the boats. He then made a more general point, which implied that differences between him and the naturalists had been simmering for some time: 'The advantage of using persons employed in the navy for these kinds of expeditions cannot be stressed enough, since such persons (being more aware of what is permitted in such circumstances) would not make impossible demands.' If d'Entrecasteaux's advice was to be followed, the era of the gentleman-naturalist on board discovery expeditions was drawing to a close.

By late January 1793 the ships were once more off Van Diemen's Land, where they anchored in Port du Sud (Rocky Bay) before moving on to Adventure Bay. The visit the year before had been in the autumn; this time, in the height of summer, the naturalists were delighted that the plants were in flower, and on one day alone they collected thirty species. Among the specimens later described and published by Labillardière were three species of the carnivorous, insect-trapping sundew plant.[58] He also found traces of Bligh's visit in the *Providence* to the bay the previous year in the form of a carving on the trunk of a tree: 'Near this tree, Captain William Bligh planted 7 fruit-trees, 1792; Messrs S. and W. Botanists.' The wording was not to Labillardière's liking; as he pointed out, the botanists (Christopher

Smith and James Wiles) were referred to only by their initials, and the implication was that Bligh had personally done the planting. It was an example of the 'same marks of deference' to the officers that he resented on the French ships.[59]

Unlike on the previous visit, the expedition's members met a considerable number of the Aboriginal inhabitants. As on other occasions in other regions, the first contacts were made by the naturalists – in this case, Labillardière and the gardener Delahaye, who in company with two unnamed crew members were botanising away from the shore when they became aware of a large number of spear-carrying natives approaching through the trees. After some hesitation both sides dropped their weapons, 'and from that moment onwards,' d'Entrecasteaux wrote, 'the most cordial relationship was established' (Pl. 22).[60] Soon meetings were on a regular footing, with the crews holding and caressing the children. Piron, the expedition's artist, sketched the Aborigines in groups and individually, and engravings from his drawings appeared in Labillardière's account of the voyage. Aboriginal nudity no longer shocked, gifts were exchanged and Labillardière put together a vocabulary of eighty-four Tasmanian Aboriginal words (Anderson on Cook's brief third voyage visit had listed just nine), while d'Entrecasteaux was embarrassed to remember that the year before he had suspected nearby Aborigines of cannibalism. Rather, as Lieutenant Elisabeth-Paul-Edouard de Rossel wrote, they seemed a 'simple and good people'.[61] The chaplain Louis Ventenat realised the significance of the encounters with the Aborigines when he observed that his shipmates were not 'like Tasman who saw no-one; or like Marion [Dufresne] who saw them only to kill them; or like Furneaux and Cook, who never saw them all together at their meal and in their huts'.[62]

Some of the Frenchmen still had reservations. Labillardière protested, in vain, at the way in which the women had to dive in the icy sea in search of shellfish while their menfolk sat idly around the fire,[63] while Deschamps went further with his remark that it was 'difficult to find a human species which is more brutish and further from civilisation than these men'.[64] D'Entrecasteaux saw the Aborigines at close quarters only once. Even so, he was keen to record his impressions, which were significantly different from the unenthusiastic observations of the few earlier French and British visitors to Tasmania's shores, and from the contemporary descriptions by the Port Jackson settlers of the Aborigines of New South Wales:

> This tribe seems to offer the most perfect image of pristine society, in which men have not yet been stirred by passions, or corrupted by the vices caused by civilization. Composed of several family groups, without property other than their wives and children, there must exist no cause for dissent, as the only chiefs are those designated by nature itself: the fathers and the old men ... Secure in finding their means of subsistence, these men must enjoy peace and contentment. Thus, their open and smiling expression reveals a happiness that has never been troubled by intrusive thoughts and unattainable desires.[65]

The idyll, real or imagined, was to be short-lived: within a few generations most of d'Entrecasteaux's 'simple and kind' people had disappeared off the face of the earth, wiped out by settlers, whalers and disease.

From this point on, ill fortune seemed to attend the expedition at every turn. La Motte du Portail, sub-lieutenant on the *Espérance*, sensed that all was not well with d'Entrecasteaux as they left Tasmania: 'The more I have observed our Chief, the more I could see in him traces of a deep feeling of sadness. Through my various conversations with Mr. Huon [de Kermadec], I know that his sorrows were caused by the dissensions among his officers. Jealousy and ambition tormented them. They tore one another apart and they destroyed the peace of mind of our leader.'[66] As the ships approached the northern tip of New Zealand, Labillardière pleaded to be allowed to land to collect specimens of New Zealand flax, whose fibres he believed would make immensely strong ships' cables; but d'Entrecasteaux, fearing hostile attacks, refused. 'No one,' complained Labillardière, 'ought to have been more sensible than the commander of the expedition of the great utility of this plant for our navy.'[67] In the Tongan Islands there was no news of La Pérouse, but violence at Tongatapu left one of the crew seriously injured, and at least two Tongans dead. As the ships reached New Caledonia, Labillardière discovered that some of the breadfruit cuttings collected at Tongatapu had died. Saltwater had been poured on them, a deliberate and malicious act by someone on board the *Recherche*. When the naturalists ventured inland they were accompanied by a large escort, for continual thieving and evidence of cannibalism had led them to query Cook's conclusions on his second voyage that the inhabitants were 'Courteous and friendly and not in the least addicted to pelfering [sic]'.[68] Labillardière took full advantage of his botanical excursions in this little-known island, and towards the end of his life published the first flora of New Caledonia. Throughout the voyage it is noticeable that, unlike some earlier botanists, a day he spent ashore collecting would sensibly be followed by a day on board ship sorting and classifying the new specimens.

Shortly before the ships left New Caledonia, Captain Huon de Kermadec of the *Espérance* died after a long, wasting illness. He was replaced by d'Auribeau from the *Recherche*, while Lieutenant Rossel became captain of the flagship. Kermadec's death had a devastating effect on d'Entrecasteaux. In his memoirs Jurien remembered how he 'lapsed into a sombre silence, exhibiting almost total disgust at any kind of food', and soon began to show signs of scurvy.[69] As the ships headed northeast from New Caledonia they again passed near the site of the La Pérouse shipwreck on Pitt's Island (Vanikoro), but did not stop at this tiny island – one among hundreds in the vast expanse of the southwest Pacific. They continued on to the Solomon Islands, where d'Entrecasteaux confirmed Fleurieu's theories about their identity – a subject of much speculation and mystification since Mendaña's discovery of them in 1568 – and then sailed along the coasts of New Guinea and New Britain into waters that had seen no European vessel since Dampier's voyage almost a hundred years earlier. The survey work was valuable and was recorded in the fine charts of Beautemps-Beaupré, but landings were few and the naturalists had no opportunity to carry on their work. On 8 July 1793, d'Entrecasteaux wrote that the ships must head for Java 'where it was becoming more and more pressing for us to arrive' as provisions were running out and the health of the crews was 'drained by the fatigues of a long and harsh journey'.[70] It was his last journal entry, and on 20 July he died. D'Auribeau took overall command of the expedition, but he himself was ill, too weak to attend d'Entrecasteaux's funeral. The commander's death was a cruel blow to an expedition already beginning to disintegrate, for, as a junior officer wrote, d'Entrecasteaux 'had for us more of the qualities of a father than of a captain'.[71]

As the ships sailed towards the Dutch East Indies more crew members fell ill with scurvy and dysentery, and only one-third were in any reasonable state of health when the expedition reached the port of Kajeli on Buru Island in early September. Lieutenant Rossel wrote how their eyes, 'long since wearied by the spectacle of arid and desert coasts, have now fallen with quiet satisfaction upon a fertile country'. In a passage that probably represented the views of all on board except the naturalists, he remembered how 'the very beauties of raw Nature, which had at first enraptured us, now struck us as a dismal monotony . . . The sense of genuine curiosity, which had aroused the desire in us to visit primitive people and to learn about their customs, had become entirely extinct.'[72] In late October the expedition reached the major port of Surabaya in Java. As Rossel had remarked, in such distant regions 'every European becomes a compatriot', and the crews confidently expected recognition of the expedition's status as a scientific venture and help from the Dutch governor. Instead they were shocked by the news that France was at war with the Dutch, that Louis XVI had been executed and a republic proclaimed, and that they themselves were prisoners of war.

The news, the first received since the expedition had left France two years earlier, brought into the open the ideological divisions that had been present on the expedition from the beginning. The royalist d'Auribeau accused Labillardière of 'spreading perfidious advice', Riche of exulting in the king's execution and both men of conspiring with republican members of the crew. The Dutch for their part wavered between a reluctance to allow ashore men who were now enemies, and a humanitarian impulse to relieve suffering among those confined to the ships. In February 1794 the Dutch occupied the two French ships, arresting either on the vessels or in their lodgings ashore all those d'Auribeau accused of republican sympathies. Labillardière was among those taken prisoner and sent overland to Samarang. His main preoccupation was keeping possession of his journal, demanded by d'Auribeau and the Dutch, and in this he was successful. In August 1794, d'Auribeau died and Rossel, the last senior officer, took command. To pay off the debts incurred by the crews since their arrival in Dutch-controlled waters, he was forced to sell the *Recherche* and *Espérance*. Soon after, Rossel and the first contingent of survivors from the expedition sailed for Isle de France in a Dutch convoy, which also carried the precious collections from the expedition. A second group, which included Labillardière, left Java for Isle de France in March 1795 on a further prisoner-exchange scheme. Labillardière arrived in France a year later, to be stunned by the news that the ship in the first convoy carrying thirty-seven cases of his natural-history specimens had been captured by a British man-of-war and taken to England.[73]

In France, it seemed, everything had changed, even in the world of science, which was normally divorced from political upheavals. Some colleagues of Labillardière had been executed during the Terror, others had been demoted or lost their posts. The Jardin du Roi's unacceptable name had been changed to the Jardin des Plantes. Even the calendar had changed. For Labillardière, all this was secondary to the need to recover his collections. The situation was far from straightforward. The exiled king of France, Louis XVIII, had laid claim to the collections, but agreed that Queen Charlotte (wife of George III) should be allowed to take her pick. Sir Joseph Banks, president of the Royal Society and adviser to the government on all matters scientific, was the main intermediary in the affair. Labillardière had already written to him while a prisoner in Samarang, complaining of d'Auribeau's behaviour and accusing him of wishing to hand over the expedition's scientific collections to the Dutch. Once in France, Labillardière again wrote to Banks, appealing to his 'love of science' for help in recovering his collection. By June 1796, Banks had come to the view that despite earlier agreements this was the correct course to follow. As he told Labillardière, it was 'an Axiom we have learned from your Protection to Capt Cook' that 'the science of two Nations may be at Peace While their Politics are at war'. Among Banks's papers is a description of the state of the collection. It had suffered considerably during its sea passage, and perhaps during its detention by British customs. Most of the boxes had been soaked up to a third of their height, so that many plants had been 'reduced to the condition of manure'. Some birds were 'in the most awful state of preservation and had to be thrown away', though most were in good condition. Parts of the entomological collection had 'crumbled to dust', while the shell collection had been so pillaged that only the most common specimens were left. On the other hand, Banks thought that the herbarium contained at least ten thousand specimens, and after the collection was returned to Labillardière he wrote to him praising 'the Glorious Collection which I had the Good Fortune to Send over to you, I Envy you the possession of Some of them which I Saw accidentaly & hope when Peace returns to Procure from you Some duplicates'.[74]

After the break-up of the expedition Labillardière's fellow naturalist, Louis-Auguste Deschamps, decided to stay behind in Samarang, where he spent several years botanising before returning to France, but the ship in which he was sailing was captured in 1803. He was released as a noncombatant but his six cases of Javanese plants were lost, and although many of his drawings survived they were never published. More fortunate was the gardener, Lahaye, who managed to preserve live plants from Java as well as dried specimens and seeds collected on the voyage, and who in time became gardener to Empress Josephine.

La Pérouse and his ships had disappeared without trace, but his journal up to the expedition's arrival at Botany Bay had survived. Sent back to France in instalments from the ships' various ports of call, it was put together and edited by Louis-Marie-Antoine Destouff Milet-Mureau. His expertise was in military rather than naval matters, and much of his editorial work consisted in eliminating references to the monarchy and *ancien régime* titles. When the book was finally published in 1797 in three volumes, it sold poorly in France, although it was quickly translated into English and several other languages. Because the manuscript journal had disappeared, the published version was regarded with some reservations because of the uncertainty about the extent of Milet-Mureau's editorial changes, but the rediscovery of La Pérouse's original journal in 1977 showed that these were minor. What had gone beyond possibility of reclamation were the collections and observations made by the scientists on the expedition, even though some remains were eventually found. In 1827 an Irish trader, Peter Dillon, discovered on Vanikoro relics from the lost expedition and heard from the islanders that ships had been wrecked on the reefs.[75] In the following years further relics were brought up – cannon, anchors, a ship's bell – and when they were sent back to France some of these were identified by the elderly Barthélémy de Lesseps, the last known survivor of the La Pérouse expedition. In time, part of the story of the fate of the expedition was pieced together from islander traditions and archaeological investigations.

The *Boussole* and the *Astrolabe* seem to have been wrecked close together on the same reef in a great storm. Many of the crews either died in the shipwreck or were killed by the islanders as they struggled ashore, but enough survived to build a small craft in which they sailed away to an unknown fate. Two men were left behind and are thought to have survived for some years. They were almost certainly alive when the *Pandora* passed within sight of Vanikoro in 1791, and if d'Entrecasteaux had followed his intention of landing on the island the following year they would have been rescued.[76] Despite all the upheavals of the revolutionary and Napoleonic period, memories of the lost expedition lasted long in France, and in 1828 King Charles X granted Dillon the 10,000-franc reward which the National Assembly in 1791 had promised for news of La Pérouse.

The survivors from the d'Entrecasteaux expedition were slow in giving accounts of their voyage. First to do so was Labillardière, who gave priority to publishing his journal rather than his botanical observations. *Relation du voyage à la recherche de La Pérouse* went through several French editions after its first publication in 1799, and was followed by English and German editions. It undoubtedly helped Labillardière's election to the Académie des Sciences in November 1800. Fears that the book would have a polemic tone because of the botanist's republican sympathies proved to be unjustified. It was a mixture of narrative detail, passages on native customs, and recommendations for the practical uses of many of the trees and plants Labillardière had come across. Accompanying it was a large number of engravings, mostly from Piron's drawings of native peoples, natural-history specimens and artefacts. Labillardière's great botanical work, *Novae Hollandiae plantarum specimen*, was published in parts between December 1804 and August 1807. Its two volumes were accompanied by 265 plates. There were a number of errors, primarily in plant location, and some confusion between Labillardière's specimens and those collected by the later Baudin expedition, but, in Edward Duyker's words, the work's 'crowning glory ... was little diminished by the fact that some of its jewels were misplaced'.[77]

What was perhaps Labillardière's most original work had to wait for thirty years after the voyage before it made its appearance. In his advice before the expedition sailed Banks had particularly recommended New Caledonia to Labillardière's attention: 'These islands have been observed only very superficially. Forster, who did not get on well with Cook, did hardly anything there, and no other naturalist has visited them.'[78] Labillardière had spent little more than two weeks in New Caledonia in mid-1793, but on several trips into the interior he made large collections of specimens. After a long delay, he published in 1824–5 his pioneering *Sertum Austro-Caledonicum* which contained descriptions of seventy-seven of the island's plants. One striking feature was that, late in life, he used in this work the natural classification system of Antoine-Laurent de Jussieu in preference to the older Linnaean method.

D'Entrecasteaux's original manuscript journal, kept to within twelve days of his death, has disappeared. For some years it was in the possession of Rossel, the last surviving senior officer from the expedition, during a period of exile in England. In 1802, when the Treaty of Amiens brought a temporary peace between the warring countries, he returned to France with the journal and other documents. There Fleurieu read the journal before its publication and complained that it was 'overburdened with erasures, additions, transpositions' and was consequently very 'painful to read'.[79] Rossel was told to make a clean copy, but there is no way of telling whether the resultant printed version was a faithful transcription of d'Entrecasteaux's manuscript original. It is not cast in the form of a seaman's log, but is a continuous narrative, with much skipping of detail and compression of tedious periods of the voyage. It is possible that d'Entrecasteaux, with his official daily log to hand, himself kept this more readable journal with a view to publication (Cook had done something similar on his second and third voyages). This still leaves uncertain Rossel's role as editor: he was an ardent royalist, but the neutral tone of the journal suggests that he made little effort to slant it towards his own political views. The *Voyage de d'Entrecasteaux envoyé à la recherche de La Pérouse* was finally published in 1808, fifteen years after d'Entrecasteaux's death. Probably because the expedition had failed in its main objective, the narrative attracted little interest either in France or in other countries. A request by d'Entrecasteaux's heirs to share in the book's profits was met with the brusque response that only eighty-seven copies had been sold. It was a sad end to an ill-fated expedition.

CHAPTER 8

'All our efforts will be focussed on natural history'

The Scientific and Political Voyage of Alejandro Malaspina

THE SUPERBLY EQUIPPED expedition commanded by Alejandro Malaspina that left Cádiz for the Pacific in July 1789 was intended to reassert the tradition of Spanish voyaging in the *Mar del Sur* which had faded from view in the glare of publicity that accompanied the voyages of Cook, Bougainville and La Pérouse. It represented at one level the philosophical and scientific interests of the European Enlightenment held by Spain's Bourbon monarch, Carlos III, at another a determination to investigate the political and economic state of Spain's sprawling Pacific empire. It would not be a voyage of discovery in the traditional sense, for, as Malaspina explained, 'The safest and shortest routes between the most distant corners of the earth had been pieced together. Any further voyage of discovery would have invited scorn.'[1] It was the oceanic equivalent of the Royal Scientific (or Botanical) Expedition to New Spain, which was approved in 1787 and resulted in the establishment of the Botanical Garden and Institute of Botany in Mexico City.[2] The land expedition – in effect a series of excursions – was not the result of a government initiative but the brainchild of an Aragonese physician and botanist, Martín de Sessé y Lacasta. In the same way, the oceanic enterprise was suggested not by any government department but by Alejandro Malaspina and his fellow naval officer José Bustamante.

Thirty-three years old at the time of sailing, the Italian-born Malaspina had joined the Spanish navy in 1774, four years later than Bustamante, who was already an experienced officer although he was not yet thirty. Malaspina was a skilled hydrographic surveyor who, together with several of his officers, had served on Don Vicente Tofiño's comprehensive charting of the coasts of Spain. A man of wide reading and radical thinking, he held political and economic opinions that were much influenced by the Enlightenment. In upbringing, education and attitude he was far removed from the stereotyped image of a Spanish naval officer still held in Britain and other Protestant countries.[3] Between 1786 and 1788, Malaspina had sailed around the world while on secondment to the Royal Philippines Company, a voyage that taught him helpful lessons for his later expedition. More immediately, his experiences and observations during the voyage led to his presenting, together with Bustamante, a 'Plan for a Scientific and Political Voyage around the World' to the Spanish navy minister, Antonio Valdés y Bazán, in September 1788. Within five weeks the 'Plan' was approved, although it is not clear whether Carlos III saw it. Seriously ill, the latter died in December 1788 and was succeeded by his son, Carlos IV. The new king held few of his father's enlightened views, but preparations for the voyage continued without hindrance. The scientific part of the voyage was modelled on the expeditions of Cook and La Pérouse; there were to be hydrographic and astronomical observations, and natural-history investigations to match those of the land expedition to New Spain. The naturalists were to collect specimens for the Real Jardín Botánico in Madrid, established in 1755. In the wider scholarly world much was made of these objectives, so that in his 1803 *Ensayo político sobre el reino de la Nueva España*, Alexander von Humboldt proclaimed that 'no government has sacrificed greater sums than Spain's to advance the knowledge of plants'. Before sailing, Malaspina wrote to Joseph Banks assuring him that 'the main aim of our voyage will no longer be to make discoveries ... All our efforts will be focussed on natural history and perhaps also on acquiring a knowledge of native languages.'[4]

However, the voyage had less-publicised aims, directed towards strengthening Spain's national interests. Among the expedition's priorities would be the making of detailed charts of Spain's overseas possessions, investigating their commercial and defensive capabilities, and suggesting measures for their improvement. In addition, it would report on the Russian settlements rumoured to have been established in California, and the new British settlement at Botany Bay – 'places of interest whether from a commercial point of view or in the event of war'.[5] The voyage, Malaspina and Bustamante calculated, would take three and a half years. Their ships would enter the Pacific around Cape Horn, and sail along the South American coast from Chile to Mexico, across the North Pacific to the Hawaiian Islands, then back to the American mainland to trace the coast north from California before calling at Canton. The second part of the voyage would take the expedition to the Spanish possessions in Guam and the Philippines, and south to New Holland, New Zealand, the Society Islands and Tonga, before returning home by way of the Cape of Good Hope. It was an ambitious programme, and one whose political objectives of report, recommendation and reform distinguished it from its British and French predecessors.

No cost was spared in preparing for the voyage. Unlike the voyages of Cook and La Pérouse, where workaday colliers or storeships had been converted for the Pacific, the corvettes or small frigates *Descubierta* ('Discovery') and *Atrevida* ('Daring') were specially built for the voyage, and carried the latest navigational and hydrographic instruments. They were identical vessels, and each had a complement of 102. The officers were carefully chosen, and most had sailed with Malaspina or Bustamante before.[6] Among the scientists, a scholarly army officer, Antonio Pineda, who was appointed director of natural history on the expedition, was described by Malaspina as having 'not only all the talent and capacity necessary in his science but also ... an admirable disposition, genuine love of the study of new things and concern for his honor which can be his only motives'.[7] He was supported by the hard-working if often underrated Luis Neé, a French-born botanist who had become a Spanish subject, and in his mid-fifties was probably the oldest member of the expedition. He was described by Pineda as 'an indefatigable man appropriate for this investigation having long experience in plant collecting in far provinces and mountains' of Spain.[8] A late appointment, not known personally to any member of the expedition, was an impressive Czech-born botanist, Tadeo Haenke, already an outstanding scholar although only in his late twenties. Apart from his botanical competence, he had interests in geology, mineralogy, anthropology and linguistics. On his journey across Europe to join the expedition he arrived in Cádiz two hours after the *Descubierta* and *Atrevida* had sailed, took passage in a merchantman to their first port of call at Montevideo, only for his ship to be wrecked off the harbour. An oft-repeated story has him swimming to safety carrying only a copy of Linnaeus, but in a letter to friends at home he described how he came ashore in one of the ship's boats without clothes, papers or books – 'everything is lost'.[9] Once ashore, he discovered that the ships had left for Cape Horn and the Pacific eight days earlier. Undaunted, Haenke crossed the pampas and the Andes on a transcontinental journey that took eight months, and finally caught up with the *Descubierta* at Valparaiso in April 1790. On his way, Malaspina reported, this remarkable young man 'had gathered close to 1,400 plants which were either new or had been poorly described'.[10]

By the time Haenke had joined the expedition, his fellow botanists Pineda and Neé were well into their stride. On Isla San Gabriel in the Río de la Plata they 'gathered such a variety of shrubs, plants and flowers that they seemed to be the fruit of the examination of an entire country rather than that of one small island'. At Port Egmont in West Falkland they botanised with 'notable success', while Pineda brought on board the skeleton of a penguin.[11] Several other members of the expedition became enthusiastic if sometimes inexpert collectors, while others tried their hand at taxidermy – not always with successful results.[12] Local collectors brought specimens to the ships, notably at Guayaquil, where the mayor of a town in the interior sent down a large number of dried and stuffed birds and medicinal plants. For his part, Haenke returned on board with a live crocodile.[13] Despite all precautions, the casualty rate among the specimens was high. There is a doleful entry in Malaspina's journal at Panama: 'Pineda took the opportunity to air and check again the invaluable collection of dried birds and animals from Guayaquil ... he had managed to find room for the cases in a deck-house, so that they would benefit from fresh, clean air. Our surprise can be imagined when we saw, upon opening the cases, that despite all our precautions the entire collection was spoilt and useless.'[14]

Pineda paid special attention to plants and trees that had a practical or medicinal value. He reported to Valdés in Madrid that on an excursion inland from Guayaquil he and Neé had found valuable drugs, and in the forests a gum-producing tree: 'It manifests the same properties as the elastic gum which is brought from Portugal and sold in Madrid, in ignorance of the fact that it is produced in our colonies. The natives ... liquify it over a fire and with it varnish linens, obtaining excellent and flexible oil-cloths, which are used in the making of rain-proof caps, boots, and hat-covers.'[15] In a report written after his return to Spain, Neé listed no fewer than fifty-five trees of the Guayaquil region whose timber and fibres might be used for shipbuilding.[16] For the botanists generally, the area 'was a paradise that offered notable scientific achievements'.[17] Again, while at Realejo, Pineda reported to Valdés on the many useful trees and plants, whose exploitation, he told the minister, 'could give vigor to this port, a part of which is now very decadent and poor'.[18]

Essential to the collecting and recording activities of the naturalists were the artists who served at various times on the expedition: José Guío and José de Pozo (the two original artists), José Cardero, Tomás de Suria, Fernando Brambila and Juan Ravenet. Some indication of their importance is given in Neé's comments on Guío (although the artist was later dropped from the expedition because of ill health and his reluctance to draw any except plant specimens). He 'is good and patient. He knows the principles of botany and is well able to identify the parts of a plant including the means of propagation. The drawings I have had from him so far are uncluttered except with what is essential for systematic classification. With an accompanying methodical description, they suffice to know the plant in question thoroughly.'[19] Pozo was a zoological artist, and the early-voyage pictures of birds and animals are his; however, Malaspina found him lazy and obstinate, and sent him home from Peru. Cardero began the voyage as a cabin boy, but his facility in drawing led to his acceptance as a member of the 'scientific corps'. By the time the expedition reached New Spain his drawings of birds, fish and animals had made an important contribution to the natural-history record, while he also contributed a number of fine views. Suria, a young artist from the Academy of San Carlos in Mexico, joined the expedition at Acapulco. Apart from being a proficient artist, he kept a personal journal for part of the voyage that adds colour to the more restrained journals of Malaspina and the other officers. On boarding the *Atrevida*, Suria was pleased to find that he had been allocated a cabin, even though he was to share it with the second pilot. He soon discovered that it did not allow him the space that he had anticipated for his artistic and literary work, or indeed any space at all: 'While stretched in my bed with my feet against the side of the ship and my head against the bulkhead ... the distance is only three inches from my breast to the deck, which was my roof. This confined position does not allow me to move in my bed and I am forced to make a roll of cloth to cover my head and, although this suffocates me, it is a lesser evil than being attacked by thousands of cockroaches.'[20] Brambila and Ravenet joined the expedition for its Pacific phase, but their main contributions were pictures of people, places and views. In addition, Pineda, Neé and Haenke made drawings and paintings to complement their collecting activities. Among the great mass of Haenke manuscripts investigated by Victoria Ibañez, 'one of the most beautiful and mysterious' is a notebook kept by the botanist of which twelve pages 'contain more than 2,500 different numbered watercolour "shades" for artists who wish to paint botanical specimens, along with some descriptions of the pigments that need to be mixed to build up some of the colour groupings'.[21] The tables formed a code that helped Haenke to reproduce accurately in his paintings the reality of the plant and other specimens he had seen in their live state before their colours faded. The system would later be used by Ferdinand Bauer on the voyage of Matthew Flinders.[22] Whatever the constraints, the artists produced work that made the Malaspina expedition the best recorded in visual terms of any in the eighteenth century.[23]

At Callao (the port of Lima) the first major consignment from the expedition was prepared for despatch to Spain in August 1790. The viceroy of Peru arranged for a shipment of charts, journals, astronomical observations, and botanical and other drawings. Nineteen boxes of plants, stuffed animals, minerals and other items were to follow. Later in the voyage substantial cargoes of journals and specimens were despatched from Acapulco and Manila. These mid-voyage shipments reveal an important difference between the Malaspina expedition and the British and French voyages to the Pacific. Because the *Descubierta* and *Atrevida* were in Spanish-controlled waters for much of the voyage, they spent about 50 per cent of their time in harbour, and another 10 per cent at anchor on open coasts. They were at sea for only about 40 per cent of the time, in contrast, for example, to the 70 per cent of the time that Cook's ships spent at sea on his second voyage – much to the frustration of Johann Reinhold Forster. Moreover, the naturalists were encouraged to make trips into the interior, and these could last several weeks if not months, and were usually accompanied by an armed escort from the ships. There was little of the tension between naturalists and ships' officers that marked the other discovery expeditions of the period. Even so, the stays on land were not an unmixed blessing, and Malaspina's journal records with increasing frequency the problems presented by his crews in port – ill-discipline, desertion and outbreaks of venereal disease.

In March 1791 the *Descubierta* reached Acapulco, near the northern limits of Spanish settlement on the Pacific coast (the *Atrevida* was already at the port). A list that Malaspina drew up at this time, twenty months into the voyage, showed that crew losses from one cause or another, but mostly from desertion, totalled 143 men. Given that the original complements of the corvettes amounted to 102 men for each vessel, this was an extraordinarily high proportion. While at Acapulco another shipment for Spain was prepared. It included 132 botanical and zoological paintings, and four chests of botanical, zoological and mineral specimens. Meanwhile, the botanists were busy collecting new specimens. Pineda was generous in his praise of Neé's endeavours, going so far as to say that he had 'discovered as many new plants as have been described by botanical writers since Hippocrates to the present day'.[24] A further indication both of the help the expedition's scientists received from local people and of the importance they attached to a specimen's utility came during the *Atrevida*'s stay at San Blas. There Pineda was presented with a list of local woods accompanied by comments on their medical uses. Iris Engstrand describes a dozen of these trees, with decoctions from their bark, pods or fruit used to treat a variety of complaints ranging from insect bites to scurvy, jaundice and rabies.[25]

Until its arrival in Acapulco the expedition had followed a predictable course, although the original timetable had slipped, and Malaspina had decided that he would abandon the projected circumnavigation and return to Spain by Cape Horn. It was the halfway stage of the voyage, but already much had been accomplished. Pineda thought that the botanical specimens collected would increase the list of the world's known plants by about a third 'if one counts those that are scarcely known or those which the great Linnaeus has described badly'. On his reckoning seven thousand plants had been collected, and five hundred species of animals and four hundred fossils examined.[26] From Acapulco, Malaspina travelled on horseback to Mexico City, the capital of New Spain, where he met the viceroy, the count of Revilla Gigedo. By this time, after almost two years away, it is clear that Malaspina was resenting the lack of information and fresh instructions from home. Letters from Valdés approved of his activities 'but gave no indication as to what our future steps should be', while the viceroy made no suggestions 'as to the best course of action for the corvettes in future'.[27] Without official orders to the contrary, preparations were made for the ships to sail across the North Pacific for a three-month visit to the Hawaiian Islands. Meanwhile, the scientific contingent split, for Antonio Pineda, accompanied by Neé and Guío, left Acapulco on a journey that was to take them to Mexico City and through much of New Spain, investigating, reporting and collecting – though the latter activity was on a minor scale because of transport constraints. While they were away, Pineda was promoted to colonel of infantry.

It was while Malaspina was still in Mexico City that he received news from Bustamante on the coast that startling new orders had arrived from Spain. These were brief but to the point.[28] The expedition was to sail north to Alaska and there search for the strait to the Atlantic reportedly discovered by Lorenzo Ferrer Maldonado in 1588. For centuries the search for the Northwest Passage had been encouraged by the accounts of apocryphal voyages, but Ferrer Maldonado's story was the most extraordinary of them all. He claimed to have made a voyage from Lisbon through Davis Strait and north of the Arctic Circle before sailing southwest for more than two thousand miles to latitude 60°N., where he reached the Pacific coast of North America through the fabled Strait of Anian. Near its entrance was a harbour capable of holding five hundred ships, and while anchored there Maldonado met a large vessel bound for the Baltic through the passage with a rich cargo of pearls, gold, silks and porcelain. The lack of interest shown at the time in this nonsensical story is entirely understandable; what is difficult to explain is why the account should have been taken seriously on its rediscovery in the Spanish archives two hundred years later. It was supported by a memoir by the French royal geographer, Jean-Nicolas Buache de la Neuville, but his credibility had been damaged by the disappointments of La Pérouse on the Alaskan coast a few years earlier during his search in Lituya Bay for the entrance to the Northwest Passage.[29]

Malaspina had no option but to return to Acapulco, and there prepare to sail north, abandoning the Hawaiian part of the programme. In an official response to Madrid he referred to some of the 'difficulties' in Maldonado's account, which in private letters he dismissed as 'apocryphal' and 'false', while, on the *Atrevida*, Bustamante castigated Buache's memoir as intended 'to mesmerize Europe'.[30] The 'scientific and political voyage' had after all turned into a conventional voyage of discovery, the very thought of which Malaspina had earlier said 'would have invited scorn'. With Pineda and Neé away in the interior of New Spain, Haenke was the only full-time naturalist on the ships. Suria and Cardero were on board as artists, but their main – and most important – contribution consisted of making records of peoples and places in Alaska rather than natural-history drawings (Pl. 24).

It was in a mood of resignation rather than eagerness that Malaspina and Bustamante sailed north from Acapulco in May 1791, while a dozen seamen attempted to desert when they heard that their destination was Alaska, not Hawaii. Following the advice of Spanish navigators who had made earlier voyages along the northwest coast, Malaspina took the ships on a long, curving track well out to sea before heading towards Alaskan shores in latitude 56°N. The snow-covered mountains awed the journal-keepers, and the cold was so intense that Suria was unable to sketch on deck, being forced to retreat below to complete his drawings. On 27 June excitement grew on board as the *Descubierta* and *Atrevida* steered towards a great cleft in the coastal range in latitude 59°15′N. Malaspina had to admit that the inlet (Yakutat Bay) resembled the one described by Ferrer Maldonado, and added that 'imagination soon supplied a thousand reasons in support of hope'.[31] In 1787 the bay had been visited by the English fur trader George Dixon, who named it Port Mulgrave, but his main preoccupation was trade, and he had not explored far into the inlet. By nightfall the corvettes had dropped anchor close to a beach and a Tlingit village. Here a portable observatory was set up, Suria and Cardero began sketching, and Haenke set off collecting ethnographic items as well as botanical specimens. As he moved among the local inhabitants, Haenke, as he later told a friend, was apprehensive but excited to feel that he was observing humankind in its original state.[32] During the expedition's stay relations with the Yakutat Tlingit remained tense. There were two serious confrontations with them, and both on shore and in their boats the crews were always armed.

On 2 July, with wooding and watering completed, Malaspina was ready to explore the inner reaches of the inlet in search of the Strait of Anian. He took two launches and fifteen days' provisions. A disappointed Haenke was left behind, and it must have been scant compensation when Malaspina assured him that he would personally collect 'everything relating to natural history' and hand it all over to him on the boats' return.[33] It took only a few hours to dispel hopes of a major discovery, for soon the water shoaled and the thunderous sound of large chunks of ice calving from a glacier could be heard. Then the end of the inlet came in sight from the launches, its low shore blocked by a glacier (the Hubbard Glacier) behind which rose the steep walls of the coastal range. A frustrated Malaspina named the inlet Puerto de Desengaño, or Deceit Bay. In his absence in the boats there had been an awkward confrontation with the Tlingit at the anchorage, and after a hasty act of possession he prepared to sail. Although the expedition members had collected a rich harvest of ethnographic and natural-history material, and Suria and Cardero had made some superb sketches and paintings,[34] the overriding disappointment was that there was no strait leading deep into the interior. Later, Malaspina reflected in his journal that a reader in the twenty-first century would be amazed to see how seriously the fictitious accounts of Ferrer Maldonado and other navigators had been taken 'in an age which we call scientific and enlightened'.[35]

From Alaska the ships sailed south in August 1791 to Nootka Sound, flashpoint of the crisis the previous year that had almost led to war between Spain and Britain. Moving around the infant Spanish settlement at Nootka, Haenke was as active as ever, noting 'details of the flora and fauna, fishes, birds, soils and beaches'.[36] It was a sign of the confusion that could accompany the scientific side of long-range discovery expeditions that the specimens he collected here were later attributed to Neé. The next month at Monterey the botanist was even more in his element, for 'a recent fruiting almost as complete as that of spring had occurred. The leafy shores of Río Carmelo had mingled plants near the sea in such a variety (their seeds no doubt carried down by the winter rains) that they seemed to come from a territory over one hundred leagues wide rather than from the small area that our excursions could cover.'[37] At Monterey, Haenke discovered and arranged for shipment to

Spain specimens of the giant Californian coast redwood as well as spec-imens of local oaks. In all, fifteen trees were described, 'useful for ship-building and for houses'.[38] On the ships' return to Acapulco, Malaspina received news from Mexico City that Pineda and Neé had covered twelve hundred miles of territory in their excursions, and were leaving thirteen cases of specimens to be sent to Madrid for the Real Gabinete de Ciencias Naturales or natural history museum. They had received every possible assistance from the viceroy and from the judge of the Royal Audiencia of Mexico, Don Ciriaco González Carajal – 'his house was rather like a lyceum and his magnificent natural history collections were at all times available to Don Antonio de Pineda'.[39] Other shipments were made from Acapulco, including a box of Haenke's specimens, but the botanist was showing signs of homesickness. On his travels inland from the port he was delighted to find that some of the supervisors of the mines of New Spain came from his native Bohemia, but he lamented that he had received no news from home since his departure two and a half years earlier.[40]

By the end of 1791, as he prepared to sail once more from Acapulco, Malaspina had finished the main part of his mission. His expedition had produced detailed charts of long stretches of the coasts of Spanish America, established the exact locations of the main ports, carried out numerous scientific experiments, collected vast numbers of natural-history specimens, and made observations and drawings of native peoples from Patagonia to Alaska. Less publicly, Malaspina had inves-tigated the political, economic and defensive state of Spain's colonies, with a view to making recommendations for change. But in terms of distance and time the voyage was only half-completed. There was the long run across the Pacific to the Philippines by way of Guam that had to be made before the ships turned south. Pineda's diary of a land journey on Guam from Umatac to Agat reveals his interest in miner-alogy. He began by listing some trees but then gave up when encoun-tering 'others with whose genera I am not familiar, but of which I will be informed by the learned botanists [Neé and Haenke] who are surveying the island'. He was more interested in fissures in the rocks from which gas and ashes emerged to form crystals, and carried out some limited experiments before heavy rain drove him from the spot. He heated the discharge ('it turns white as snow . . . and becomes phosphorescent towards the tips') and mixed it with borax ('it forms a white porcelain').[41] Neé, more conventionally, continued to collect plant specimens on the island, sixty of which he later described in his 'Observaciones Botanicas – Islas Marianas'.

Malaspina spent almost nine months in the Philippines in 1792, while Bustamante in the *Atrevida* visited Macau on the coast of China. Malaspina's time in the Philippines was spent in making coastal surveys as his naturalists headed inland on arduous and sometimes dangerous excursions. Neé left the *Descubierta* at the port of Sorsogon near the southern tip of the main island of Luzon for an overland trip to the capital at Manila that lasted three months. Towards the end he had five men to carry his collections and baggage. His instructions 'did not limit my investigations to [botanical collecting] alone. I was also to include the cultivation and plantations of cocoa, mulberries, indigo, pepper, sugar and cotton ... and, as far as possible, the advancement of the branches of zoology and conchology.' For reasons that may have been as much commercial as scholarly, shell collecting had become something of an obsession on the *Descubierta*. At Palapag on the island of Samar, after the officers bought all the shells the inhabitants had to offer, they scoured the beaches themselves, so that 'within three days not a stone remained unturned on the nearby shores, nor was there a single species of the shells of these waters that we did not find'. When the ship reached Manila it was characteristic of Haenke's enthusiasm that he went ashore even before the ship anchored. His journey from the capital to the north of Luzon took more than three months, during which he faced danger from unassimilated tribesmen on land and from pirates at sea. As he warned in one part of his account, 'any travellers who undress, drop their guard or fall asleep, are almost sure to have their throats cut by dawn'. Even so, during his journey he collected about two thousand dried plants, enough, he wrote, to 'provide a lifetime's work for the most diligent and practised of botanists'.[42]

Antonio Pineda also set off in late April 1792, on a collecting expedition which took him east and north of Manila. He was accompanied by Ravenet and by a local naturalist. After experiencing 'intolerable' heat that dragged 'the traveller down into a sort of lethargy, broken only by annoying skin rashes and bites of insects', Pineda unwisely decided to scale the Santa Inés Mountains, but 'after climbing to the top by hauling himself up on tufts of grass, he spent the night in the open with no other shelter than a blanket and, above all, with nothing to eat but cold food, which he could not stomach. The damage to his health from these hardships was to be beyond recovery.' Despite his weak condition, Pineda insisted on continuing the journey, even when his notes, 'drafted in some confusion, bearings incorrectly given or duplicated in contradiction, began to show the first signs of the writer's fatal condition. His memory weakened, he confused species with each other, and, his body lying exhausted on a litter or portable bed (carried by Indians), he was left powerless but for his will.' Although racked by a high fever, he refused rest, 'never ceasing to describe the country, and plan his itinerary'. On 20 June he reached the small town of Badoc, where he fell into a coma and died three days later (Pl. 25). For Malaspina, the death of his friend and colleague was a shattering personal blow as well as a grievous loss to the expedition. In a tribute to Pineda, he gave some indication of the range of his interests when he lamented: 'His ideas, as ambitious as they were viable, about the land and inhabitants of almost the entire continent of the Americas subject to the monarchy, the comparative exploitation of its minerals, the analysis of its languages, the administration, situation and customs of our colonies, although partially described in his notebooks, have largely perished with him.' The loss was the greater because some of his notebooks, probably owing to the sensitive nature of their contents, seem to have been kept in cipher.[43] The job of putting his notes in order was left to his brother, Arcadio Pineda, but he seems to have made little progress with this daunting task. In a final tribute to Pineda, the expedition's officers met the costs of a memorial erected in his honour in Manila.

After calling at Mindanao, the large island south of Luzon, the *Descubierta* and *Atrevida* followed a semicircular track into the Pacific. At the end of February 1793 they reached Cook's Doubtful Sound on the southwest coast of New Zealand's South Island, where a chart of the area was made that was to remain in use for many years. Bad weather prevented landings either there or at Dusky Sound, and Malaspina decided to sail to New South Wales and the British convict settlement at Port Jackson. Before leaving Spain, Malaspina had suggested that New South Wales should be visited because of its potential political and commercial significance. His month's stay at Port Jackson reflected this attitude. The courtesies exchanged by the Spanish officers and their British hosts no doubt represented a genuine sense of companionship between men many thousands of miles from home, but Malaspina's journal also revealed that the ceremonies enabled him to cast 'a discreet veil over our national curiosity'.[44] In effect, Malaspina was engaged in unobtrusive espionage in a colony that had made remarkable progress in the five years since its unpromising beginning. In his memoir 'A Political Examination of the English Colonies in the Pacific', Malaspina described the threat the new colony posed to Spanish interests in the Pacific, possibly even serving as a base for an invasion of Chile or Peru in wartime. To ward off this danger, he suggested the establishment of commercial links between Spanish America and the colony, which would turn a potential enemy into a satisfied trading partner.[45] There was no realistic prospect that the Spanish government would adopt such a policy, but the recommendation showed the unconstrained nature of Malaspina's thinking on political and economic matters.

For Haenke and Neé, the region was an exciting new world. After four months at sea Haenke, as he wrote to Joseph Banks, was enthralled by the sight of Port Jackson: 'When, gradually, we penetrated further and further into the harbour, the appearance of a land new and strange to me, with a wealth of trees and bushes, caused me, just as it should, very great joy, and also aroused hopes of a rich harvest.' Soon after landing, the botanists found a space near the expedition's observatory where they could care for the plants they had collected. These, Haenke wrote, 'surpassed all our expectations'. They made several excursions away from the settlement, inland towards the more fertile area of Parramatta and south to Botany Bay. This last excursion probably prompted Haenke's letter to Banks (though he had also written to him from Lima), for it was the spot where Banks and Solander 'added such a number of Plants to the treasury as to be judged worthy of being known by the name of the beloved Science of Botany'.[46] A reminder that Haenke collected more than plants came in his list of zoological specimens, including three kangaroo foetuses, a dried kangaroo, four dried possums, a black swan, twenty-three parrots and a shark.[47]

On leaving Port Jackson, Malaspina sailed directly to Vava'u, the northernmost group of islands in the Tongan archipelago. It had not been visited by Cook during his calls in the archipelago during his second and third voyages, and Malaspina's stay there had an even clearer political purpose than it had at Port Jackson – it was a follow-up to the visit made by the Spanish navigator Francisco Mourelle in 1781, and towards the end of his ten days at Vava'u Malaspina carried out his second act of possession on the voyage. Like Yakutat Bay, site of the first act of possession, Vava'u had seen few European visitors, so the observations, paintings and vocabularies brought back by the expedition are of special value. Although Haenke continued to collect plants, he devoted most attention to birds and fishes, finding 'great numbers and species that had not yet been fully recorded in the descriptions of natural history already published'.[48] The brief interlude at Vava'u had more in common with the previous voyages of the Cook era than any other part of the voyage, and remarks on the expedition's stay by Bustamante could have been taken from the ecstatic pages of Hawkesworth or Bougainville: 'Nothing can compare to the beautiful variety of scenery ... In these delightful places the dullest imagination could not resist the sweet and peaceful sensations that they inspire. Here our minds were gently drawn to philosophical reflections on the life and happiness of these people ... their tranquil existence in the midst of abundance and pleasure.'[49]

Not all was peace and harmony at Vava'u. There were outbreaks of petty thieving and some misunderstandings with the island's ruling chiefs. More serious for the continuation of the voyage was the unruly behaviour of the crews, many of whom, Malaspina wrote, resented any discipline, however mild. By this stage there were problems in working the ships, for the crews were divided between worn-out sailors who had been on board for the whole voyage and inexperienced youths from the Philippines and elsewhere. Some indication of the strain Malaspina was under can be seen in an angry outburst in his journal two weeks into the voyage from Vava'u to the South American mainland.[50] The trust that he had tried to build up with his crews had 'vanished like smoke'. More worryingly, there was a breakdown in relations between him and his officers, who were demanding 'more rest and fewer obligations'. The situation was, he said, one in which 'discipline was seen as tyranny, caution as fear, and a normal desire for calm taken as a sign of weakness'. He even hinted at the possibility of mutiny. The rift was also a sign of Malaspina's own deterioration, certainly physically, and perhaps mentally and emotionally too. Not only had he been in sole command for almost four years, but before sailing he had enjoyed only a few months' rest after his circumnavigation in the service of the Royal Philippines Company. Malaspina's strictures did not apply to Neé and Haenke. On the contrary, when the ships arrived at Callao he praised them as 'tireless, intelligent and useful'. Rather than keeping them on board for the long homeward passage around Cape Horn and into the Atlantic where there would be few opportunities to botanise, he gave them permission to leave the expedition and strike out overland across those parts of the South American continent that 'remained unknown to the physical sciences, particularly botany'.[51] Haenke left the ships at Callao, Neé farther south at Concepción; both were ordered to rejoin the expedition at Montevideo.

At Callao, Malaspina left the *Descubierta* and took his instruments and books to the nearby retreat of La Magdalena, away from his squabbling officers. There, he wrote, he could 'shake off the hateful guise of commanding officer and attend quietly to the restoration of my own much weakened health'.[52] His state of mind would not have been helped by the fact that he had not received any official letters or instructions from Madrid since October 1791, more than two years earlier. He may well have felt betrayed by his closest associates and abandoned by his superiors. To add to Malaspina's worries, during the expedition's stay at Callao news was received of the outbreak of war with revolutionary France, for this might have serious consequences for the lightly-armed corvettes. It explains Malaspina's decision that the *Descubierta* and *Atrevida* should make separate passages to Montevideo, both to increase the amount of surveying that could be done, and to reduce the risk of the expedition and all its work being wiped out by a chance encounter with enemy ships. Pineda's journals and papers were to be stored on the *Atrevida*, where his brother, Arcadio Pineda, and the ship's surgeon, Pedro González, would continue to sort and classify them. From the wording of Malaspina's instructions, it seems as if little had so far been done in this regard, for he ordered Pineda and González 'to distinguish and separate the material that is purely scientific, and which goes into the minute detail which is necessary for natural history, from the pleasant, entertaining and no less instructive parts that include the philosopher's contemplation of the immense variety of nature'.[53]

So serious were the combined ravages of sickness and desertion among the crews that Malaspina and Bustamante were able to sail only after receiving fresh seamen from naval ships already at Callao. Of all Malaspina's problems the most intractable was the turnover of crew. The records show that during the voyage about 215 men served on board the *Descubierta* at different times, and about 240 on the *Atrevida*, making the total number of those who served on the two corvettes (each normally crewed by about a hundred officers and men) more than 450.[54] Fatalities among the 250 or so 'losses' suffered by the expedition seem to have been comparatively low at twenty named individuals, although there would probably have been further deaths among those crew members left behind in port hospitals whose fate is not known. What is certain is that Malaspina and his officers were faced with continual problems of training and discipline because of the changes among the crews, and his journal reflects his growing irritation with the situation.

With weakened crews, and worried by the prospect of meeting hostile ships, Malaspina dropped some of the more ambitious parts of the survey of the coasts of South America that he had originally had in mind. He carried out survey and scientific work in West Falkland (Pl. 26), while Bustamante did the same in East Falkland, before the ships sailed for their rendezvous at Montevideo. There Neé rejoined the expedition. Malaspina was reassured to see him 'because I now saw that he would soon be safe from the many risks to which his blind love of botany exposed him almost daily'. He had collected more than four hundred new plant specimens as he crossed the cordillera and the pampas, as well as rocks from the mountains that were 'now to add a great deal to our lithological investigations'. Haenke's efforts had been even more impressive. He was still in Upper Peru (Bolivia), and Malaspina gave him leave of absence for another year: 'As well as his detailed research in the fields of botany and lithology, and the excellent new collection of birds made on his long journey, he had also added greatly to general knowledge by his analysis of many mineral springs and the famous cinnabar [mercury sulphide] mine in Guancavelica.'[55] Haenke was never to see Europe again, but settled near Cochabamba, where he died in 1817. He ignored all orders to return to Spain, but was careful to keep in touch with his royal paymaster, receiving regular payments for the rest of his life.[56]

The last stage of the voyage from Montevideo to Cádiz was an anti-climax as the *Descubierta* and *Atrevida*, together with a royal frigate, escorted a convoy of clumsy merchant ships across the potentially hostile waters of the Atlantic. Despite the frustrations of the slow crossing, Malaspina and his officers continued to make observations and check their instruments, and it was wholly in character that the final sentence in Malaspina's last entry in his journal, that for 21 September 1794, concerned the rating of the ship's chronometers. The expedition had been away for five years and two months, and brought back with it 'one of the largest stocks of news and information ever assembled by a single expedition'.[57]

Within two days of his return Malaspina wrote to his friend Paulo Greppi, expressing his eagerness to join the Spanish army campaigning in Roussillon. On 7 October he wrote in more sober mood to Valdés (still navy minister) that he 'would have rushed at this very instant to serve in this war if the pressure to collate the mass of papers from our recent tasks did not stop me'.[58] This should not take long, he told the minister; an optimistic assessment when the nature of the task is considered. Malaspina intended his published account of his voyage to be on a scale that would dwarf the narratives of his predecessors in the Pacific, even the lavish three-volume account of Cook's last voyage. It would answer all accusations that Spain kept its discoveries secret. Rather, the publication of the results of the expedition would 'draw aside at last the thick curtain of mystery' that had concealed Spain's overseas possessions. In the first instance there would be seven volumes, an atlas of seventy charts, harbour plans and coastal views, and a folio of seventy drawings. Later, Malaspina hoped that there would be additional volumes based on the work of Antonio Pineda, Tadeo Haenke and Luis Neé. This monumental work would be more than a matter of factual record, for it would include an exposure of 'the slow but deleterious effects of present laws' on Spain's overseas empire, together with detailed recommendations for change and reform.[59]

In December 1794, Malaspina was received at court together with Bustamante and other officers from the expedition, but felt that he had been fobbed off with superficial ceremony. A letter to Greppi shows that deference and tact were not among his strong points: 'One single day would have been sufficient to explain my system. I have seen everything. I have been everywhere. I had hoped that no matter the chaos of the present system it would be realised that there is but a small step from the wrong route to the right one, from absurdity to sane philosophy.'[60] Nor did Malaspina confine himself to the state of the overseas empire, for at this time he presented Valdés with a memorandum setting out his views on the terms of a peace treaty with France. In commenting on so sensitive a matter, this serving naval officer without any diplomatic experience enraged the powerful first minister, Manuel Godoy, who referred to the memoir's 'lack of principles and moderation', and advised Valdés that Malaspina should burn it and be told to mend his ways.[61] Malaspina was treading on dangerous ground, although he was again received at court by the king and queen in March 1775 and promoted to *brigadier* (admiral).

Despite this sign of official favour, Malaspina foolishly became involved in an unsuccessful conspiracy aimed at replacing Godoy and other ministers. In November a warrant was made out for the arrest of 'Brigadier of the Royal Navy Don Alejandro Malaspina'. It was the last time that he was accorded his rank; from now on he would be simply and brutally referred to as 'the criminal Malaspina'.[62] After a hurried hearing of his case (in his absence) by the Council of State, Malaspina was stripped of his rank and sentenced to ten years' imprisonment in the grim fortress prison of San Antón at La Coruña. In his statement to the Council, Godoy remarked: 'In my opinion this subject has lost his mind.'[63] The officers who had been helping him prepare his account of the voyage were ordered to stop work and to surrender all papers relating to the expedition. In prison Malaspina occupied himself in reading and writing, but at times was subject to delusions, for example claiming that when he was arrested he was being considered to replace Valdés as navy minister.[64]

Of the seven volumes relating to his expedition planned by Malaspina, only one was published, and this dealt with a subsidiary part of the expedition (the surveys carried out by Dionisio Alcalá Galiano and Cayetano Valdés on detachment in the Strait of Juan de Fuca in 1792). The volume was a belated attempt to counter the publication of George Vancouver's account of his voyage to the northwest coast, which included his rival survey of the strait; but although the Spanish account included a handsome atlas, its appearance ten years after the event, and four years after the publication of Vancouver's narrative, was too little, too late.

The volume contained no reference to the expedition as a whole, and no mention of Malaspina other than an occasional allusion to the unnamed 'commander of the corvettes'. He had been removed from the historical record, and so, for the most part, had his expedition, although gradually the work of some of its members found its way into print.

The shipments of natural-history specimens from various ports of call ensured their arrival and preservation at the Real Gabinete de Ciencias Naturales and the Real Jardín Botánico in Madrid. In all, seventy cases had been sent, and at the end of the voyage Luis Neé listed the provenances of the dried plants he had sent back to Spain:

On the Pampas of Buenos Aires	507
Ports Deseado and Egmont [Falklands]	235
Chiloé and Chile	1,167
Peru	1,609
Panamá	449
New Spain	2,940
Philippines	2,400
Botany Bay	1,155
Friendly Island [Tonga]	160
	[Total: 10,622]

The addition of more than five thousand specimens of grasses, mosses, algae and fungi brought the total number to just short of a colossal sixteen thousand.[65] Many of the more important of Neé's botanical discoveries were included by Spain's leading botanist, the director of the Real Jardín Botánico, Antonio José Cavanilles, in the fifth volume of his *Icones et descriptions plantarum* (6 volumes, Madrid, 1791–1804). Although some of the plants were wrongly attributed because of incorrect labelling, Cavanilles estimated that half of those he described were new to science, and they formed the nucleus from which the herbarium of the Real Jardín Botánico was established.[66] Plants grown from Neé's seeds adorned botanical gardens in Madrid and other parts of Spain, and were used in exchange for specimens from other countries. Neé himself published four articles on specific plants in *Anales de ciencias naturales* between 1801 and 1803, but his work was handicapped by the fact that eight of the thirteen folders of plant descriptions and field notes he had sent back to Spain never arrived. His comprehensive list showed that he had collected 1,155 plants at Botany Bay, but his notes have descriptions of only sixty.[67] Neé had hoped to publish a full account of his collections as 'Historia general de las plantas', but Spanish officialdom was resistant to pleas that national honour demanded publication, and he died in 1807 before he could make much progress.[68]

Other parts of the expedition's vast collections received little attention. Assiduous though Tadeo Haenke had been in collecting, he was less careful in sending his natural-history and ethnographic specimens back through official channels to Spain. Many he treated as part of his personal collection, and they were shipped to Europe by a Bohemian trading company together with fifteen thousand field-work notes to find their home in the Náprstek Museum in Prague.[69] After Haenke's death the Czech botanist K.B. Presl began a major publication based on the collections in Prague, *Reliquiae Haenkeanae*, but lack of funds halted the project in 1835 after he had issued only six out of the intended twenty parts. Only in recent years has there been widespread scholarly recognition of the importance of Haenke's work, with biographies, editions of his writings and an international conference in Prague, followed by a volume of essays.[70]

Study of Antonio Pineda's collection, especially strong in zoological specimens, was seriously handicapped by his premature death. As Malaspina lamented, 'if only fate had allowed us to bring Pineda himself back to his homeland in safety, how much would Spain had benefited from his work in prospect – in studying the collection with that breadth of knowledge he had studied to embrace.'[71] Pineda's papers included diaries, scientific observations, logbooks and personal documents. After the return of the expedition his brother reported that his natural-history papers alone made up five quarto volumes each of four to five hundred pages. The intention to publish these collapsed following Malaspina's disgrace, and Pineda's documents were dispersed throughout a number of archives in Spain and other countries.[72]

The hydrographic work of the expedition fared better, and when, in 1809, Espinosa y Tello, who had served on the expedition, published his *Memorias sobre las observaciones astronómicas, hechas por los navegantes españoles en distintos lugares del globo*, he not only included details from

Malaspina's journal, but dared to mention Malaspina by name. Oddly, the first appearance of Malaspina's journal in print was in Russian. A Russian diplomat in Madrid obtained a copy of the Spanish original, which was translated first into French and then into Russian. It was then issued in six parts between 1824 and 1827 in a Russian naval periodical at St Petersburg, where its existence was recently revealed by the Russian scholar Alexsey Postnikov.[73] With its text twice removed from the original and its appearance made in serial form, this was a less than ideal first publication, and the journal had to wait almost a hundred years before it appeared in Spanish in 1885, in a heavily edited version by Pedro Novo y Colson. In Oskar Spate's words, the editor 'resurrected and promptly re-buried' Malaspina's account 'in a folio of 681 pages, 573 of them double-columned, with no index and not even a table of contents'.[74] The journal's original draft, in Malaspina's handwriting and including his corrections, was not published until 1990 as part of a magnificent nine-volume set that when completed included Bustamante's journal, natural-history observations by Neé, Pineda and Haenke, astronomical and hydrographic surveys, and political and anthropological writings by Malaspina and other members of the expedition. It brought to fruition, two hundred years after the expedition's return to Cádiz, the plan for a comprehensive scholarly publication such as Malaspina had envisaged in his abortive 'Introducción'.

Malaspina was released from prison in 1803, following the intercession of Napoleon Bonaparte on his behalf, but on the strict condition that he was never to set foot in Spain again. The remaining years of his life were spent at Pontremoli, near his birthplace in Mulazzo, where he died in April 1810. Although he seems to have had reasonable means during his retirement, among the sad details of this period of his life is a document dated September 1806 recording that he had sold his sextant, that most personal and treasured of a naval officer's possessions.[75] There was no rehabilitation, no restoration of his rank and no resumption of work on the ambitious edition of the voyage that he had planned so carefully. An expedition that had set new standards in terms of hydrographic, astronomical and natural-history observations slipped from sight, and for long Alejandro Malaspina was the forgotten man among the Pacific navigators of the eighteenth century.

CHAPTER 9

'When a botanist first enters so remote a country he finds himself in a new world'

The Australian Surveys of Nicolas Baudin and Matthew Flinders

WHATEVER THE AURA of prestige that surrounded Sir Joseph Banks in his later career when as baronet, president of the Royal Society and adviser to the king and cabinet, he became in effect unofficial minister for the sciences, in some ways the most important years of his life were those when as 'Mr Banks' he sailed with Cook on the *Endeavour* voyage. The importance of that experience on the development of his character and outlook was shown in later years when he became patron of Cook's men and promoter of enterprises associated with Cook's discoveries. The decision to locate the convict settlement at Botany Bay owed much to his recommendation, and the First Fleet carried fruit trees, plants and seeds that, in Banks's words, should do well in 'a Climate similar to that of the South of France which Botany Bay probably is'.[1] In return, successive governors sent him specimens of plants and animals in what has become known as the 'Antipodean Exchange'. It was no surprise, then, that when a young naval officer, Matthew Flinders, returned to England from New South Wales in 1800 with a plan for a comprehensive survey of the coasts of Australia and the land's natural resources he approached Banks in the first instance. Flinders had sailed with Bligh on the second breadfruit voyage to Tahiti, and had then joined George Bass, a ship's surgeon, in a number of small-boat surveys along the uncharted coastline north and south of Port Jackson. By navigating Bass Strait, the two men proved that Van Diemen's Land (named Tasmania in 1856) was an island, and that the strait provided a short cut on the long voyage from the Cape to Sydney. Undaunted by the problems between captain and naturalist that had surfaced on Vancouver's voyage, Banks threw his weight behind another ambitious project that combined hydrography and natural history.

In a letter to Banks written at Spithead on 6 September 1800, Flinders summarised his work in Australian waters before proposing a comprehensive coastal survey that he hoped would determine whether a north–south strait separated the New Holland of the Dutch discoveries from the New South Wales of 'the great captain Cook'. If one were found, it would provide a short route between India and the new settlement at Sydney, with 'almost incalculable advantages'. In language designed to appeal to Banks, Flinders wrote: 'It cannot be doubted, but that a very great part of that still extensive country remains either totally unknown, or has been partially examined at a time when navigation was much less advanced than at present. The interests of geography and natural history in general, and of the British nation in particular, seem to require, that the only remaining considerable part of the globe should be thoroughly explored.'[2] Banks had submitted a similar plan to the Admiralty two years earlier, but at the height of the war with Revolutionary France it had been rejected. In a new proposal he now argued that no worthwhile import from the young colony of New South Wales had yet been found to compensate for the settlement's cost. Banks concluded: 'It is impossible to conceive . . . that such a country, situate in a most fruitful climate, should not produce some native raw material of importance to a manufacturing country as England is.'[3]

In 1800 nervousness about French ambitions influenced ministers in favour of the expedition, for that summer the French government had requested a passport allowing safe passage for a scientific expedition to Australian waters commanded by Nicolas Baudin. As was customary, this was granted, despite Banks's suspicions that the stated French intention to survey the northwest coast of New Holland was a blind, since that 'part of the Coast [is] better known to navigators than another'.[4] Banks had some grounds for his misgivings, for although the initiative for the proposed expedition came from Baudin, backed by the scientists of the Institut National, there was a clear implication of larger intentions in the statement of the first consul, Napoleon Bonaparte, when he ordered 'a voyage which will have for its main object the exploration of the south-west coast of New Holland where Europeans have not yet penetrated'[5] – in effect, the coasts of modern South Australia and western Victoria. If a great strait divided this region from the British-claimed territory of New South Wales, then a French settlement could follow, to turn the old dream of France Australe into a reality.[6] Flinders represented an important motive for his voyage and Baudin's when he wrote that 'there is a great geographical question to be settled'.[7] Further evidence that commercial benefits might be expected from his voyage came in the decision of the East India Company to give him and his officers £1,200 for their table 'to Encourage the men of Science to discover such things as will be useful to the Commerce of India & you to find new passages'.[8]

By the end of November 1800 the Admiralty had selected a vessel for the voyage, a sloop-of-war converted from a collier. Flinders was delighted that, like Cook, he would be sailing in a converted North Country collier, but the *Investigator*, as the 334-ton vessel was renamed, had been weakened during the conversion work and gave continual problems on the long voyage. She was copper-sheathed, but underneath the copper the wood was already rotting.[9] While Flinders attended to the fitting out of the vessel, Banks supervised the scientific side of the enterprise. As far as that aspect was concerned, his old friend Evan Nepean wrote from the Admiralty, 'Any proposal you make will be approved. The whole is left entirely to your decision.'[10] There were to be five civilian supernumeraries, among whom the most famous in later years would be Robert Brown, a young army surgeon with a passion for botany. In 1793 the botanist James Edward Smith, the first president of the Linnean Society, had outlined both the opportunities and the challenges that Brown and his fellow naturalists could expect. In the few years since the founding of the convict settlement in New South Wales considerable collections of plants and animals had been sent back to England, where they were eagerly awaited. One of the first books on the new colony, by its surgeon-general, John White, included '65 plates of Non-descript Animals, Birds, Lizards, Serpents, Curious Cones of Trees, and other Natural Productions'.[11] In his book, *A Specimen of the Botany of New Holland*, Smith explained: 'When a botanist first enters on the investigation of so remote a country as New Holland, he finds himself as it were in a new world ... The whole tribes of plants, which at first sight seem familiar ... prove, on a nearer examination, total strangers ... not only the species themselves are new, but most of the genera, and even natural orders.'[12] If the plants of Australia offered a challenge to the naturalist, some of its animals were still more perplexing, as the arguments among zoologists about the classification of the kangaroo and platypus showed. All in all, the more that was known about the natural history of the new land, the more doubts emerged about Linnaeus's insistence on the universal Chain of Being with its corollary that 'truly may it be said, that there is nothing new under the sun'.[13]

Also sailing on the *Investigator* was Peter Good, a gardener with experience of caring for plants at sea; Ferdinand Bauer, an Austrian-born botanical artist; William Westall, a nineteen-year-old landscape artist; and John Allen, a miner. Their status and salaries were decided by Banks, who also approved Flinders's instructions. Lessons had been learned from the Vancouver expedition in that the scientific contingent signed an agreement requiring them 'to render voluntary obedience to the Commander of the ship in all orders', and to hand their journals and other observations to the Admiralty on their return. An even more sensitive issue was touched upon by Banks when he passed on to Nepean his ideas about the management of the plant cabin that was to be erected on the *Investigator*'s quarterdeck, reminding him how 'the King's interests & that of Science which were Embarked together in a Plant Cabbin on board Vancouvers Ship was abusd'.[14] A more tactful arrangement on the *Investigator* allowed Flinders and his officers an unobstructed quarterdeck on the outward voyage, for the prefabricated frame was not to be erected until the ship reached Port Jackson (Pl. 27). Among others on the expedition were Flinders's brother Samuel as second lieutenant, and his cousin John Franklin (the future Arctic explorer) as one of the midshipmen.

During the preparations a backstage drama was being played out, for in an attempt to reconcile career and marriage Flinders foolishly promised his new wife, Ann, that she should sail with him in the *Investigator*, an understanding that was totally unacceptable to Banks and the Admiralty when they discovered it. For Flinders, the choice was clear. In language reminiscent of Cook, he wrote that his ambition was to make such a thorough investigation of New Holland 'that no person shall have occasion to come after me to make further discoveries'.[15] Before marrying Ann, he had warned her: 'Half my life I would dedicate to thee, but the whole I cannot.'[16] Many naval officers might have written the same. In July 1801, Flinders, promoted to commander, sailed from Spithead, leaving behind his wife of three months. It would be more than nine years before they saw each other again.

By the time Flinders left England, Baudin had reached Isle de France (Mauritius) with his two ships, whose names, the *Géographe* and the *Naturaliste*, were as appropriate for his voyage as the *Investigator* was for that of Flinders. Baudin's expedition was larger and more elaborate than the British one, and carried no fewer than twenty-two civilian supernumeraries, many more than Baudin thought necessary. They included botanists, zoologists, mineralogists and artists. With their equipment and baggage, they added to the overcrowding on the ships, which were already having to cope with the unnecessary addition (as Baudin saw it) of fifteen well-connected midshipmen. Among the documents relating to the voyage that have survived are comprehensive lists of the natural-history supplies taken on the voyage. They range from flowerpots to drying and drawing paper, watering cans to tweezers, and pruning knives to insect needles (twenty thousand of the latter).[17] The lists give some indication of the care with which preparations for the voyage were made. Baudin's instructions were drawn up by the experienced Claret de Fleurieu, but his advice on the relationship between captain and naturalists would have left Baudin little wiser as to how he should act: 'On all occasions, and by every means at his disposal, he will facilitate . . . the research of the naturalists. So long as he sees no inconvenience in so doing, he will prolong his stay in those places which promise a more valuable harvest for Natural History . . . But on no account must he lose sight of the fact that the monsoons are in control, and that a few too many days spent on one call may condemn him to six months of inactivity.' He should also make due allowance for 'men unaccustomed to the sea', but not to the extent of harming discipline.[18]

Fleurieu's instructions were supplemented by recommendations prepared by the Muséum National d'Histoire naturelle, whose director was Antoine-Laurent de Jussieu, and by the Institut National.[19] Among the others who had a hand in the instructions were Georges Cuvier, founder of the science of comparative anatomy, his collaborator Geoffroy Saint-Hilaire and the pioneering evolutionist Jean-Baptiste Lamarck, scientists with European reputations. All three were professors at the renamed Jardin des Plantes, whose enlarged premises at the time contained natural-history collections secured during Napoleon's campaigns. With its twelve professors and dozens of assistants and technicians, new museum buildings and lecture halls, the Jardin was the largest and best equipped centre of natural history in Europe. Cuvier's museum in the Jardin contained more than sixteen thousand animal specimens, and provided the foundation for his replacement of the hierarchical Great Chain of Being thesis with an arrangement that divided the animal world into four sub-kingdoms: vertebrates, arthropods, molluscs and marine creatures such as jellyfish and sea anemones. His later research and publications on fossil bones pointed the way towards Darwin's work.[20] One novelty as the Baudin expedition prepared to set sail came in the form of a set of suggestions for anthropological investigations presented to the expedition by Joseph-Marie Dégerando, a young member of the newly-founded Société des Observateurs de l'Homme.[21] Although circumstances on the Baudin expedition did not allow most of Dégerando's careful recommendations to be put into effect, they anticipated much of the standard practice of anthropologists in the nineteenth century.

Nicolas Baudin seemed an appropriate choice for the venture. He had commanded botanical voyages while in the service of Joseph II of Austria, bringing back collections for the magnificent imperial gardens at Schönbrunn Palace. Of his four different ships on those voyages, three bore the same, appropriate, name – the *Jardinière*. Closer to home, in 1798 he had brought to France from the Caribbean a large collection of natural-history specimens that were displayed in Paris in a triumphal procession celebrating Napoleon Bonaparte's victories in Italy. Baudin was interested in all branches of natural history, and a zoologist on the Australian voyage later remembered how the captain's personal journal 'was a huge hard-bound volume . . . [it] contained a multitude of drawings in which molluscs, fish and other objects of natural history were painted with an incomparable perfection and trueness to life', although he regretted the lack of anatomical detail.[22] The artists who illustrated Baudin's personal journal, Nicolas Petit and Charles-Alexandre Lesueur, were engaged by him in an unofficial capacity, and appear on the ship's muster roll as 'assistant gunners'. The captain's affection for natural history did not extend to most of its practitioners, and the voyage was marked by a series of disputes between him and the scientists on the *Géographe*. Significantly, there were fewer problems on board the consort vessel, the *Naturaliste*, commanded by the affable Emmanuel Hamelin. It did not help shipboard relations that, although appointed *capitaine de vaisseau* (post captain), Baudin's early years at sea had been spent in the merchant navy, and that he had then made his name, not in the French navy, but sailing under a foreign flag. Nor would his mood have been improved by a last-minute order requiring him, in addition to collecting specimens for the Institut National, to bring back a special collection – 'living animals of all kinds ... and especially of birds with beautiful plumage' – for Mme Bonaparte, wife of the first consul.

Early signs of friction were evident when the ships reached Tenerife in November 1800, less than two weeks into the voyage, when Baudin complained in his sea-log that the scientists 'had pestered me from the moment we dropped anchor to allow them to go ashore, and I had been obliged to give my permission in order to be rid of them'.[23] The expedition took five months to reach Isle de France, with the slow-moving *Naturaliste* continually forcing Baudin on the *Géographe* to shorten sail. On the overcrowded ships life was one of boredom interrupted by frequent quarrels – between botanists and hydrographers, artists and zoologists, and the 'professional' naturalists on board as opposed to the 'amateur' naturalists represented by the ships' surgeons.[24] As the winds remained contrary, Baudin wrote, 'most men had become so peevish and depressed that they lost their tempers with each other over the smallest things'.[25] During the stay at Isle de France, twenty officers and scientists left the expedition, and forty sailors deserted. Some of the savants were in ill health, or claimed to be (when Baudin visited some of them in hospital he found only empty beds, for they were 'on some outing or dining elsewhere'); others, and some of the junior officers, objected to Baudin's authoritarian attitude; the senior botanist left because he intended to write a natural history of Madagascar; many sailors were offered better wages to work in the colony, while others were offered a generous advance to sail on a corsair.[26] Among those who left the expedition was the naturalist Bory de Saint Vincent, who in 1804 published an account of the first leg of the voyage in which he accused Baudin of becoming 'invisible for us' (i.e. the scientists).[27] All three official artists departed, leaving the pictorial record of the voyage in the hands of Petit and Lesueur. For Baudin, not all was loss. The expedition's chief astronomer, Frédéric Bissy, remained behind at Isle de France and was replaced by Pierre-François Bernier from the *Naturaliste*, 'who in every respect is worth a hundred thousand times as much'.[28] Although some of the losses among the seamen were made up locally, the ships were under-manned as they left Isle de France in April 1801 for New Holland.

The delays on the first leg of the voyage led Baudin to make a change to his instructions, which ordered him to sail to Van Diemen's Land, and then back along New Holland's largely uncharted south coast to its western extremity at Cape Leeuwin. Instead, Baudin headed first to the area of the cape, and then north along Australia's west coast, already visited by Dampier and the Dutch. The decision, François Péron wrote in his account of the voyage, 'caused general distress',[29] although at the time the French had no knowledge that Flinders was about to begin surveying the continent's south coast. Péron, a protégé of Georges Cuvier, was attached to the expedition as an assistant zoologist, but had abilities and ambitions far in advance of his humble status. Some indication of these came in a memoir he presented to the Institut National before sailing – 'Observations sur l'anthropologie ou l'histoire naturelle de l'homme'. He was, his biographer has written, 'a sure and self-confident naturalist',[30] with interests in anthropology, oceanography and mineralogy as well as zoology. He also had a difficult and irritating personality, something of which was revealed in his own self-analysis: 'Irresponsible, scatter-brained, argumentative, indiscreet, too absorbed in my own opinions, incapable of ever giving way for any reason of expediency, I can make enemies and alienate my best friends.'[31] He was on frosty terms with Baudin from the start of the voyage, when as a late addition he was not welcomed by the captain, who already considered himself excessively burdened with four zoologists. For his part, Péron, who had lost an eye in the revolutionary wars, was unlikely to be well disposed towards a captain who had spent years in the service of France's traditional enemy. Their antagonism was revealed to the reading public when Péron was entrusted with writing the official account of the voyage after the expedition's return. Several times on the voyage Baudin dismissed Péron's enterprise, as he saw it, in collecting specimens as irresponsible recklessness. In an episode reminiscent of the misadventures of Claude Riche on d'Entrecasteaux's voyage, Péron became lost on an island in Shark Bay and spent an uncomfortable twenty-four hours in the bush. Baudin's irritated reaction to Péron's escapade recalls d'Entrecasteaux's admonition to Riche: 'I firmly promised him [Péron] that when he went ashore again, I would send someone with him who would keep him constantly in sight.'[32] Despite this, Baudin's journal included in full Péron's report of 118 paragraphs on the natural history of the expedition's earlier landing place at Geographe Bay. A more serious matter than these personal differences was the loss of the longboat, essential for close inshore work. In the event, landings were infrequent on the arid shoreline, but from the ships the naturalists saw plenty of sharks, sea snakes, huge jellyfish and turtles, and Lesueur proved to be a superb artist of all forms of marine life, down to minuscule zoophytes (Pl. 29).

On the way north from Geographe Bay to Shark Bay what could be seen of the land was 'disagreeable and dreary', and Baudin entered in his sea-log that, 'as this coast appears to be of no interest for navigation and even less for Natural History, I did not think it necessary to stop'. He continued to be at odds with several of his officers, and at Shark Bay could not help noting that 'several people who are sick when it comes to performing duties on board ... gathered up sufficient strength to run around all day hunting while the sun was at its hottest'. Sub-lieutenant Antoine Picquet was a special cause of irritation, and Baudin's patience finally snapped when he dismissed Picquet from his watch: 'Your innumerable impertinences, your conduct on duty and your impolite replies every time that I have reminded you of your responsibilities, have at last exhausted my patience.'[33]

Péron, never one to understate the perils and hardships he experienced, left a dramatic account of his misfortunes while collecting shellfish on Bernier Island in Shark Bay:

> At the moment that I was most occupied in carefully detaching them from the rock, a rough wave broke over the reef with such force that I was washed against the boulders nearby and rolled over their fearsome surfaces. In the same instant my clothes were all ripped to shreds, and I found myself with cuts everywhere and covered in blood ... summoning

all my strength to escape from the wave that would take me back on to
the reefs as it retreated, I clung tightly to the tip of a rock and so managed
to prevent the last misfortune, which would undoubtedly have killed me.

Péron limped back to the shore camp, where he said his colleagues
wept to see his condition, and even Baudin 'appeared touched by my
woeful state'.[34] Others among the naturalists fared better. The zoologist
René Maugé collected ten new species of birds on Bernier Island, while
the gardener Anselm Riédlé found seventy plants, most of which he
thought were unknown to botanists.

During the weeks of the survey several of the crew of the *Géographe*
fell ill with scurvy, while during the stay of almost three months at the
Dutch port of Kupang near the southwest tip of Timor many went
down with 'bloody flux' (dysentery) and malaria. These included Baudin,
whose cheery greeting to Hamelin when the *Naturaliste* arrived at
Kupang after losing company with the *Géographe* near Shark Bay was to
inform him that he was to take command of the expedition in the event
of his (Baudin's) death. Among those who died at Kupang was Riédlé, a
serious loss to the expedition. When Baudin finally left Timor for Van
Diemen's Land in November 1801 he did so without two of his officers.
One had been wounded in a duel, while Picquet, as insolent as ever, had
been discharged into Dutch custody. Relations with Picquet form an
obsessive theme in Baudin's personal journal, with confrontations with
the recalcitrant officer described in terms that vary from 'this indecent
scene' to 'a scandalous scene'.[35] Without explanation, Baudin stopped
keeping this journal after the stay at Timor, and confined his observations to his sea-log. A more worrying matter than difficulties with individual officers were the deaths of a dozen crew members from dysentery
in the first weeks back at sea. As Baudin described it, 'The symptoms of
this terrible disease are so frightening that the moment one is struck
down by it one feels dead already.'[36]

After two months and the deaths of eleven more men, the ships
reached D'Entrecasteaux Channel, Van Diemen's Land, in mid-January
1802. The coastline had been charted by the meticulous Beautemps-
Beaupré on d'Entrecasteaux's expedition, so Baudin was able to concentrate on relations with the indigenous Tasmanians, while the naturalists
searched for specimens. The shore excursions during the expedition's two-

month stay were the most rewarding of the whole voyage, but for Baudin the behaviour of his officers continued to perplex and irritate. When he checked on the progress of the replacement longboat that had been sent ashore to fill the water casks, he found that its crew were engaged in the more appealing pastime of shooting swans. An observatory was erected on a small island, but when Baudin visited it he discovered that the site was more like a tavern, that several kitchens had been set up, and 'the day before the whole company had nearly been burnt alive, along with the tents and the instruments, for the fires in two different kitchens had run into each other and the grass had caught alight'.[37] Nor were relations with Hamelin altogether harmonious, for Baudin was critical of the failure of his fellow commander to inform him of problems with an insubordinate officer. The death of the assistant zoologist on the *Naturaliste* led to further problems, for when his chests were examined they were found to contain little of value, even though he was known to have collected a large number of shells, guaranteed to fetch a high price from collectors at home. On investigation it was found that many items had been sold, even though – as Baudin reminded Hamelin – they were government property. In places Baudin's daily entries in his log read more like those of a surgeon than a ship's captain as he reported on the condition of his sick officers and naturalists. He was particularly attentive to René Maugé, the senior zoologist, whose sickbed he visited every two hours on most days. His death in early February, Baudin wrote, was 'an irreparable loss for the expedition ... alone, he did more than all the scientists put together'.[38] Maugé's death, following that of Riédlé, was a hard blow for Baudin to bear, for the pair were 'the most valuable people on the expedition and my best friends'.[39]

As the ships left D'Entrecasteaux Channel, Baudin described the area's Aboriginal inhabitants: their number (the largest group seen was fifty-five, including women and children), appearance ('nearly every one of them has spindly legs and weak arms, they nevertheless seem strong and vigorous'), dwellings ('the most miserable things imaginable'), boats ('more like a sort of buoy than anything else') and behaviour (friendly except for the occasional bout of stone-throwing and a more serious attack when a spear was launched). Apart from the important but chance discovery by Péron of two funerary structures containing ashes which pointed to ritual cremation (Pl. 28), there was little that had not already been reported by Labillardière on the earlier visit by the d'Entrecasteaux expedition. The relatively brief encounters on the beach made it impossible for Péron or any other member of the expedition to obtain the detailed information on 'language, clothing, kinship, family relations, sex, marriage, divorce, law, education, politics, war, cannibalism and illness' demanded by Dégerando.[40] On neither of the French expeditions did the naturalists venture far inland, nor were they able to observe the Aborigines in their normal living places. Péron managed to persuade seventeen reluctant Aboriginal men to have their strength tested on a dynamometer, an exercise that satisfied him that they were weaker than French crew members. He also added some words to the Aboriginal vocabulary collected by Labillardière, but like many of his manuscripts his original list has disappeared. The main method of communication on both sides was by gesture, and there is only one recorded instance of an Aboriginal word being used by the French. When Péron felt threatened by spear-carrying Tasmanians on Maria Island, he pointed to his companion's gun and called out the word *mata* – 'dead'.[41] The written descriptions of the Aborigines were supplemented by the drawings of Nicolas Petit. He drew not only portraits of individual Aborigines, but illustrations of their artefacts and a series of sketches of their social or ritual gatherings, including one of the earliest known drawings of a corroboree or Aboriginal dance ceremony.[42]

As the ships continued their survey work they became separated once more. Hamelin in the *Naturaliste* explored the mainland coast until, in mid-April, with little food left and many of his crew sick, he changed course and headed for Port Jackson. Baudin, meanwhile, was sailing west along the mainland towards Port Phillip, assuming that he was the first to survey that stretch of coast. Disillusionment came on 8 April 1802 near Bass Strait when a sail was sighted, thought to be the *Naturaliste* until the vessel ran up the British flag: 'As they spoke us first, they asked what the ship was. I replied that she was French. They then asked if Captain Baudin was her commander. I was very surprised, not only at the question, but at hearing myself named as well. When I said yes, the English ship brought to.'[43] She was Flinders's *Investigator*, which sailing from west to east had already surveyed most of the south coast of Australia. The two captains engaged in restrained mutual courtesies and exchanged hydrographical information. In his log Baudin made no mention of the disappointment he must have felt at this encounter, but when the two ships met again at Port Jackson, Flinders was given an indication of the fraught atmosphere on the *Géographe* when Henri de Freycinet, a young officer, remarked to him, 'Captain, if we had not been delayed so long picking up shells and catching butterflies in Van Diemen's Land you would not have discovered the South Coast before us.'[44]

Baudin's log continued to record in painstaking fashion his disciplinary problems. When he ordered his midshipmen to take turns at the wheel because of the shortage of fit seamen, all save one refused on the grounds that such a task was 'dishonouring'.[45] With food running short and scurvy increasing among the crew, Baudin decided to give up surveying 'the thankless coast of New Holland' and to make for Port Jackson. As they passed the coast of Van Diemen's Land, Baudin could not resist a jibe at Péron, 'who makes fresh discoveries, or always thinks he does', when he claimed to have discovered a river no one else had seen. At the beginning of June only four men from each watch were fit to take the deck, and when the *Géographe* reached Port Jackson on 20 June she needed the help of a party of English seamen to make her anchorage.

The expedition remained at Port Jackson for five months, during which time the ships were refitted, the crews began to recover their health and Baudin purchased a small schooner, the *Casuarina*, for close inshore work. With Britain and France at peace under the terms of the Treaty of Amiens, relations between Baudin and Captain Philip Gidley King, governor of New South Wales, were warm enough for them to plan to meet again once they were back in Europe. Baudin agreed with Hamelin that the *Naturaliste* should return to France, laden with the natural-history specimens collected thus far. Péron, by this time the only zoologist left on the *Géographe*, ranged freely around the colony collecting further specimens, but he also spent much time packing those obtained earlier on the voyage. In all, the *Naturaliste* took thirty-three large cases of zoological specimens and seventy tubs of live plants back to France, as well as kangaroos, black swans, emus and other exotic creatures. Although Baudin observed the terms of neutrality under which he sailed, Péron carried out some rather amateurish espionage investigations into the resources and defences of the new colony. His findings seem to have aroused little interest when delivered to Isle de France the following year.[46]

The French ships sailed from Port Jackson on 18 November 1802; three weeks later the *Naturaliste* parted company with the *Géographe* at King Island in Bass Strait and began her long homeward voyage. Helped by the *Casuarina*, commanded by Louis de Freycinet (younger brother of Henri), and a new longboat, Baudin carried out useful survey work along the south coast of the continent. In his shallow-draught schooner Freycinet was able to survey the deep inlets of Spencer Gulf and Gulf St Vincent, already investigated by Flinders. However, once at sea, familiar problems reappeared. During the night of 24 November, off the Furneaux Islands a current drove the *Géographe* dangerously near the coast. 'I had an anchor got ready,' Baudin wrote, 'and all the crew were on duty, except for the officers who, according to their usual custom, spent the night in their beds.'[47] During a survey of King Island in December the large dinghy went ashore, 'carrying the scientists, their knowledge and their baggage, for these gentlemen never move without pomp and magnificence. The cooks with their utensils, the pots, the pans and the saucepans cluttered up the boat so much that not every one could fit in.' In despair Baudin retired to his cabin, 'extremely dissatisfied that the whole lot of them had not left on the *Naturaliste*'.[48] Extravagant living was not an accusation that could be levelled at the hyperactive Péron, whose collecting efforts were helped by Lesueur. Both men used their personal allocation of alcohol to preserve their zoological specimens, On King Island, in a chapter that his biographer has described as a 'landmark in Australian ecological writing', Péron wrote with dismay of the mass killing by English sealers of the huge elephant seals. 'They will not now,' he lamented, 'escape the mercantile greed which appears to have sworn the annihilation of their race.'[49] Giant kangaroos and dwarf emus were taken alive at Kangaroo Island off the Gulf of St Vincent, but two of the kangaroos in their rain-soaked pens on deck died within days. To keep the remaining five kangaroos healthy, Baudin made a botanist and a midshipman give up their cabins, meeting their protests with an apology for not realising that they preferred their 'own comfort and a few temporary advantages to the greater success of the expedition and whatever may serve our country'. Far to the west at King George Sound, visited and named by Vancouver in 1791, the botanists collected a rich haul of live plants, including the insect-trapping pitcher plant, while Péron found 160 species of shells, most of them new to him. The plants were distributed throughout various parts of the ship as well as in Baudin's cabin – the latter to forestall 'the lamentations that I shall have to listen to when it becomes absolutely necessary to take the cabins of those still in occupation for housing the additional objects of Natural History that we may collect during the remainder of the expedition'.

Among the officers, Louis de Freycinet, in command of the *Casuarina*, was a frequent source of Baudin's ire on the grounds that 'your pleasures ... always take precedence over your duties'. He was, Baudin added, 'insubordinate on principle'. As a cause of anger and frustration, however, Freycinet had to concede pride of place to Péron, 'the most thoughtless and most wanting in foresight of everyone aboard'. Going ashore in Shark Bay in the longboat as the expedition surveyed parts of the continent's west coast missed on its first visit in 1801, Péron became so engrossed in collecting shells that he once more lost his way. Without food or water for forty-eight hours, he and his two companions barely survived their ordeal, at one stage submerging themselves in the sea up to their chests to escape the burning rays of the sun. When the longboat finally returned, its crew brought with them a large number of shells, and despite Baudin's orders to the contrary the naturalists paid the sailors in rum for them, 'so that in the evening sixteen people were fighting in every corner of the ship and I was obliged to strike several of them in order to obtain peace ... however, any prohibition is worthless for scientists like ours'.[50] Since shells were much prized as collectors' items, it pained Péron that most of those from the *Naturaliste* had been bought by the English at Port Jackson.[51] While at Shark Bay, he cleared up a minor mystery stemming from William Dampier's improbable report in 1699 that he had found the head of a hippopotamus, complete with teeth, in a shark's stomach in the bay.[52] On the beach Péron came across the partly decomposed body of a dugong whose teeth, he realised, resembled those of its fellow herbivore the hippopotamus, although differently placed in the jaw.[53]

After several narrow escapes from shoals and rocks during the voyage north from Shark Bay, Baudin lost confidence in the charts of his predecessors: 'Those geographers who have attempted to trace the direction of the coast, and even the details of it, would have done much better to omit their work than to lead navigators into error or expose them to certain destruction ... Everything that they have done is faulty.'[54] Distrusting his officers, Baudin spent long hours on deck, until on 12 April he recorded that he was so tired that he was scarcely able to stand. On 7 May the *Géographe* reached Timor, where fresh provisions were obtained, but at the cost of outbreaks of dysentery. After a month's stay in Timor, Baudin was determined to resume the survey, this time of the northwest coast of Australia. The death soon after sailing of the astronomer Bernier was a sad blow, both to Baudin personally and to the efficiency of the expedition's survey work. The twenty-three-year-old had all the qualities necessary for a successful scientist on a long voyage, Baudin wrote, except a strong constitution. The numbers of the sick grew daily, some suffering from venereal disease, while Baudin had a persistent cough and was spitting blood. The emus on board were force-fed with pellets of rice, and the kangaroos were given wine and sugar from Baudin's personal supply. On 7 July, Baudin finally decided to abandon the survey and head for Isle de France. The ship had only a month's ration of biscuit and two of water, and twenty of the crew were on the sick list. The news was greeted with general relief but also with incredulity, as Baudin recorded in his log: 'Throughout the whole voyage no one has ever known where I was going or what I wanted to do. Several of those who spent the whole night on deck went frequently and consulted the compass for fear they had not heard aright.'[55] Sick and alone, Baudin kept to his cabin for the four-week voyage to Isle de France, 'no more than the shadow of a captain on a ghost ship'.[56]

The *Géographe* reached Isle de France on 7 August 1803, and Baudin died there on 16 September, probably from tuberculosis. His funeral was attended by the officers and scientists of the expedition, but if the later remarks of a midshipman on the voyage were correct this was a matter of form only. Baudin had shown 'great strength of spirit in his last days', but 'his funeral was a dismal event; he was universally detested'.[57] Baudin's and other deaths left only half the original number of officers and savants to complete the expedition. On her homeward voyage the *Naturaliste* had been captured in the English Channel, but on appeal to Joseph Banks was allowed to continue her voyage to Le Havre, which she reached in June 1803 with her precious cargo of twenty live animals and birds and 133 cases of plants and other specimens. The *Géographe* was even more crowded, but finally reached Lorient in Brittany in March 1804, bringing back an extraordinary cargo of natural-history specimens, as Péron's account of the voyage reveals: 'Apart from a host of cases of minerals, dried plants, shells, fishes, reptiles and zoophytes preserved in alcohol, as well as stuffed or dissected quadrupeds and birds, we also had seventy large boxes filled with plants in their natural state (including almost two hundred different species of useful ones), about six hundred types of seeds in several thousand packets.'[58] Also brought back to France were almost a thousand zoological illustrations by Lesueur, although he had made no botanical drawings. Outstanding were his watercolours of an astonishing variety of marine fauna, all drawn with great attention to detail.[59] Especially valuable were his drawings of molluscs and zoophytes, long neglected by naturalists because of their 'weird, changing shapes, difficult to describe, draw and conserve; colours most often dull, dark and disagreeable; a soft, viscous consistency, disgusting to the touch; a stale or even foul odour . . . rapid, almost instantaneous decomposition'.[60] Fortunately, Péron had the advantage of the advice of Georges Cuvier, under whom he had studied for four years, and the collaboration of Lesueur: 'Whatever I applied myself to describing with care, he drew or painted with that accuracy and skill that has earned him so much honourable approbation. All our work and all our observations were carried out upon living animals.'[61]

The *Géographe* was a floating menagerie, carrying seventy-two animals, 'rare or absolutely new'. These included two kangaroos, as well as a panther and tiger sent on board at Isle de France by the governor for Mme Bonaparte (soon to become Empress Josephine). Equally troublesome gifts were presented to the expedition at the Cape of Good Hope, including two apes and two ostriches. From Lorient, road trains took the animals and the collections to Paris, where plants and trees in three hundred tubs were received by the Muséum National d'Histoire naturelle. Antoine-Laurent de Jussieu described the shipment as the largest and most valuable ever received in France, while other professors claimed that Péron and Lesueur had made known 'more new creatures than all the recent travelling naturalists put together'.[62] Ministers were pressed to authorise an early publication of the voyage, but in the midst of a European war this was not likely to be a priority, and although

Péron and Lesueur spent their time cataloguing and describing, they did so without official support or finance: 1805, the year of Trafalgar and Austerlitz, was not the time for the French government to pay attention to such matters, and reports of dissension on the voyage cast a pall over the whole venture. Napoleon's supposed comment that Baudin 'did well to die, on his return I would have had him hanged'[63] is probably apocryphal but it summed up much official opinion.

In 1806, two years after the return of the *Géographe*, the emperor approved publication of the voyage at government expense, to be written by Péron and Louis de Freycinet, while Lesueur would be responsible for the illustrations – Petit had died in an accident soon after the expedition's return. Napoleon was swayed by the praise for the expedition by the scientists of the Institut Impérial (formerly the Institut National), and was perhaps influenced by Josephine, who had been presented with animals and live plants brought back on the ships for the gardens at Malmaison, her château on the outskirts of Paris (Pl. 31). She partly compensated for the lack of botanical drawings on the voyage by commissioning France's best-known flower illustrator to paint many of the Australian plants sent to Malmaison. As far as the official account of the voyage was concerned, only the first volume of *Voyage de découvertes aux Terres Australes* (1807) appeared in Péron's lifetime, dealing with the voyage up to the visit to Port Jackson in June 1802. There would be no separate part, as Péron had planned, on the zoology of the voyage. The volume was accompanied by an atlas, which contained no maps but a series of fine engravings by Lesueur and Petit of Aborigines, animals and coastal profiles.

The publication, issued in English translation in 1809, and Louis de Freycinet's atlas of 1811 containing twenty-six charts and plans outraged British officialdom with their apparent disregard of Flinders's surveys and their lavish scattering of Bonapartist names along Australia's southeast coast. Adding insult to injury, the latter was shown as 'Terre Napoléon' on Freycinet's charts. An article in the *Quarterly Review*, probably written by John Barrow, the influential second (permanent) secretary to the Admiralty, stated that 'a strong suspicion arises that the whole has the effect of a premeditated design to snatch the merit of the discovery from its rightful possessor by setting up a [French] claim, at some future day, to this part of New Holland'.[64] Because of Flinders's detention with his journal and charts on Isle de France, Péron and

Freycinet were able to claim priority in the exploration of Australia's south coast. If they were reluctant to recognise Flinders's achievements, they were even less ready to acknowledge those of their captain. Most of Baudin's original names for coastal features, some reflecting their physical appearance, others commemorating distinguished naval officers or local wildlife were unceremoniously discarded in favour of names that paid homage to the all-encompassing Napoleonic regime.[65]

In many ways Péron's account was an impressive piece of work, blending natural-history and anthropological observations with a lively narrative of the voyage, but it was distorted by its author's pathological hatred of Baudin. This seems to have been shared by Freycinet, who in a recently discovered letter of October 1806 says of the forthcoming account that 'the name of that despicable commander will not figure in it. His memory and his name, forever blackened, will not dishonour the splendour of the fine work conducted in spite of him.'[66] Baudin was only once mentioned by name, when his death was recorded on 16 September 1803 in almost contemptuous terms: 'M. Baudin ceased to exist.'[67] References to 'le Commandant' were invariably critical. They formed a mirror image of Baudin's feelings for Péron, but unfortunately for Baudin's posthumous reputation, while his log with his side of the story remained unpublished, its derogatory references to Péron were seen by the naturalist before he wrote his account. Péron died in 1810 at the age of thirty-five, like Baudin from tuberculosis (Pl. 30). He was never to publish a comprehensive study of his findings on the zoology of New Holland. An epitaph on a proposed (but unbuilt) memorial to him read: 'He withered like a tree laden with the finest fruits, which succumbs to its excessive fertility.'[68]

After Péron's death the second volume of the *Voyage* was completed by Louis de Freycinet, but it was not published until 1816. Pulling together the various observations of the scientists and other expedition members had not been an easy task, 'each of us,' as Freycinet wrote, 'having worked in accordance with his own view.'[69] Deaths, delays and official indifference prevented full justice being done to the scientific results of the Baudin expedition. Since no senior botanists had returned with the ships, the botanical specimens were entrusted to Jacques-Julien Houtou de Labillardière. This seemed a logical decision since the botanist was in the midst of publishing his *Novae Hollandiae plantarum specimen*, containing the results of his labours on the d'Entrecasteaux expedition.[70] Unfortunately, he did not consistently distinguish the specimens collected on Baudin's expedition from his own, and the two collections were amalgamated in his own personal herbarium which after his death went to Italy (today they are in the Botanical Institute in Florence). As far as zoology was concerned, for decades Péron's specimens and notebooks and Lesueur's superb drawings lay neglected in the storerooms of the Paris natural-history museum before their transfer in the later nineteenth century to the more accommodating environment of the Muséum d'Histoire naturelle in Le Havre, where work continues to this day to identify and publish the specimens from the Baudin expedition.[71] Even more frustrating was the fate of the ethnographic items brought back on the *Géographe*. On the ship's homecoming, 206 Aboriginal and other artefacts were sent to Empress Josephine, but after her death the whole collection disappeared and has never been rediscovered.

That Baudin had character defects as commander of a major discovery expedition there can be no question. By his own admission he was secretive and demanding, rarely able to see another's point of view in shipboard disputes. His journal and log show that he took an almost perverse delight in anticipating differences with his officers and scientists. On the other hand, it is clear that many of Baudin's officers were prejudiced against him from the beginning. A contemporary English observer noted 'a general want of discipline which pervaded the whole body of the officers', a weakness that he attributed to 'the levelling principles at that time predominant in France'. A more recent commentator has suggested that the tensions between Baudin and his officers were 'inspired not by the egalitarian ideas of the Revolution, but by a pride in family and class persisting from an earlier period'.[72] Baudin's deteriorating health did not help his judgement, but he showed remarkable stoicism while in the final stages of a fatal illness, and only near the end gave up the objectives of his voyage. By any standards he accomplished much. Although the knowledge that on the south coast Flinders had surveyed much of it before him was a cruel and unexpected disappointment, his achievements in charting the coast of Victoria were substantial.[73] It is true that during the long voyage his crews suffered terrible losses through sickness, but his two ships survived and brought back unrivalled collections of natural-history specimens. Finally, his strict control of the use of firearms by his crew ensured that there were no casualties among the native peoples encountered. 'History is written by the survivors', however, and Baudin's name is not attached to any major feature of the Australian coastline.[74] Perhaps even more than Malaspina, he was the forgotten explorer of his time.

Matthew Flinders in the *Investigator* reached Cape Leeuwin in December 1801, six months after Baudin's landfall. Despite the poor condition of his ship, he had made excellent time from Cape Town. His instructions ordered him to sail to longitude 130°E. (near the present Western Australia–South Australia border) before beginning his survey of the coast eastwards as far as Bass Strait. Instead, Flinders departed from his instructions and immediately began a five-month survey of the south coast from Cape Leeuwin. This 'slight deviation', as Flinders called it, had unforeseen consequences; for although Baudin had left Europe ten months before Flinders, his slow passage down the Atlantic, his decision to chart first the west coast of New Holland, and the three-month stay at Isle de France, put him well behind Flinders in terms of charting those long stretches of the south coast unseen by d'Entrecasteaux or Vancouver. By the time Baudin reached Tasmanian waters in January 1802, remaining there until March, Flinders had surveyed (and named) the Great Australian Bight and reached Kangaroo Island off Spencer Gulf. His longest stay was at King George Sound, where relations with the Aborigines were relaxed enough for Brown and the ship's surgeon to measure the body parts of one of the men, and to collect a list of Aboriginal words. Despite the early loss of the expedition's astronomer, who had returned home from the Cape, the charts made by Flinders of some of the less-frequented stretches of the Australian coast were accurate enough to be in use until the Second World War.[75]

In the Archipel de la Recherche, so named by d'Entrecasteaux, Flinders remained at anchor for five days to allow the botanists to collect plants. Peter Good saw himself as being more than a gardener. Before sailing, he had made it clear to Banks that he wanted 'the honor of being recorded as the introducer of such plants and seeds as I shall be able to collect',[76] and there were occasions when Good went ashore botanising while Brown remained on board arranging and listing his plants. Brown was not an easy man to please, but he clearly had a high opinion of Good, and wrote of him to Banks that 'a more active man in his department could hardly [I] believe have been met with but has not sufficient facilities on board for keeping plants'.[77] Clearly, Brown regretted that the plant frame had not yet been erected on the *Investigator's* quarterdeck.

As on Baudin's expedition, the naturalists and artists seemed unable to combine their collecting activities with even the most rudimentary sense of direction. On Christmas Eve 1801 in Torbay Inlet, Brown's party became lost. In extreme heat, with little water, 'Mr Bauer, quite exhausted, lay down & seemd unable to proceed . . . He felt the strongest inclination to sleep every time he stopt, wch was nearly every 40 paces.' The little group took four or five hours to return to the beach, where they had to spend the night.[78] A few weeks later Brown himself became separated from his party:

> In intolerable heat . . . I at length reachd the beach & rather staggerd than walkd to a small hollow in the rocks where I was shaded indeed from the direct rays of the sun but there being no current of air the heat was still distressing & neither the ship nor the island under which she anchored was in sight. Either from fatigue or the Zig Zag direction I had taken to reach the beach I had lost all ideas of her direction . . . After walking about ½ mile I returnd to the same hole, rested again, and set off in the opposite direction. Soon after having climbd up from the beach I got sight of the Ship apparently about 4 miles distant.[79]

One and a half hours later he was picked up by a boat from the ship. The coast to the east was inaccessible and dangerous, and in one tragic incident the ship's cutter disappeared with the loss of the ship's master, a midshipman and six seamen. Two wide openings to the north raised hopes that they might lead to a strait running into an inland sea or even to the distant Gulf of Carpentaria, but they proved to be deep inlets, Spencer Gulf and Gulf St Vincent. At the aptly named Kangaroo Island dozens of the marsupials were shot, to provide Flinders and his crew with 'a delightful regale . . . after four months privation from almost any fresh provisions'.[80]

On 8 April 1802, near the mouth of the Murray River, Flinders sighted Baudin's ship, the *Géographe*, sailing west after her stay in

Tasmanian waters. Neither captain had sighted the mouth of Australia's largest river. Only on their second meeting the following day did Baudin learn Flinders's name, and realise that he was the explorer of Bass Strait. Flinders aptly summed up the Frenchman's reaction: 'He expressed some surprise and congratulation; but I did not apprehend that my being here at this time, so far along the unknown part of the coast, gave him any great pleasure.'[81] Robert Brown accompanied Flinders on board the *Géographe*. He had been disappointed in his efforts to collect plants in the heat of the Australian summer, so was interested to see how the French botanists had fared: 'C. Baudin informed us that a very considerable collection of natural curiosities had been made & if I mistake not said that he had on board about 100 boxes of them. I saw no living plants on board. A row of flower pots which stood in his own cabin were filled with earth but none of them seemed to contain any plants.'[82] Both surveying and collecting activities on the *Investigator* were coming to an end as autumn gales swept through Bass Strait, and Flinders was relieved to reach Port Jackson on 9 May, with his crew 'in better health than the day we sailed from Spithead'.[83] From there Brown wrote to Jonas Dryander, Banks's librarian, with a summary of his botanising along the continent's south coast: 'Of absolutely new species I cannot reckon more than 300, for of 750 observed, 120 have previously been found in this neighbourhood or along the east coast by Sir Joseph; 140 more by Menzies at King George's Sound. I reckon that Billardiere may ... have found 140 more.'[84]

The *Géographe* arrived during Flinders's stay at Port Jackson. She was, Flinders reported, 'in miserable condition ... there being not more out of one hundred and seventy, according to the commander's account, than twelve men capable of doing their duty'.[85] News had just arrived of the Treaty of Amiens between Britain and France, and relations were more relaxed than during the earlier meeting at sea. Flinders showed Baudin one of his charts of the south coast, and underlined the limits of the French discoveries. Pointedly, he mentioned in his published account that Péron was present at the meeting, so indicating that the Frenchman had no excuse for his later claims. Another admission of British priority came in Henri de Freycinet's complaint about the time spent ashore by the French naturalists.[86] In a letter to his wife, Flinders wrote: 'Had we found an inlet which would admit us into the interior of New Holland,

I should have been better pleased, but as such did not exist, we could not find it; several important discoveries, however are made, of islands, bays and inlets.'[87] As well as repairing and refitting the *Investigator*, Flinders supervised the erection of the greenhouse on the quarterdeck, but found that when the structure was filled with boxes of earth its weight would be too much for the weak upper works of the ship. The height of the frame was reduced by a third, while the live plants collected on the voyage were placed in the garden of the colony's governor. Brown wrote to Banks that because of the delay in erecting the greenhouse there had been severe problems in transporting the living plants he had collected, 'so that of upwards of 70 species, mostly taken up in good condition, scarcely more than 10 of them, unfortunately the least interesting, have been brought alive here'.[88] Good's care in storing seed paid dividends, and a supply he sent back to England at this time in a whaler was successfully acclimatised at Kew less than twelve months later. In his own letter to Banks, Flinders wrote enthusiastically: 'It is fortunate for science that two men of such assiduity and abilities as Mr Brown and Mr Bauer have been selected: their application is beyond what I have been accustomed to see.'[89]

Flinders sailed north from Port Jackson on 22 July 1802, accompanied by a small, sixty-ton brig, the *Lady Nelson*, intended for inshore surveying. Another addition to the expedition was Bongaree, an elderly Aborigine who it was hoped would act as an interpreter amd mediator. As they reached the Great Barrier Reef, Flinders searched the coast for relics of the La Pérouse expedition, but found nothing that would indicate a shipwreck. The botanists collected five hundred new species, while the coral reefs both entranced and alarmed Flinders – 'glowing under water with vivid tints of every shade betwixt green, purple, brown and white, equalling in beauty and excelling in grandeur the most favourite *parterre* of the florist . . . but whilst contemplating the richness of the scene, we could not long forget with what destruction it was pregnant.' The *Lady Nelson* proved to be a slow-sailing and clumsy vessel, and when her keel was damaged Flinders sent her back to Port Jackson. After a hazardous passage through a maze of reefs the *Investigator* negotiated the Torres Strait at the end of October and entered the Gulf of Carpentaria. Any satisfaction at this achievement was tarnished by news that the chronometers had twice been allowed to stop, and that the ship was gaining water at more than ten inches an hour. The master and carpenter reported that her state was so rotten that 'in twelve months time there will scarcely be a sound timber in her'. The approaching monsoon left no possibility of escaping from the Gulf of Carpentaria for at least three months, so Flinders settled to a charting of the gulf's little-known shores. The survey work was tedious and dangerous, and the state of the *Investigator* was not reassuring: 'In a strong gale, with much sea running, the ship would hardly escape from foundering; so that we think she is totally unfit to encounter much bad weather.' Flinders noted frequent landings by 'the botanical gentlemen' – Brown and Good, and probably Bauer and Westall – and in one short trip on shore Brown counted about two hundred plants, twenty-six of them new to him. Flinders in turn was excited by the discovery on an offshore island of dozens of Aboriginal cave paintings, some of which were copied by Westall (Pl. 33). Most intriguing was

the representation of a kangaroo, with a file of thirty-two persons following after it. The third person of the band was twice the height of the others, and held in his hand something resembling the *whaddie*, or wooden sword of the natives of Port Jackson, and was probably intended to resemble a chief. They could not, as with us, indicate superiority by clothing or ornament, since they wear none of any kind, and therefore, with the addition of a weapon, similar to the ancients, they seem to have made superiority of person the principal emblem of superior power.[90]

Some days later a clash occurred between a boat party and a group of Aborigines in which one seaman was wounded, another died of sunstroke and an Aborigine was killed. The corpses of the dead men were treated rather differently. The seaman was buried at sea with the customary honours; the Aborigine's body was dissected, and his head preserved in spirits to be taken back to England for scientific examination.

By March 1803 more than twenty of the *Investigator*'s crew were suffering from scurvy, and Flinders decided to head for Timor to obtain fresh provisions. The ship stayed there little more than a week, but it was enough for dysentery to take hold among the crew. From Timor, Bauer sent a letter to his brother Franz that illustrated some of the problems faced by artists on the discovery voyages. In a complaint that would have been familiar to Parkinson and other artists, he wrote that his supply of paper 'has gone mouldy because of the damp and warmth of the cabin and is covered with spots of mould and can no longer be painted on'.[91] The return passage to Port Jackson was made by way of the west coast of Australia, so making Flinders the first to circumnavigate the continent. It was a miserable time, in a ship so decayed that in any severe gale 'she must have been crushed like an egg and gone down'.[92] After an unsuccessful attempt to locate and chart the notorious Trial Rocks off the northwest coast of Australia, the sickly state of the crew forced Flinders to give up any further efforts at detailed surveying. Port Jackson was reached at the beginning of June 1803. Five of the crew had died at sea, followed by another four deaths, including that of the invaluable Peter Good, after the ship anchored at Port Jackson. There the *Investigator* was surveyed and found to be totally unseaworthy, with timbers so rotten that Flinders could push a stick through them. A possible replacement vessel, the *Porpoise*, was found to be in not much better condition, but it was decided that Flinders should return to England in her with his charts, in the hope that the Admiralty would send him out with another ship to complete the survey. A greenhouse was set up on the vessel's deck, containing some of the plants collected in the previous twelve months.

His new command marked the beginning of a chapter of unrelieved misfortunes for Flinders. Seven hundred miles north of Port Jackson the *Porpoise* and an accompanying ship, the *Cato*, were wrecked on an uncharted reef. Facing a desperate situation, Flinders and the captain of the *Cato* managed to sail back to Port Jackson in a cutter to organise a rescue expedition. They arrived at Government House while Philip Gidley King was at dinner with his family: 'A razor had not passed over our faces from the time of the shipwreck, and the surprise of the governor was not little at seeing two persons thus appear whom he supposed to be many hundred leagues on their way to England.'[93] Once he had recovered from the shock, King quickly organised three vessels to pick up the shipwrecked men. In his published account Flinders related an engaging story about his brother, Samuel, who had been left in charge on Wreck Reef. When the sails were sighted, he 'was in his tent calculating some lunar distances, when one of the young gentlemen ran to him, calling "Sir, Sir! A ship and two schooners in sight!" After a little consideration, Mr Flinders said he supposed it was his brother come back, and asked if the vessels were near? He was answered, not yet; upon which he desired to be informed when they should reach the anchorage, and very calmly resumed his calculations.'[94] Matthew Flinders was in a twenty-nine-ton schooner, the *Cumberland*. She was, he said, smaller than a Thames ferry, but he nevertheless decided to return to England in her after the rescue had been completed. It was a reckless decision, with miserable consequences for Flinders.

Crossing the Indian Ocean, the *Cumberland* was in such bad condition that Flinders decided to look for help at Isle de France. Unknown to him, war had broken out again between Britain and France, and the governor of the island, the Bonapartist General Decaen, treated Flinders as a possible spy rather than as an officer engaged in a scientific expedition. Flinders did not help matters by his arrogant attitude, and by the fact that his safe conduct related only to his original ship, the *Investigator*. Despite all efforts to secure his release, he remained interned on Isle de France for six years.[95] By the time Flinders returned to England in 1810, the first volume of Péron's account of Baudin's voyage had been published both in French and in English translation, while news of his own voyage had made little impact. Particularly galling was that, although he constructed his general chart of Australia in 1804 during his detention on Isle de France (Pl. 34), its finished version was not published until 1814, whereas Louis de Freycinet's chart, with its French toponyms and its confrontational use of the term 'Terre Napoléon', was issued in 1809. For any naval officer, the conferring of place names on new geographical features was one of his most cherished prerogatives, and the circumstances in which Flinders saw his being overwritten were particularly hurtful. As he wrote to Banks in 1809 on hearing of the French publications, 'they search at Paris to deprive me of the little honour with the scientific world which my labours may have procured me'.[96]

When Flinders sailed from Port Jackson in the tiny *Cumberland*, Robert Brown was left behind with that part of his collection not lost on the *Porpoise*. He had written to Banks bemoaning the fact that all the plants on the wrecked vessel had been destroyed, an 'irreparable' loss, 'for altho' I possess duplicates of almost all the specimens, yet those sent were by far the best'.[97] Fortunately, his boxes of seeds had been saved.

Relations between Brown and Flinders appear to have been good, but in a grumpy letter to Banks in August 1803 the botanist told a familiar story of shipboard tensions, although how far Flinders was aware of these is not clear: 'I am on the whole disappointed ... Our excursions have nowhere extended to more than a few miles from the shore. The interior of New Holland, therefore, is as completely unknown as ever ... wherever the ship has anchor'd I have had at least opportunities of landing, though very frequently my time has been so limited that but little could be done.' Especially irritating was the lack of watertight boxes for his specimens. Workmen in the colony had made some, but they were so poorly constructed that one fell to pieces as it was hoisted on board the *Investigator*. The importance Flinders attached to this matter seemed to Brown 'very small', and he suggested that 'a few words' from Banks would help. Instead, he received a tart reply from Banks, praising Flinders for giving Brown 'a variety of opportunities of landing and botanising', and wishing that Cook had been so helpful on the *Endeavour* voyage.[98]

Astonishingly, the cut-down hulk of the *Investigator* was made fit for the sea. The removal of her upper deck left her at half her previous tonnage, and she was re-rigged as a brig (with two masts instead of her original three). With great misgivings, Brown agreed to sail to England in her, with his herbarium, 'the least perishable part' of his collection, stowed away in the ship's after-hold rather than have it 'mouldering away in the storehouses of the colony'. Living plants were left at Parramatta in the care of George Caley, one of Banks's many helpers who collected natural history specimens for him in farflung parts of the world. After a nerve-wracking voyage around Cape Horn, the *Investigator* reached Liverpool in October 1805 without calling at any port. Brown wrote that 'this extensive Voyage was performed in four months & sixteen days in the crazy low cut down Investigator, perhaps the most deplorable Ship in all the World'.[99] An eyewitness as the ship came into Liverpool described 'her sides being covered with barnacles and seaweed, and her sails, masts, and rigging presenting the usual signs of a vessel that has been abandoned ... [On board] a sight was presented still more astonishing – plants we had never before beheld, black swans and other curious birds and animals surrounded us on every side.'[100] Also on board was Ferdinand Bauer with more than two thousand natural-history sketches. From Liverpool, four thousand dried plants (seventeen hundred of them previously unknown to science) were taken by road to London, where they filled all available spaces in Banks's herbarium at 32 Soho Square. Brown and Bauer arrived in London on 5 November 1805, the day before news of Trafalgar and Nelson's death reached the Admiralty; unsurprisingly, the two men received little attention. After his return Brown continued working on the collection, and in 1810 he published his *Prodromus florae Novae Hollandiae*, which described 464 Australian genera and a thousand species. Unindexed, without illustrations and expensive, the book sold only twenty-six copies, but it was described by Joseph Hooker in 1859 as 'the greatest botanical work that has ever appeared'.[101] In it Brown set a pattern for future botanical publications when he adopted Jussieu's 'natural' system of classification rather than the 'artificial' or sexual system of Linnaeus, a move helped by the ability of Bauer to reveal the entire morphological structure of a plant in his drawings. Unlike Péron, whose commitment to his general account of the Baudin voyage prevented him from giving full attention to the expedition's scientific investigations, Brown produced a work of pioneering importance in Australian botany.[102] It was done with scant help from the Admiralty, and was never completed in the form Brown had originally contemplated. The huge number of exotic specimens collected by Brown and other botanists of the period seems, perversely, to have had led to a lessening of public interest in their efforts. In 1808, Banks wrote to Caley in Parramatta: 'I cannot say that Botany continues to be quite as fashionable as it used to be. The immense number of new Plants that have every year accumulated seem to deter the people from making Collections as they have little hopes of making them perfect in any branch.'[103] After Banks's death in 1820, Brown supervised the removal of the collection to the British Museum, where under his direction as keeper it became the first national botanical collection open without charge to the public. Up until the time of his death in 1858, he continued to work on the backlog of specimens from the *Investigator* voyage, including those intended for the nation, still in the bundles originally stored in the herbarium at 32 Soho Square.[104]

After the return of the *Investigator*, Ferdinand Bauer worked on his botanical and zoological sketches to produce a series of superb paintings. Of his finished watercolours, 236 were on botanical subjects, only fifty-two on zoological, a proportion that, taken together with Brown's publications, shows that botany was the main preoccupation of the naturalists on the Flinders expedition. By contrast, on the *Géographe*, Péron and Lesueur had concentrated on zoology and anthropology. To overcome the problem of lack of time in the field, Bauer made only pencil sketches on the spot, and later coloured them by referring to a complex system of codes that represented a thousand different variations of colour and enabled him to designate the subtlest shades (Pl. 32). It was a version of the method used by Haenke on Malaspina's voyage, and it is significant that both men were pupils of the Dutch botanist Nikolaus J. von Jacquin in Vienna.[105] Unable to find a publisher for his work, Bauer returned to Austria in 1814. Another fifteen engravings of his drawings were included in his *Illustrationes florae Novae Hollandiae* (1813–17), but not until the late twentieth century was there any attempt at a comprehensive publication of the drawings and paintings by the artist described by the art historian Bernard Smith as the 'Leonardo of natural history illustration'.[106]

Flinders arrived back in England in October 1810. Despite the joy of the reunion with Ann (and the birth of a daughter in 1812), the last four years of Flinders's life were ones of unremitting toil as he struggled to complete his account of the voyage, secure the support of the Admiralty for its publication, and cope with financial difficulties and chronic illness. The days when he hoped that his name would come second to 'the immortalized name of Cook' were far distant.[107] Although he was promoted to post captain, the First Lord of the Admiralty felt unable to backdate the promotion by more than a few months, whereas Flinders had hoped it would have come into effect from the time of his detention on Isle de France in 1804. The Admiralty refused to cover the full cost of the published account of his expedition, which was not the easiest of voyage accounts either to write or to read. As Flinders admitted in the preface, 'a polished style was not ... attempted ... Matter, rather than manner, was the object of my anxiety.'[108] Eager to establish his priority in charting the coasts of Australia, Flinders included a heavy weight of hydrographical data, while his resentment at his treatment by General Decaen on Isle de France took up much of the latter part of the two-volume work. Its size was swollen by four technical appendices, including an eighty-page essay by Brown, 'General Remarks on the Botany of Terra Australis' (in English, unlike his *Prodromus*), while the accompanying atlas featured engravings of Bauer's natural-history drawings and William Westall's landscape scenes. Flinders spent much time and effort checking and redrawing the charts, especially his magnificent 'General Chart of Terra Australis or Australia', whose title reflected both old and new perceptions of the continent.

By early 1814, Flinders was too ill to leave his house, and he died on 19 July; hours earlier, an advance copy of his *Voyage* had arrived at his bedside, but he never recovered consciousness to see it. Its publication coincided with momentous events in the wars against France and the United States, and fewer than half of its 1,150 copies were sold. A final disappointment for Flinders in his last months was the title settled upon by Banks for the book – *A Voyage to Terra Australis* – over his own preference for calling the whole continent 'Australia'.[109] A few years after Flinders's death, the governor of New South Wales, Lachlan Macquarie, decided to use his preferred name in his despatches in the hope that it would become 'the Name given to this Country in future'. It was an appropriate memorial for an explorer whose achievements were not fully recognised until many years after his death.

CHAPTER 10

'Like giving to a blind man eyes'
Charles Darwin on the Beagle

IN THE YEARS after the Napoleonic Wars, the Royal Navy dominated the world's oceans. With no enemy fleets to engage, the shrunken peacetime navy's main responsibilities were law-and-order operations, anti-slavery patrols and surveying. For many young officers, denied hopes of promotion through battle, an alternative route to advancement, and a focus of interest on routine and often tedious voyages, was the pursuit of scientific investigations in fields ranging from natural history and geology to geomagnetism and meteorology.[1] Encouraged by John Barrow, second secretary to the Admiralty from the war period to 1845, and by Francis Beaufort, hydrographer of the navy from 1829 to 1855, many officers developed scientific interests that were not directly related to their professional duties. In the period from 1804 to 1855, four hundred naval officers and naval surgeons were members of scientific societies, and 136 of them published papers on scientific subjects. Many surgeons were appointed to vessels on specific voyages primarily for their knowledge of botany, zoology or geology. There was no official policy of supplying naval vessels with naturalists, but most ships bound for distant parts carried an officer or surgeon with an interest in collecting. They did not entirely replace civilian naturalists, but they far outnumbered them. There were few salaried naturalists in British universities or other institutions, for botany as an academic discipline lacked a broad philosophical base. With its emphasis on the mundane tasks of collecting and classifying plants, for long it 'was regarded as the preserve of hobbyists – women flower painters, self-taught artisans and minor clergy'.[2] In 1818 the Regius professor of Greek at Cambridge characterised natural history as 'a pursuit which demanded little exertion of the highest powers of intellect'.[3] In these circumstances of shortage of funds and lack of scholarly interest, the navy performed a useful function in organising the collection of data and specimens from regions otherwise out of reach; and under Sir Francis Beaufort, the navy's official hydrographer, the Admiralty became a clearing house for such data. An intriguing example of its international reputation comes in a letter from Alexander von Humboldt to Beaufort asking for information about the distribution of Sargasso weed. Beaufort replied that his search of ships' logs had revealed little, but that 'had I earlier known that you were intent on that subject, I would have specially pointed the attention of our seamen to it'.[4]

In the period after the Napoleonic Wars, the French sent more discovery expeditions to the Pacific than did the British, but they too carried few civilian scientists on board. The first voyage of the post-war era was commanded by Louis de Freycinet, whose experiences on the Baudin expedition had alerted him to the problems of dealing with non-naval personnel. The surgeons on Freycinet's ship, the *Uranie*, were both interested in natural history, as was the pharmacist, but they were the only naturalists on board. The wreck of the *Uranie* in the Falklands in 1820 on the homeward voyage destroyed many of the natural-history specimens on board, but an impressive number survived, including 3,000 botanical specimens, 1,300 insects, almost 450 mammals, birds, reptiles and fish, and 1,300 insects. Although the expedition made no geographical discoveries of note, its researches in natural history and anthropology were of exceptional value, and Freycinet paid tribute to them in his account of the voyage, which eventually ran to five volumes.[5]

Freycinet's expedition set the pattern for much of what followed. On the three Pacific voyages of Dumont d'Urville, the surgeons usually doubled as naturalists, and brought back vast collections of botanical and zoological specimens which went to the Muséum National d'Histoire naturelle, sometimes exceeding its capacity to deal with them. His voyage of 1826–9 on the *Astrolabe* was the most important, for, in addition to amending the existing charts, d'Urville confirmed the wreck-site of La Pérouse's ships on Pitt's Island. Increasingly, the French voyages to the Pacific had political and commercial motives. Showing the flag, protecting traders and missionaries, and eyeing possible vantage points for future annexation became an important, if sometimes understated, part of the expeditions' instructions.[6] But there was another problem. In the account of his second voyage d'Urville wrote that he was 'haunted' by Cook. The successors of Cook, whatever their nationality, were faced with the fact that his three voyages had determined the main features of the Pacific. There was still detail to be confirmed, charts to be corrected, little-known stretches of coast to be surveyed, dozens of small islands to be located, more work to be done on peoples, flora and fauna; but the heroic age of Pacific exploration had passed. Increasingly, expeditions in search of significant discoveries sailed to the ocean's extremities, into the Arctic in the hope of finding a Northwest Passage, or south towards the ice-covered vastness of Antarctica. It is not surprising that when d'Urville was 'tormented by dreams' about Cook, 'they always tended to bring me towards the [South] Pole'.[7] Important exploration voyages of the mid-century years were made to the Southern Ocean and the Antarctic, notably d'Urville's third voyage, the turning to the far south of James Clark Ross after his earlier Arctic journeys had located the North Magnetic Pole, and the ambitious United States Exploring Expedition commanded by Charles Wilkes.

The other extremity of the Pacific saw some of the most notable scientific expeditions of the immediate post-Cook era in the shape of Russian voyages. The first major Russian government expedition to the North Pacific after Cook's third voyage was commanded by Joseph Billings, who had sailed on the *Discovery* in the humble position of able seaman and 'astronomer's assistant'. In form it was almost a repetition of Bering's Great Northern Expedition, with an overland trek to Kamchatka followed by a voyage towards the American coast in locally-built ships. Billings came to Russia 'with the intention of putting a finishing hand to Capn Cook's & Clark's discoveries in the Eastern Ocean', although the priority of the Russian government was control of the fur trade in the waters between Siberia and Alaska.[8] In all, the expedition in its various forms lasted from 1785 to 1792, and brought back a mass of information – hydrographic, botanical, ethnographic – but its main practical importance was in asserting Russian sovereignty over the region.

A better-known venture that was a precursor of later Russian voyages to the Pacific was the expedition of Adam Johann von Krusenstern (1803–6), the first Russian circumnavigation. Krusenstern came from a distinctive group in Russia, the Baltic German nobility, as did two officers on Krusenstern's ship, the *Nadezhda*, Fabian von Bellingshausen and Otto von Kotzebue, who in time commanded their own expeditions to the Pacific. Both Krusenstern and his fellow commander, Yuri Lisianski, had served in the Royal Navy, their ships were built in Britain and they were equipped with British navigational instruments. Unlike Billings, they followed Cook's routes in the Pacific, often calling at places he had visited to pay tribute to the great explorer. In Simon Werrett's words, 'Cook was a kind of Napoleon, and the voyage of exploration a kind of battlefield' where Krusenstern and the others could make their own reputations.[9]

Sailing with Krusenstern were the German naturalists Georg Heinrich von Langsdorff and Wilhelm Gottlieb Tilesius, but they were forced to carry out their scientific tasks on a voyage that began in confusion, continued in acrimony and ended in disappointment as responsibility was disastrously divided between Krusenstern and Nicolas Petrovich Resanoff, a court official with no experience of the sea. On the outward voyage Langsdorff and Tilesius experienced the familiar frustrations of naturalists on discovery voyages. At Santa Catarina off the coast of Brazil, Langsdorff noted that they received 'minimal support', although 'the wealth and variety' of the animal and plant life 'could occupy hundreds of natural scientists for years'. At San Francisco he complained that his seal and bird skins were thrown overboard, and live specimens were set free as soon as he turned his back, while requests for help on boat excursions were turned down on 'the pretense that the men had more important business and our expedition had not been undertaken for purposes of natural history'.[10] Matters were not helped by the fact that Langsdorff and Tilesius were at odds with each other. In the preface to his book, Langsdorff was effusive in his praise of the artistic and scientific abilities of his 'friend and travelling companion', but the recently published diary of Hermann Ludwig von Löwenstern, fourth officer on the *Nadezhda*, tells a story of disputes and rivalries between the two naturalists. The fractious relationship between them was only one of several feuds on the ship, for, as Löwenstern noted, 'there is no place else where men can become so estranged from each other as on a ship. Little annoyances build up; vexation grows bigger and bigger; the necessity of having to deal with the other men causes you to begin to wish that they would land at the other end of the world.'[11]

In the years between the outset of Krusenstern's voyage and Charles Darwin's departure from England in the *Beagle* in 1831, almost thirty Russian expeditions left the Baltic for the Pacific, about half of them naval, the remainder fitted out by the Russian American Company. Most had the primary purpose of supplying the Russian posts on the northwest coast of America, but several made outstanding contributions to the hydrography and natural history of the Pacific. After the end of the Napoleonic Wars, Von Kotzebue, the young officer in the Russian navy who had sailed with Krusenstern on the *Nadezhda*, was chosen to command a new expedition to the North Pacific. It was financed by Count Nicolai Rumiantsev, a director of the Russian American Company, which at this time occupied trading posts from Alaska to California. Its main purposes were to protect the Company's interests and to discover whether a Northwest Passage existed through the Bering Strait, but it also had wider scientific objectives. In his introduction to Kotzebue's published account of his voyage in the *Ryurk*, Krusenstern wrote: 'The crossing of the South Sea twice, in quite different directions, would certainly not a little contribute to enlarge our knowledge of this great ocean, as well as the inhabitants of the very numerous islands scattered over it; and a rich harvest of objects of natural history was to be expected.' To this end two naturalists – Adelbert von Chamisso, born in France but educated in Germany, and the surgeon-zoologist Johann Friedrich Escholtz – sailed on the *Ryurk*.[12] The expedition spent time in Alaska, California and the Pacific Islands, and returned with a rich haul of collections and natural-history observations. Chamisso brought back about 2,500 plants which he presented to Berlin's botanical gardens, of which he became curator. His insights into coral and his observations on the life cycle of medusa jellyfish later influenced the marine biologist T.H. Huxley, while the former's ethnographic investigations into the cultures and languages of the peoples he encountered on the voyage marked him out as an exceptional observer.

During his voyage on the *Ryurk* and a later one on the *Predpriyatie*, Kotzebue carried out valuable hydrographic work, much of it recorded in Krusenstern's magnificent atlas of the South Pacific. Glynn Barratt's comment that in the Tuamotus 'the Russians strove to remove the many layered errors adhering, like so many barnacles, to European knowledge of the archipelago', applies more generally to Russian hydrographic activity in the Pacific, where islands 'were numerous and wretchedly charted, bore varieties of names and were a hazard to the shipping of the world'.[13] Another notable venture was the first Russian Antarctic expedition, commanded by Fabian von Bellingshausen (1819–21), who also visited Tahiti and New South Wales. He brought back natural-history specimens and drawings, but the expedition's contribution suffered from the absence of the two German surgeon-naturalists appointed to the *Vostok* who missed the ship at Copenhagen. One of them, Karl-Heinrich Mertens, compensated by sailing with Fedor Petrovich Litke on the *Seniavin* in 1826–9, the last major Russian scientific expedition to the northwest coast of America. Its naturalists brought back a large haul of zoological and botanical specimens, including seven hundred bird specimens, and these were deposited in the St Petersburg Academy of Sciences. Litke was more sympathetic than most captains to the scientists on board and their work, but he was at pains to emphasise the constraints under which he operated: 'The commanders were not denied the right to engage in scientific research, [but] they could only do so in passing, and when the main part of the voyage permitted.'[14]

A revealing below-deck view of these voyages was provided by Adelbert von Chamisso, whose natural-history observations had been included in Kotzebue's official account of the voyage of the *Ryurk*, published in 1821, but not to Chamisso's satisfaction: 'What I wrote was defaced and made incomprehensible in many places by innumerable sense-distorting typographical errors.'[15] In 1836 he published his own account of the voyage, and it revealed something of the difficulties he encountered. He had never been to sea before, and on reporting on board the *Ryurk* he was given no indication of his rights and duties except a warning from Kotzebue that 'as a passenger on board a warship I might not make any demands as none could be fulfilled'.[16] His complaints about his accommodation echoed those of the Forsters on Cook's second voyage. The cabin he shared with others was twelve feet square, and only a small part of it was his: 'My berth and three of the drawers under it comprise the only space in the ship that belongs to me ... in the narrow space of the cabin four people sleep, six live, and seven eat.' The expedition's artist, Louis Choris, had first claim on the cabin's table, followed by the other occupants, so Chamisso was able to use it only for 'fleeting moments'.[17] At first Kotzebue was 'amiable and kind', but Chamisso soon realised that on board ship the captain was 'a more unlimited monarch than the tsar'. As the voyage progressed Kotzebue became increasingly irritated by Chamisso's incessant questioning, 'so I, the oldest in years, was rebuked for being the youngest in mind and heart'. Unfamiliar with shipboard routine, Chamisso found life 'monotonous and empty, like the surface of the water and the blue of the sky above: no stories, no events, no news', and matters were not helped by the fact that 10 p.m. each evening saw lights out.[18]

Chamisso's collecting efforts suffered from the same problems of carelessness and, sometimes, malice experienced by other naturalists on long voyages. Ashore in Brazil, his shipmates sleeping in his tent used his plants as pillows; these were destroyed when the tent collapsed in a storm. Off Cape Horn, Chamisso was allowed to put his specimens in the crow's-nest to dry – an odd place one might think – but when the ship was cleaned 'my little treasure was pitched overboard without notice'. In Hawaii helpful islanders protected his plants with a taboo, but once on board a collection of four days disappeared. When revising his account for publication, Chamisso reflected that the days of the civilian scientist on discovery voyages had passed. On her homeward voyage, the *Ryurk* met Louis de Freycinet's *Uranie* at Cape Town en route to the Pacific. Chamisso noted that the French officers were also the expedition's scientists, and he reflected: 'In France and England no titular scholars are taken on voyages of discovery anymore.'[19]

Unknown to Chamisso, the year of the publication of these remarks saw the return to England of a survey vessel with a civilian supernumerary on board whose presence proved a dazzling exception to his lament. As far as non-service naturalists and other supernumeraries sailing on naval ships were concerned, the Admiralty had a relaxed policy, permitting them on voyages where space and the commanding officer allowed. Sometimes the initiative came from the officer himself, as shown by a

letter of March 1842 from Beaufort to Sir William Hooker, the distinguished botanist who had just been appointed director of the Royal
Botanic Gardens at Kew: 'My friend Capt. Vidal who is about to survey
the Azore islands has been good enough to offer a cabin to any botanist
or geologist who has a mind to explore them.'[20]

It was in this context that the celebrated partnership between naval
officer and naturalist had taken place on the voyage of HMS *Beagle*
some years earlier. Appointed in 1831 to command the *Beagle* on an
expedition to South America intended to complete the coastal surveys
begun on an earlier voyage by the same vessel, Captain Robert Fitzroy
decided that he would welcome both the expertise and the company of
a naturalist. As the Rev. John Stevens Henslow, professor of botany at
Cambridge, explained, Fitzroy wanted someone 'more as a companion
than a mere collector & would not take anyone however good a Naturalist
who was not recommended to him likewise as a *gentleman*'. Undoubtedly
playing a part in Fitzroy's request was the memory of the *Beagle*'s earlier
survey voyage to the same part of the world, when the ship's captain,
Pringle Stokes, committed suicide off the stormbound coast of Patagonia.
After briefly considering, and then rejecting, the possibility of going on
the voyage himself, Henslow wrote to Charles Darwin, a twenty-two-
year-old Cambridge graduate with a growing interest in natural history,
at that time unenthusiastically contemplating a career in the Church,
and urged him to accept the position, 'not on the supposition of yr. being
a *finished* Naturalist, but as amply qualified for collecting, observing, &
noting any thing worthy to be noted in Natural History'.[21]

A meeting between Darwin and Fitzroy went well, the latter making
no effort to conceal the hardships of life at sea. As Darwin reported,
Fitzroy suggested that they should eat together each day, but 'thought it
his duty to state every thing in the worst point of view ... I must live
poorly, no wine & the plainest of dinners.'[22] For Darwin, an attractive
feature of the proposed expedition was the Admiralty's plan that after
Fitzroy's Patagonian survey the *Beagle* would sail into the Pacific and
return home by the Cape of Good Hope, so completing a voyage around
the world. As he wrote in an excited letter to Henslow, thoughts of
hunting foxes near his family home in Shropshire had dramatically
changed to hopes of chasing llamas in South America.[23] At a more
practical level Darwin was reassured by the news that all the *Beagle*'s

officers from her previous voyage were sailing on her again, together with two-thirds of the crew.

The autumn of 1831 passed in a flurry of activity as Darwin equipped himself for a voyage expected to last for three, perhaps even four, years. Fitzroy, technically one of the ablest officers in the Royal Navy, possessed a large number of instruments and books, most bought at his own expense. The ship's instruments included no fewer than twenty-two chronometers, six of them purchased by Fitzroy. Darwin in turn brought on board a whole library of books, and instruments ranging from a simple rain-gauge to a portable microscope (Pl. 39). The money for these came from his father, Robert Waring Darwin, a successful doctor and shrewd investor, who met his son's considerable expenses during the voyage. Among Darwin's books was Alexander von Humboldt's *Personal Narrative of Travels to the Equinoctial Regions of the New Continent*, which he had first read at Cambridge, and there is no doubt that the opportunity to see at first hand the extraordinary natural life of South America revealed by Humboldt was an important factor in Darwin's decision to sail with Fitzroy. Darwin arrived in Plymouth in mid-September, only to find that work on refitting the *Beagle* was still in progress, and even when it was completed bad weather prevented the ship from sailing until late December. Originally a ten-gun brig of a type that had a poor safety record, the *Beagle* was rigged as a three-masted barque for her survey voyages while the raising of her upper deck gave her crew more headroom and helped her seaworthiness. Only ninety feet long, she was to carry seventy-four people; these included three Fuegians being returned to their homeland together with a young missionary. For work purposes Darwin shared the poop cabin with the mate and assistant surveyor, John Lort Stokes, while a young midshipman, Philip Gidley King, slept in the cabin for at least part of the voyage. A chart table took up much room, and lack of space in his cramped quarters was as much a worry for Darwin as it had been for Chamisso on the *Ryurk*. The corner of the ten-by-eleven-foot cabin farthest from the door was his, but was 'most wofully small – I have just room to turn round & that is all'.[24] Soon he was 'in a panic on the old subject want of room', but Fitzroy came to his rescue, being 'such an effectual & good natured contriver that the very drawers enlarge on his appearance'.[25]

Everything was new and strange. Darwin's first night on board was a sleepless fiasco because of his unfamiliarity with his hammock, into which he tried to insert himself legs first. It was just as well that for weeks after his arrival the *Beagle* lay quietly at Devonport, for Darwin felt that the slightest task on the gently swaying ship took extra effort and time, whether it was removing a book from the shelves or picking up a piece of soap from the washstand. Even so, he was determined to put his leisure hours to good effect, 'collecting & observing & reading in all branches of Natural history that I possibly can manage. Observations in meteorology. – French and Spanish, Mathematics & a little Classics, perhaps not more than Greek Testament on Sundays ... If I have not energy enough to make myself steadily industrious during the voyage, how great & uncommon an opportunity of improving myself shall I throw away. – May this never for one moment escape my mind.'[26] It was a retrospective indictment of his abortive spell studying medicine in Edinburgh and his lacklustre years reading theology at Cambridge.

On 27 December 1831 the *Beagle* finally sailed, and the voyage began quietly for Darwin except for the seasickness that bothered him throughout. Stokes later remembered how this malady frequently interrupted Darwin's use of the chart table: 'After perhaps an hour's work he would say to me, "Old fellow, I must take the horizontal for it" ... and stretch out on one side of the table. This would enable him to resume his labours for a while when he had again to be down.'[27] On the uneventful passage to the Cape Verde Islands, Darwin busied himself trawling the surface of the water with a plankton net made out of bunting, only the second known example of such a device being used. In his diary he noted of the plankton: 'Many of these creatures so low in the scale of nature are most exquisite in their forms & rich colours. – It creates a feeling of wonder that so much beauty should be apparently created for such little purpose.'[28] Darwin soon settled into his confined quarters, and even found advantages in them. As he told his father: 'Everything is so close at hand, & being cramped, make one so methodical, that in the end I have been a gainer.'[29] More important was Darwin's proximity to the ship's officers as they made their surveys, recorded their observations and wrote up their daily logs. The methodical recording of events on a naval vessel had its counterpart in Darwin's diary, much of it kept daily, and running to 770 pages. Revised by Darwin on his return, it was first published as *Journal and Remarks, 1832 to 1836*, making the third volume of the official *Narrative* of the expedition published under Fitzroy's name in 1839.[30] In addition he kept detailed notes of his geological and zoological observations, and wrote frequent letters to his family and friends at home. Towards the end of his published *Journal*, Darwin included hints for collectors, insisting that they 'Trust nothing to memory' – this an admission that some of his own recording on the voyage had been less than perfect. All specimens should be numbered with date and location, with labels both on the specimens themselves and on their containers. And so the advice went on, down to the minutest detail: 'A series of small numbers should be printed from 0 to 5000; a stop must be added to those numbers which can be read upside down (as 699 or 86).'[31]

Darwin's *Journal and Remarks* belied its uninspiring title. It was written in a vivid, personal style easily accessible to readers of all kinds. In his diary he revealed how overwhelmed he was by the tropical exuberance of the voyage's first landing, at St Jago in the Cape Verde Islands: 'It has been for me a glorious day, like giving to a blind man eyes.'[32] On the shoreline his journal shows that he was fascinated by the octopuses:

> Although common in the pools of water left by the retiring tide, these animals were not easily caught. By means of their long arms and suckers, they could drag their bodies into very narrow crevices; and when thus fixed, it required great force to remove them . . . These animals also escape detection by a very extraordinary, chameleon-like, power of changing their colour . . . I was much amused by the various arts to escape detection used by one individual, which seemed fully aware that I was watching it. Remaining for a time motionless, it would then steadily advance an inch or two, like a cat after a mouse, sometimes changing its colour; it thus proceeded, till having gained a deeper part, it darted away, leaving a dusky train of ink to hide the hole into which it had crawled.

The passage on the octopus ends with a sentence that provides an unexpected glimpse of life in the cramped quarters shared by Darwin, Stokes and King: 'I observed the one which I kept in the cabin was slightly phosphorescent in the dark.'[33]

Darwin's indeterminate status on the *Beagle* was reflected in the flippant if affectionate titles given to him by Fitzroy and the other officers – 'Philos' (or ship's philosopher) on some occasions, 'Fly-catcher' on others. His apparently inconsequential diary entry for 1 April 1832 is revealing in more ways than one:

> All hands employed in making April fools. – at midnight all the watch below was called up in their shirts; Carpenters for a leak; quarter masters that a mast was sprung. – midshipmen to reef top-sails . . . The hook was much too easily baited for me not to be caught: Sullivan cried out, 'Darwin, did you ever see a Grampus: Bear a hand then'. I accordingly rushed out in a transport of Enthusiasm. & was received by a roar of laughter from the whole watch.[34]

A helpful relationship during Darwin's time on the *Beagle* was his friendship with the first lieutenant, John Clements Wickham, who was responsible for much of the day-to-day running of the ship. It even survived the casual way in which Darwin dumped his specimens on the *Beagle's* spotless decks when he came on board from his collecting expeditions. One of his daughters remembered Darwin describing how Wickham, 'a very tidy man who liked to keep the decks so that you could eat your dinner off them – used to say If I had my way, all your d—d mess would be chucked overboard, & you after it old flycatcher.'[35] Darwin seemed to be on good terms with all except Robert McCormick, the ship's surgeon, who had assumed that he was to be the ship's naturalist and given priority in his collecting activities. Angered by what he saw as the favours granted to a civilian interloper, McCormick left the *Beagle* at Rio de Janeiro, only four months into the voyage. He was replaced by the ship's assistant surgeon, Benjamin Bynoe, who was also an enthusiastic field botanist.

Darwin's relations with Fitzroy were respectful, though perhaps not as close as might be expected between two men who, when both were on board, ate together three times a day. In a letter home Darwin wrote: 'His ascendancy over every-body is quite curious: the extent to which every officer & man feels the slightest rebuke or praise would have been, before seeing him, incomprehensible ... His greatest fault as a companion is his austere silence: produced from excessive thinking: his many good qualities are great & numerous: altogether he is the strongest marked character I ever fell in with.'[36] The two men had a mutual interest in geology, but Darwin shared neither his captain's evangelical beliefs nor his High Tory politics, and on occasion they had strenuous differences of opinion. During the expedition's stay in Brazil, Darwin's challenge to Fitzroy's defence of slavery caused such offence that it seemed as if his time on the expedition had come to an end. Apologies restored the situation, and matters were helped by the fact that the two men did not spend as much time together as might be imagined. While Fitzroy was engaged in his time-consuming hydrographic surveys, Darwin was often ashore, sometimes many miles away, collecting and observing. Unlike most of his shipboard predecessors, he was on land for three-fifths of the voyage of almost five years. As is often the case, distance seemed to give a warmer edge to the relationship. In August 1833, Fitzroy wrote from the *Beagle* to Darwin, then on one of his inland excursions, in humorous and affectionate terms:

My dear Philos

Trusting that you are not entirely expended, though half-starved, occasionally frozen, and at times half drowned, I wish you joy of your campaign ... I do assure you that whenever the ship pitches (which is *very* often as you *well* know) I am extremely vexed to think how much *sea practice* you are losing, and how unhappy you must feel upon the firm ground ... Take *your own time* – there is abundant occupation here for *all the Sounders*, so we shall not growl at you when you return.

In his return letters Darwin's usual ending read: 'Adios, dear FitzRoy / yr faithful Philos. / C.D.'[37]

Darwin's diary and his letters home, especially those to Henslow and to his three unmarried sisters, Caroline, Susan and Catherine, tell us much about his activities, and something about his state of mind. There were moments of nostalgia. As he wrote to one of his sisters, 'Although I like this knocking about. – I find I steadily have a distant prospect of a very quiet parsonage, & I can see it even through a grove of Palms.'[38] More often he tried to convey his excitement at the opportunities that seemed to present themselves at every turn. To his Cambridge friend and second cousin, William Darwin Fox, he wrote, 'My mind has been since leaving England in a perfect *hurricane* of delight and astonishment'; to Henslow, 'you will think me a Baron Munchausen among Naturalists ... today I have been out & returned like Noahs ark. – with animals of all sorts'; to his sister Caroline, as he anticipated the arrival of a parcel of natural-history books from England, 'No schoolboy ever opened a box of plumcake so eagerly as I shall mine'; to his youngest sister, Catherine, 'There is nothing like geology; the pleasure of the first days partridge shooting or first days hunting cannot be compared to finding a fine group of fossil bones.'[39] Whether ashore or afloat, Darwin continued to collect every conceivable type of natural-history specimen – bird skins, beetles, fossils – and sent crates and casks of them to England (two hundred animal and bird skins in one crate alone). In this respect he was more fortunate than most naturalists on the eighteenth-century voyages, who struggled to find long-term storage for their specimens on board ship.

In scientific terms Darwin's interest was focused increasingly on the geology of the regions he visited. He wrote to Fox from Rio de Janeiro that, although he had collected a host of land and marine animals as well as insects, 'Geology carries the day; it is like the pleasure of gambling, speculating on first arriving what the rocks may be; I often mentally cry out 3 to one Tertiary against primitive; but the latter have hitherto won all the bets.'[40] He told Henslow that his hammer and Humboldt's *Personal Narrative* were his constant companions; and while 'I formerly admired Humboldt, I now almost adore him'.[41] To Darwin, Humboldt's breadth of vision was both a model and a challenge, although it is doubtful whether he was aware of some of Humboldt's more stringent criticisms of the naturalists of his day. In 1806, Humboldt had complained to the Berlin Academy that 'travelling naturalists' were 'concerned exclusively with the descriptive sciences and with collecting, and have neglected to track the great and constant laws of nature'. A year later he returned to the charge: 'The investigations of naturalists are usually limited to objects which comprehend a very small part of botany. They occupy themselves almost exclusively with the discovery of new species, descriptions of their external form, and the characters by which they are grouped in classes or families.'[42] With his long spells ashore, sometimes travelling on horseback hundreds of miles inland, Darwin did much to meet Humboldt's other criticism of seagoing naturalists, that while they 'have given precise notions of the coasts of countries, of the natural history of the ocean and islands, their expeditions have advanced neither geology nor general physics as travels into the interior of a continent should have', and that their coastal investigations had done little to 'reveal the history of the earth'.[43]

As the voyage continued, a more immediate influence on Darwin was Charles Lyell, professor of geology at King's College, London, a copy of whose first volume of *Principles of Geology* was given to him by Fitzroy. Lyell argued that changes in the earth's surface occurred gradually, over a long period of time; as the subtitle of his book put it, it was *An Attempt to Explain the Former Changes of the Earth's Surface, by Reference to Causes Now in Operation.* The captain's passing on to Darwin of so radical a work was evidence that in his youthful years he was not the narrow-minded believer in biblical assertions of the age of the earth imagined by some pro-Darwin writers. The second and third volumes of Lyell's work were sent to Darwin on publication, and confirmed his conviction that geology owed more to Lyell 'than to any other man who ever lived'. Impressed as Darwin was by Lyell's overriding thesis of the slow elevation of the land out of the sea over millions of years, he was aware that the geologist had never been to the Americas. During his travels along the coasts and in the interior of Chile, Darwin searched for evidence that would confirm or modify Lyell's theories, all the while insisting that 'when seeing a thing never seen by Lyell, one yet saw it partly through his eyes'.[44]

In England the long-suffering Henslow had to deal with the crates of specimens as they arrived in Cambridge, some in poor shape. As Darwin confessed to him when sending a shipment from Buenos Aires, 'I am afraid you will groan or rather the floor of the Lecture Room will when the casks arrive – Without you I should be utterly undone.'[45] Between ports that offered facilities for shipping specimens home, Darwin faced difficulties familiar to anyone who had read the journals of Banks and the Forsters. Off Cape Horn the *Beagle* was battered by some of the worst gales Fitzroy had ever experienced, and Darwin lamented: 'I have suffered an irreparable loss from yesterday's disaster, in my drying paper and plants being wetted with salt-water. – Nothing resists the force of an heavy sea; it forces open doors and sky lights, & spreads universal damage.'[46] Nor did shore excursions at this stage of the voyage bring much compensation, for of the half-naked Fuegians he wrote, 'I believe if the world was searched, no lower grade of man could be found.' (Pl. 35)[47] In tribute to his shipboard companion, Fitzroy gave the names Darwin Sound and Darwin Mount to two prominent natural features in Tierra del Fuego, but to his sisters Darwin wrote that he had become 'thoroughly tired' of the region east of the Horn. Even discovery of 'a live Megatherium would hardly support my patience' – and at the thought of reaching 'the glorious Pacific' he was 'ready to bound for joy'.[48] The letters home are a reminder that, unlike previous voyagers, for most of his time away Darwin remained in regular communication with friends and relatives in England through navy ships and merchant vessels sailing to and from the ports of South America. In Montevideo in April 1833, for example, he received letters from home dated 12 September, 14 October, 12 November and 15 December 1832. Only when the *Beagle* struck out across the Pacific on her homeward voyage was the steady flow of letters interrupted and did Darwin find little opportunity of sending crates of specimens to England.

Darwin spent much of 1833 inland, journeying dangerously from the Rio Negro to Buenos Aires across a land ravaged by revolution and a war of extermination against the local Indian peoples. If his diary is to be trusted, he greeted the turmoil with a fair degree of sang-froid. At one call at Montevideo he wrote: 'During our absence, things have been going on pretty quietly, with the exception of a few revolutions.'[49] When staying in Maldonado on the north bank of the Rio de la Plata, Darwin paid local youths to find specimens of birds and animals for him, so when he rejoined the *Beagle* at the end of June 1833 he did so with 'all my Menagerie'. Ruefully, he noted that he had 'become such a complete landsman that I knock my head against the decks & feel the motion though in harbour'.[50] At this time Darwin benefited once more from the financial support of his father in England when he engaged a member of the ship's crew, Syms Covington, to act as his servant at a cost of £60 a year. In addition to providing personal services for Darwin, Covington proved a great help in collecting and preserving specimens.

Making full use of his social status and privileged position on a ship of His Majesty's Navy, Darwin found doors opening for him in the towns and landed estates of South America as an English gentleman whose habits might be eccentric, but whose letters of introduction showed that he had influential contacts. In December 1833 he noted that during the past four months he had slept only a single night on the *Beagle*. Looking forward to geologising in the Andes, he told Henslow: 'I have not one clear idea about cleavage, stratification, lines of upheaval, I have no books which tell me much & what they do I cannot apply to what I see. In consequence I draw my own conclusions, & most gloriously ridiculous ones they are.'[51]

After delays to enable Fitzroy to complete his charts, the *Beagle* passed through the Strait of Magellan in June 1834. The next month, as the ship reached Valparaiso on the Pacific coast, Darwin felt he had entered a new world after the cloud and rain of Tierra del Fuego: 'the sky so clear & blue, the air so dry & the sun so bright, that all nature seemed sparkling with life.' Even the British merchants in the port seemed to Darwin superior to their counterparts elsewhere, discussing scientific topics with him, and surprisingly enquiring 'what I thought of Lyells Geology'.[52] By this time Darwin's geological and zoological notes ran to six hundred small quarto pages, in addition to his personal journal, sections of which he forwarded home for safekeeping. In August, Darwin made an excursion to the Campana or Bell Mountain, and it proved one of the most exhilarating experiences of the voyage:

> We spent the day on the summit, and I never enjoyed one more thoroughly. Chile, bounded by the Andes and the Pacific, was seen as in a map. The pleasure from the scenery, in itself beautiful, was heightened by the many reflections which arose from the mere view of the grand range, with its lesser parallel ones, and of the broad valley of Quillota directly intersecting the latter. Who can avoid admiring the wonderful force which has upheaved these mountains, and even more so the countless ages which it must have required, to have broken through, removed, and levelled whole masses of them?[53]

Darwin's joy at being on the Pacific slopes was dampened by an illness that kept him in bed for a month in October 1834, and still more by worries about Fitzroy's mental state. To Catherine he confided: 'The Captain was afraid that his mind was becoming deranged (being aware of his hereditary disposition)' – the latter a reference to the grisly suicide of Fitzroy's famous uncle, Lord Castlereagh – and he told her that at one stage Fitzroy resigned his command before being persuaded to change his mind.[54] A severe volcanic eruption and earthquake centred on Concepción in early 1835 took the minds of Fitzroy and Darwin off their personal troubles, and provided them with first-hand evidence to support Lyell's theories (Pl. 36). Lyell had argued that volcanic activity and earthquakes resulted in an elevation of the land surface, and the careful measurements by Fitzroy in Concepción harbour and on two offshore islands showed that the land had risen between eight and ten feet.

From this time on Darwin was absorbed by his search for evidence of elevation of the land. In April 1835 there came a dramatic moment as he crossed the Uspallata range and at the height of seven thousand feet came across the stumps of almost fifty petrified coniferous trees. The sight took Lyell's thesis to another dimension:

It required little geological practice to interpret the marvellous story, which this scene at once unfolded; though I confess I was at first so much astonished that I could hardly believe the plainest evidence of it. I saw the spot where clusters of fine trees had once waved their branches on the shores of the Atlantic, when that ocean (now driven back 700 miles) approached the base of the Andes ... I now beheld the bed of that sea forming a chain of mountains more than seven thousand feet in altitude.[55]

To Henslow, Darwin claimed: 'I can now prove that both sides of the Andes have risen in the recent period to a considerable height – Here the shells were 350 ft above the sea', following this a month later with a report of his latest mountain excursion which he introduced with the words: 'Some of the facts, of the truth of which I in my own mind feel fully convinced, will appear to you quite absurd & incredible.'[56] To his sister Susan he wrote, 'I literally could hardly sleep at night for thinking over my days work' in the mountains where 'the strata of the highest pinnacles are tossed about like the crust of a broken pie'.[57]

After more than four years away Darwin increasingly looked forward to his return to England – a welcome but still-distant prospect, as he told his cousin Fox, at this time living on the Isle of Wight: 'If a dirty little vessel, with her old rigging worn to shreds, comes into harbor September 1836 you may know it is the Beagle. You will find us a respectable set of old Gentlemen, with hardly a coat to our backs.'[58] On land Darwin continued to travel light. As he told Catherine, apart from his bed, he carried only a few cooking utensils. He always slept in the open air, for the houses were full of fleas.[59] At Lima he viewed with mixed feelings the next stage of the *Beagle's* voyage across the Pacific: 'I am very anxious for the Galapagos Islands, – I think both the Geology & Zoology cannot fail to be very interesting. – With respect to Tahiti, that *fallen* paradise, I do not believe there will be much to see.'[60]

In September 1835 the *Beagle* reached the Galapagos, isolated islands on the Equator of recent volcanic origin, best known from Dampier's visit to them in 1684. Darwin's stay in the Galapagos Islands has been seen by some writers as the climactic moment of the voyage, but there is little evidence of this in his diary, and no letters written by him from the islands survive. Darwin assumed that the islands had received their flora and fauna from the South American continent, six hundred miles distant, in the form he saw them during his visit. Only after he had left the islands far behind did he realise there might be problems with this assumption. Four months after his visit he wrote to Henslow from Sydney, but his report on the Galapagos was brief. The geology he found 'instructive and amusing', with numerous volcanic craters worth investigating, and as far as the natural history of these islands remote from any mainland was concerned, he wrote: 'I worked hard. – Among other things I collected every plant, which I could see in flower ... I shall be very curious to know whether the flora belongs to America, or is peculiar. I paid also much attention to the Birds, which I suspect are very curious.'[61] His diary adds little to this, with only one offhand mention of the islands' finches – later to hold a key significance in the development of his theories – when he described them swarming around the water holes on Albemarle Island.

The islands' most remarkable, and useful, inhabitants were the tortoises, huge, ungainly creatures, some so heavy that six or more men were required to lift them, and invaluable as a food supply to callers at the Galapagos from the buccaneers onwards. During Darwin's visit Nicholas Lawson, the British resident and acting governor (for Ecuador) who lived on Charles Island, told him that the differences between the tortoises were such that he could tell immediately from which island a particular one had come: 'It is said that slight variations in the form of the shell are constant according to the Island which they inhabit – also the average largest size appears equally to vary according to the locality.'[62] But there Darwin left the subject and, assiduous collector though he was, failed to bring home any tortoiseshells.

He had more to say about the islands' birds when he wrote up his notes from the Galapagos six months later on the homeward leg of the voyage, and noted that he had collected specimens from four of the islands, and that although those from Chatham and Albemarle Islands appeared to be the same, those from the other two islands were different. Puzzled by this, he suspected that these specimens, and the island tortoises, were 'only varieties', but if not, then 'the zoology of Archipelagoes will be well worth examining, for such facts would undermine the stability of Species'.[63] His botanical collection from the Galapagos again showed a concentration of plants in certain islands.[64]

Later realisations of the significance of the natural life of the Galapagos Islands for the development of Darwin's thinking have tended to obscure the importance of the remaining months of the *Beagle*'s voyage as the expedition visited Tahiti, New Zealand, New South Wales, Van Diemen's Land, King George Sound, the Cocos (Keeling) Islands, Mauritius and Cape Town. In Australia, Darwin puzzled over the strangeness of much of the animal life, and entered a query in his diary that would have profound implications if followed through: 'An unbeliever in everything beyond his own reason might exclaim, "Surely two distinct Creators must have been [at] work ..."'[65] The Pacific and Indian Ocean islands in particular interested him, with their isolated and confined habitats, their coral reefs and their evidence of the pressures of environmental change.[66] His observation of the reefs led him to disagree with Lyell's supposition that because corals were found only in shallow water the reefs had been built on the submerged rims of volcanic craters; instead, he suggested that they were formed by action of the tiny coral polyps incessantly building their way to the surface from a gently subsiding island. At the Cocos Islands, Darwin wrote that he regarded the coral reefs as one of the wonders of the world: 'We feel surprised when travellers relate accounts of the vast piles & extent of some ancient ruins; but how insignificant are the greatest of them, when compared to the matter here accumulated by various small animals ... we must look at a Lagoon Isd as a monument raised by myriads of tiny architects.'[67] To Caroline he repeated his rejection of Lyell's thesis when he wrote: 'The idea of a lagoon island, 30 miles in diameter being based on a submarine crater of equal dimensions, has always appeared to me a monstrous hypothesis.' In the same letter he gave a foretaste of the problems that awaited him as an author. At sea, he told his sister, he was happily kept busy sorting out his notes, but 'I am just now beginning to discover the difficulty of expressing one's ideas on paper. As long as it consists solely of description it is pretty easy, but where reasoning comes into play, to make a proper connection, a clearness & a moderate fluency, is to me . . . a difficulty of which I had no idea.'[68]

After calling at Tahiti the *Beagle* anchored at Sydney ahead of schedule, and then sailed before letters from England could arrive. From Van Diemen's Land, Darwin wrote to his cousin Fox expressing his longing for home in a passage that contained more than his personal feelings:

> I hate every wave of the ocean, with a fervor, which you, who have only seen the green waters of the shore, can never understand. It appears to me, I am not singular in this hatred. – I believe there are very few contented Sailors. – They are caught young & broken in before they have reached years of discretion. Those who are employed, sigh after the delights of the shore, & those on shore, complain they are forgotten & overlooked . . . I thank my good stars I was not born a Sailor. – I will take good care no one shall ever persuade me again to volunteer as Philosopher (my accustomed title) even to a line of Battle ship.[69]

Darwin reached his family home in Shrewsbury just before breakfast time on 5 October 1836 after a voyage lasting four years and nine months. He was fit and well. Still only twenty-seven, almost six feet tall and weighing ten and a half stone, he was far removed from the heavily bearded, patriarchal figure of his later years that became a familiar image in mid-Victorian England. The following day one of his first acts was to write to Fitzroy. The letter reflected the warmth that at this stage existed between the two men:

> I thought when I began this letter I would convince you what a steady & sober frame of mind I was in. But I find I am writing the most precious

nonsense. Two or three of our labourers yesterday ... got most excessively drunk in the honour of the arrival of Master Charles. – Who then shall gainsay if Master Charles himself chooses to make himself a fool. Good bye – God bless you – I hope you are as happy, but much wiser than your most sincere but unworthy Philos.[70]

After his return to England, Darwin took up residence in London, where he met some of the leading men of science: Charles Lyell, of course, Robert Brown of the British Museum, who had sailed with Flinders on the *Investigator*, the astronomer John Herschel (whom Darwin had first met in Cape Town), the zoologist Richard Owen, the geologist Roderick Murchison and others. Later there would be Joseph Dalton Hooker (son of Sir William), back from his voyage as botanist on James Clark Ross's Antarctic expedition in the *Erebus* (1839–43), who became one of Darwin's closest confidants as he slowly evolved his theories on natural selection, and T.H. Huxley, the biologist on the *Rattlesnake*'s Pacific survey voyage of 1846–50. Both young when they left England, they latter pair sailed as assistant naval surgeons and had few of Darwin's privileges or financial advantages. As a new arrival on the London scientific scene, ready to show and share his collections, and eager to expound his ideas, Darwin was much in demand. In addition to the crates of specimens shipped to Henslow in the early part of the voyage, Darwin brought back with him fifteen hundred specimens preserved in spirits, and three thousand dried specimens, including finches from the Galapagos. He gave his bird and mammal collection to the Zoological Society of London, though not without misgivings – they 'seem to think a number of undescribed creatures rather a nuisance', he complained.[71] Crucial was his collaboration with the ornithologist and taxonomist John Gould, who, despite problems arising from Darwin's carelessness in labelling his specimens, insisted that the Galapagos finches that he and Fitzroy had collected represented different species, identifiable by variations in beak shape and size adapted to their diet. The same seemed to be true of the Galapagos tortoises and iguanas, and, Gould said, of two different species of the South American rhea (or 'ostrich').

Gould's influence on Darwin can be seen in the published version of his *Beagle* journal, which appeared in its first edition in 1839, and

represented a stage in his gradual and cautious rethinking on the immu-
tability of species. We have seen that, while still on the *Beagle*, he
wondered whether the differences between the tortoises and the finches
of the Galapagos Islands might 'undermine the stability of Species'; but
such thoughts he kept to himself. In his published journal of 1839 the
attention of the reader would have been seized by Darwin's introductory
remark to his section on the Galapagos Islands: 'The natural history of
the archipelago is very remarkable: it seems to be a little world within
itself; the greater number of its inhabitants, both vegetable and mineral,
being found nowhere else.'[72] In a long list of birds found on the islands,
Darwin followed Gould's analysis in picking out the finches as 'the
most singular of any in the archipelago'. They were similar 'in many
points' except that 'a nearly perfect gradation of structure in this one
group can be traced in the form of the beak, from one exceeding in
dimensions that of the largest gros-beak, to another differing but little
from that of a warbler'.[73] There, for the time being, Darwin left the
subject, and turned to the famous Galapagos tortoises. Unfortunately,
Darwin admitted, during his weeks in the Galapagos 'it never occurred
to me, that the production of islands only a few miles apart, and placed
under the same physical conditions, would be dissimilar. I therefore did
not attempt to make a series of specimens from the separate islands. It
is the fate of every voyager, when he has just discovered what object in
any place is more particularly worthy of his attention, to be hurried
from it.' After some more observations on the islands' birds that seemed
to be heading for a general, and potentially startling, conclusion, Darwin
finished, rather lamely: 'there is not space in this work, to enter on this
curious subject.'[74] For most readers, his book would have represented an
engaging account of the adventures of a young man in remote parts of
the world, as indicated by its more popular title, *Voyage of the Beagle*.

In private, Darwin wrote a 'sketch' of almost 230 pages spelling out
his radical ideas on the evolution of species, but after its completion in
1844 those ideas were mentioned only to a few scientific friends, and
then in the most hesitant and convoluted terms. To Joseph Hooker, for
example, he wrote that he was 'engaged in a very presumptuous work' in
which 'gleams of light have come, & I am almost convinced (quite
contrary to opinion I started with) that species are not (it is like confessing
a murder) immutable'.[75] For Darwin, always sensitive to criticism and

bad publicity, any thoughts of swift publication were swept away by the storm of controversy whipped up by the appearance, also in 1844, of the anonymous *Vestiges of the Natural History of Creation*. Written in a lively, journalistic style, the book anticipated Darwin's emerging theories on evolution. Shocked and repelled by the uproar that greeted *Vestiges*, Darwin retreated to the safer task of revising his *Journal and Researches*. In its second edition, published in 1845, he took a cautious step forward when he speculated further on the variation of beaks in the Galapagos finches: 'Seeing this gradation and diversity of structure in one, intimately related group of birds in this archipelago, one might really fancy that from an original paucity of birds in this archipelago, one species has been taken and modified for different ends.'[76]

However, it is easy to exaggerate the influence of the Galapagos finches on Darwin's thinking (Pl. 37). He did not collect, or even see, specimens of all the thirteen species from the islands, and it has been pointed out that, 'owing to the complexities of the evidence, he did not mention the finches in any of his four notebooks on transmutation of species, or even in the *Origin of Species*'.[77] Darwin's published journal, the most popular of his writings from the *Beagle* voyage, was supplemented by weightier tomes intended for a specialist readership, among them dissertations on coral reefs (1842), volcanic islands (1844) and the geology of South America (1846). Unlike many of his predecessors, Darwin made strenuous efforts to ensure that his observations and conclusions on a variety of natural-history subjects were published. Even so, his zoology notes have only recently been edited and published,[78] while a scholarly edition of his yet more copious geological notes still awaits publication.

In 1842, Darwin and his growing family moved to Down House in Kent, in a quiet rural setting, yet only sixteen miles from central London. Importantly, the village of Downe had a post office, and Darwin made full use of the new and efficient postal service that had developed with the coming of the railway. At Down House, removed from the time-consuming obligations of London society, and with a substantial private income, Darwin forged links by correspondence with members of the scientific community at home and abroad. These included not only established scholars but a host of humble collectors who were persuaded to send or exchange specimens. While his essay on transmutation remained hidden away in a drawer, accompanied by a letter to his wife,

Emma, asking her to arrange its publication in the event of his death, Darwin was diverted from making plans for a full-scale book on the subject by a chance reunion with one of the tiniest of living organisms. More than ten years after the *Beagle*'s return, Darwin opened his last bottle of specimens from the voyage, and began studying a conch shell studded with tiny holes that he had found on one of the beaches of the Chonos Archipelago off the coast of Chile. Under his new chromatic microscope, tiny soft-bodied creatures were visible through the holes – barnacles, he thought, but without any of the characteristics commonly associated with barnacles whose shells were cemented to rocks. At the time he had simply referred to the shell-less creature as 'undiscovered', and only in 1846 did he once more examine and name it 'Mr Arthrobalanus'.[79] He intended to spend a year studying barnacles; in the end his researches into their little-known lifestyle lasted more than eight years as he begged and borrowed specimens from naval officers, collectors and institutions around the world. There were stalked barnacles, acorn barnacles, female barnacles and hermaphrodite barnacles, parasitic barnacles, barnacles with and without legs, barnacles of all shapes and sizes. Soon there were several hundred specimens in his study at Down House. The work was exhausting and difficult, for the tiny creatures could only be examined, drawn and dissected under a microscope (Pl. 38).

In the end his work on the minuscule world of barnacles resulted in four volumes, two large and two comparatively slim, published between 1851 and 1854. In his final years spent on barnacles he discovered that 'Mr Arthrobalanus' was actually female, one among many surprises and revelations in the course of his researches. These delayed further work on the epoch-making book he had in mind – 'The barnacles will put off my species book for rather a long period,' he admitted – but they brought him widespread academic recognition and a gold medal from the Royal Society. As Huxley declared, the volumes showed him 'to be as able an observer of nature on the small scale as on the large scale'.[80] More than that, his research strengthened his belief in transmutation and convinced him of 'a gradual sequence of adaptive changes in successive barnacle species', all in the interests of survival. Although in his four volumes he had been careful to avoid general conclusions about the stability of species, his research had shown him how a multitude of apparently trivial variations could result almost insensibly in a new and distinct species. It was time to return to the general work that his essay of 1844 had promised, for Darwin was far from alone in his movement away from belief in the permanence of species. The most immediate spur to Darwin was the totally unexpected arrival at Down House in June 1858 of a paper, 'On the Tendency of Varieties to Depart Indefinitely from the Original Type', sent to him by a little-known collector, Alfred Russel Wallace, in faraway Ternate. To Darwin's dismay, the theories on transmutation in Wallace's short paper seemed identical to his own, built up over twenty years of painstaking, secretive work; but Wallace was not the only one advancing ideas of evolution by natural selection. In the preface to the first edition of his *Origins of Species*, Darwin acknowledged ten authors who had published material questioning in some way the fixity of species; by the time of the fourth edition in 1866 he had become aware of no fewer than thirty-four.[81]

Published in November 1859, Darwin's *Origin of Species* changed his life.[82] Amid the acclaim and criticism that the book brought him, the voyage of the *Beagle* remained fresh in his memory. He rarely ventured from Down House where, often in ill health, at times a semi-invalid, he organised his family and scholarly life with Emma's help in the way made familiar to him from his years on the voyage. It seems altogether appropriate that a prominent feature of Down House today is a reconstruction on an upper floor of the cabin on the *Beagle* that Darwin shared for almost five years with Stokes and King. At home his telescope from the voyage was always close at hand, and Darwin used it from his study window to examine the garden for wildlife activity. He went back to his specimens and notes from the voyage, asking London Zoo for bodies of rabbits descended from those he had found on St Jago in the Cape Verde Islands, or retrieving his ethnological notes on the Fuegians he had encountered on the voyage. These latter would feature in his final major work, *The Descent of Man*. In 1862 he invited three of his shipmates from the *Beagle* to visit him and enjoyed an evening of reminiscences about the voyage.

Less happy was his relationship with Fitzroy. A sign of trouble ahead had come as early as 1837 when Fitzroy reacted furiously to Darwin's draft preface to Volume III of the *Voyage of the Beagle*, which he thought was condescending in tone and failed to acknowledge the help Darwin had received from him and the other officers. Darwin's apology was accepted and the preface was rewritten; but the publication in 1839 of the *Narrative* of the expedition under Fitzroy's name marked a more ominous parting of the ways with its inclusion of a final chapter, 'A Very Few Remarks with Reference to the Deluge' – an uncompromising twenty-five-page assertion of belief by Fitzroy in the biblical description of the creation. Darwin's discovery in the Andes of petrified trees six or seven thousand feet above sea level was simply proof to Fitzroy of the catastrophic Flood of the Bible when 'the fountains of the great deep were broken up, and the windows of heaven were opened'.[83] Darwin did not attempt to remonstrate with Fitzroy, and made a serious effort to remain on good personal terms. In 1840 he told him: 'I think it far the most fortunate circumstance in my life that the chance afforded by your order of taking a naturalist fell on me.'[84] Six years later he wrote to Fitzroy on the latter's return from his short-lived governorship of New Zealand. He invited him and his family to stay at Down House, but advised him that, although his health had improved, he was 'a different man in strength and energy to what I was in the old days, when I was your "Fly-catcher", on board the Beagle'.[85]

Relations collapsed with the publication of the *Origin of Species*. Darwin sent Fitzroy a presentation copy, but received in reply a letter that, although it began, 'My dear old friend', went on to protest: 'I, at least, *cannot* find anything "ennobling" in the thought of being a descendant to even the *most* ancient *Ape*.'[86] More troubling, the next year Fitzroy went public with his opposition. At a noisy session at Oxford of the annual congress of the British Association for the Advancement of Science which discussed the *Origin of Species*, a distressed Fitzroy, waving (according to some accounts) a copy of the Bible, shouted that he regretted the publication of Darwin's book, and that he had often warned him 'for entertaining views that contradicted the first chapter of Genesis'.[87] In 1865, Fitzroy committed suicide. His own, often overlooked, contribution to the voyage of the *Beagle* was highlighted in a tribute to him by the hydrographer to the Admiralty at the time: 'No naval officer ever did more for the practical benefit of navigation and commerce than he did . . . The Strait of Magellan, until then almost a sealed book, has since, mainly through his exertions, become a great highway for the commerce of the world . . . His works

... will be his most enduring monument, for they will be handed down to generations yet unborn.'[88]

When Darwin heard the news of Fitzroy's death he wrote a sad note to Bartholomew Sulivan, whom he had known as a young lieutenant on the *Beagle*, and who by this time was an admiral: 'I once loved him sincerely; but so bad a temper & so given to take offence, that I gradually quite lost my love & wished only to keep out of contact with him.'[89] The same was not true of his memories of his voyage on the *Beagle* generally. In 1873, more than forty years after he first boarded the ship, in an answer to a questionnaire about his personal life, Darwin confirmed that, although he thought that from the beginning he had an 'innate taste' for natural history, it was 'strongly confirmed and directed by the voyage in the Beagle'. Three years later, in his autobiographical 'Recollections', he wrote even more emphatically that the voyage had been 'by far the most important event in my life'.[90]

CONCLUSION

I N A PUBLISHED review of the account by his shipmate John MacGillivray of the survey voyage of HMS *Rattlesnake* (1846–50), the marine biologist T.H. Huxley summed up the frustrations of generations of shipboard naturalists.[1] All began their adventures, he wrote, with high expectations: 'Each of us has been his own Columbus – to each there has been a time when the idea of a voyage of discovery filled us with inexpressible longing – when we believed, that beyond the world we knew, there lay a southern cloud-land full of strange wonders and overflowing with adventure. But we have grown older and wiser.' For Huxley, there was one constant in the experience of travelling by sea. The sailor's hard physical work, 'in his constant battle with the elements, is as far apart from the speculative acuteness and abstraction necessary to the man of science as ever'. No landsman could understand 'the little world enclosed within the timbers of a man-at-war', Huxley insisted, and to illustrate the lot of the naturalist at sea he turned not to the writings of his fellow countrymen but to the reminiscences of Adelbert von Chamisso on Kotzebue's voyage for the Russian Admiralty thirty years earlier. Chamisso had complained that life on board ship was monotonous and tedious; there were no occurrences, no news. The captain was an oppressive authority 'whom one can neither remove nor avoid'. Even when he was sympathetic to the work of the naturalist, there was the problem of the junior officers, men who had been in the navy since the age of thirteen and regarded with 'singular disrespect' anything that lay outside their normal routine:

Not that there is any active opposition – quite the reverse. But it is a curious fact, that if you want a boat for dredging, ten-chances to one they are always actually or potentially otherwise disposed of; if you leave your towing-net trailing astern, in search of new creatures, in some promising patch of discoloured water, it is, in all probability, found to have a wonderful effect in stopping the ship's way, and is hauled in as soon as your back is turned; or a careful dissection waiting to be drawn may find its way overboard as a 'mess'.

In Huxley's view, Chamisso summed up the problems of the naturalist on board ship: 'He will begin full of joy, hope, and a desire to work; but too soon, he will learn that his chief business consists in getting out of the way, in taking up as little room, and in allowing his existence to be as little known as possible.'

Huxley's approval of Chamisso's censure stemmed from his own experiences on the *Rattlesnake* when he observed, as if it were a general rule, that 'exploring vessels will inevitably be found to be the slowest, the clumsiest, and in every respect the most inconvenient ships which wear the pennant'. The vessel sailed 'in such a disgraceful state of unfitness, that her lower deck was continually under water', while the Admiralty refused to supply the expedition with a single work of natural history. His criticisms told part of the story, but only a part. Many of the discovery vessels carrying naturalists on board had been carefully chosen and were well equipped, often – it must be said – at the expense of the naturalists themselves. A more legitimate complaint was that their painstaking work was often undervalued, ignored, sometimes even scorned, by crew members. The close relationship between Banks and Cook, or Fitzroy and Darwin, was unusual. On most of the voyages described here, there was a running conflict over priorities between naval officers and civilian naturalists. None of the voyages was primarily directed towards natural history. Most were concerned with geographical discovery and its subset – the charting of unknown or little-known coasts. The naturalists on board had only a subsidiary role. Their own ambition might be to increase the world's knowledge of previously unknown plants and animals; but for the government agencies that fitted out the expeditions it was the utility of such discoveries that was important. The complaints by a succession of naturalists about their accommodation and treatment were a reflection of their relative unimportance once on board ship. Given this, their achievements were praiseworthy. Almost all the naturalists whose fortunes are followed here had little or no experience of the sea, and their years in small, cramped sailing ships brought discomfort, sickness and sometimes danger. In his *Supplément au voyage de Bougainville* the first question Diderot's interlocutor asked of the navigator's years at sea was, 'Did he suffer much?'[2] For naturalists on long voyages, boredom and frustration were ever-present companions. Chamisso wrote that life was 'monotonous and empty, like the surface of the water and the blue of the sky above: no stories, no events, no news', echoing Johann Reinhold Forster who on Cook's second voyage sixty years earlier had observed that 'after having circumnavigated very near half the globe we saw nothing, but water, Ice & Sky'.

Despite all constraints, naturalists on the discovery voyages brought back rich hauls of specimens. Many of these perished on the way, but a different fate awaited some of the collections – a neglect both physical and scholarly. The limited facilities in the museums and botanical gardens of Europe could be overwhelmed by the sudden and often unannounced arrivals of shiploads of specimens, many with no known provenance and with inadequate labelling. One is reminded of Joseph Banks's observation after Flinders's voyage that 'the increasing number of new Plants seem to deter the people from making Collections as they have little hopes of making them perfect in any branch', or Darwin's complaint that he received little thanks from the Zoological Society of London for his gift of the birds and mammals he had collected on the voyage – 'rather a nuisance,' its members thought. Nor is it possible to make an accurate comparison of the number of specimens – botanical, zoological, geological – brought back by the Pacific discovery expeditions with those gathered by shorebound collectors who were the overseas agents of interested naturalists or institutions in Europe. Banks, with well over a hundred distant collectors in his employ, was probably the most energetic and wealthy individual involved in such activity, but he was only one of many. In the first half of the nineteenth century the flood of specimens reaching the museums and gardens of Europe had far-reaching effects on the disciplines of botany, zoology and geology. In the voyages that followed the pioneering work of the elder Forster and

Humboldt, naturalists such as Brown and Darwin paid increasing attention to the geographical distribution of their specimens. In this they were helped by Antoine-Laurent de Jussieu's 'natural' system of classification, which led to more precise divisions of natural life into species, genus, sub-family, family, order, division and kingdom. The way forward was pointed by Darwin, who suggested to Joseph Hooker in 1845 that geographical distribution was 'that almost keystone of the laws of creation'.[3]

The hardships of the seagoing naturalists, mostly told in their own words and from their own perspective, can be exaggerated. In contrast to their lot, Humboldt described the 'unbelievable difficulties' he faced in his land travels across Spanish America between 1799 and 1804: 'Our progress was often held up by having to drag after us for five and six months at a time from twelve to twenty loaded mules, change these mules every eight to ten days, and oversee the Indians employed on these caravans. Often, to add new geological specimens to our collections, we had to throw away others collected long before.'[4] At a later date we need only compare the ordeals endured by Alfred Russel Wallace in the malarial jungles of the Amazon basin and Borneo with the situation of his contemporary Darwin, working on his specimens in his snug cabin on the *Beagle*, to gain some sense of proportion. Difficult though most of the naturalists found their adaptation to the priorities and disciplines of shipboard life, their vessel provided a workplace, a storage space and, above all, a sanctuary unknown to their colleagues on land struggling across inhospitable terrain far from help. The fate of the expeditions of Bering and La Pérouse is a reminder of the dangers faced by voyagers in uncharted waters, but they were the exceptions. The relationship between naturalists and the ships that carried them, if not in comfort, then at least in relative safety, many thousands of miles across the globe was a more positive and fruitful one than they sometimes admitted. Even Darwin, despite the miseries of seasickness and the frequent spells of longing for home that he experienced, confessed in the final pages of his journal to the inestimable advantage of an ocean voyage to the naturalist: 'The map of the world ceases to be a blank; it becomes a picture full of the most varied and animated figures.'[5]

NOTES

Introduction

1. See Stephen Greenblatt, *Marvelous Possessions: The Wonder of the New World* (Chicago, 1991), pp.76, 78; J.H. Parry, ed., *The European Reconnaissance* (New York, 1968), p.168.
2. See David Goodman, 'Philip II's Patronage of Science', *British Journal for the History of Science*, 16 (1983), pp.49–66.
3. Richard Drayton, *Nature's Government: Science, Imperial Britain, and the 'Improvement' of the World* (New Haven and London, 2000), p.12.
4. Zelia Nuttall, ed., *New Light on Drake* (1914), p.303.
5. Johannes Heniger, 'Dutch Contributions to the Study of Exotic Natural History in the 17th and 18th Centuries', in William Eisler and Bernard Smith, eds, *Terra Australis: The Furthest Shore* (Sydney, 1988), pp.59–66.
6. Drayton, *Nature's Government*, p.18.
7. *Oxford English Dictionary*.
8. See pp.39, 123 below.
9. William Bligh, *A Voyage to the South Sea* (1792), p.5.
10. See p.119 below.
11. Michael Dettelbach, 'Global Physics and Aesthetic Empire: Humboldt's Physical Portrait of the Tropics', in David Philip Miller and Peter Hanns Riel, eds, *Visions of Empire: Voyages, Botany and Representations of Nature* (Cambridge, 1996), p.260.

Chapter 1 The 'rambling voyages' of William Dampier

1. O.H.K. Spate, *The Pacific since Magellan*, II, *Monopolists and Freebooters* (Canberra, 1983), p.vii.
2. Philip Ayres, *The Voyages and Adventures of Captain Barth. Sharp and Others in the South Sea* (1684), preface.
3. Part of the wording of a memorial brass placed in St Michael's Church, East Coker, in 1908.
4. Anton Gill, *The Devil's Mariner: A Life of William Dampier* (1997), p.36.
5. Diana and Michael Preston, *A Pirate of Exquisite Mind: The Life of William Dampier* (2004), p.42.
6. William Dampier, *Voyages and Descriptions*, II, *Voyages to Campeachy* (1699), p.80.

7. William Dampier, *A New Voyage round the World*, with introduction by Albert Gray (1937), p.4.
8. Dampier, *New Voyage*, p.157; British Library: Sloane MS 3236, fo.188.
9. Dampier, *New Voyage*, p.310.
10. Ibid., p.332.
11. Dampier, *Voyages and Descriptions*, II, p.136.
12. Dampier, *New Voyage*, p.347.
13. See Joel H. Barr, 'William Dampier at the Crossroads: New Light on the "Missing Years", 1691–1697', *International Journal of Maritime History*, 8 (1996), pp.97–117.
14. Dampier, *New Voyage*, pp.327, 330.
15. British Library: Sloane MS 3236, fo.232v.
16. Dampier, *New Voyage*, p.21.
17. For details of Dampier's route and landfalls along the Australian coast in 1688, see Leslie R. Marchant, *An Island unto Itself: William Dampier and New Holland* (Carlisle, WA, 1988), pp.101–21.
18. Dampier, *New Voyage*, pp.312–13, 315.
19. J.C. Beaglehole, ed., *The Endeavour Journal of Joseph Banks 1768–1771* (Sydney, 1962), II, p.50.
20. References from British Library: Sloane MS 3236, fo.222.
21. Ibid., fos 1, 116.
22. William Dampier, *A Voyage to New-Holland, &c. In the Year 1699* (2 vols, 1703, 1709), preface.
23. See British Library: N TAB 2026 (25).
24. Joseph C. Shipman, *William Dampier: Seaman-Scientist* (Lawrence, KS, 1962), p.8; Gill, *Devil's Mariner*, pp. 232–3.
25. *The Works of the Learned* (February 1699), p.98.
26. Thomas Sprat, *History of the Royal Society* (1667), p.86.
27. Anon., *Account of Several Late Voyages & Discoveries*, with introduction by Tancred Robinson (1694), p.v.
28. See Hans Sloane, *A Voyage to the Islands Madera, Barbados, Nieves, S. Christophers and Jamaica, with the Natural History* (2 vols, 1707, 1725).
29. Dampier, *New Voyage*, p.204.
30. Ibid., p.205.
31. Dampier, *Voyage to New-Holland*, preface.
32. For the descriptions of the anteater, hummingbird, spiders and alligator, see Dampier, *Voyages to Campeachy*, pp.60–1, 65–6, 64, 77–8.
33. Basil Ringrose, *Bucaniers of America . . .* (1685), II, p. 38.
34. British Library: Sloane MS 54, fo.7v.
35. Dampier, *New Voyage*, pp.77–9.
36. Charles Darwin, *Journal and Remarks 1832–1836*, Vol. III of *Narrative of the Surveying Voyages of His Majesty's Ships Adventure and Beagle* (3 vols, 1839), p.465, and see pp.250–1 below.
37. British Library: Sloane MS 3236, fo.233.
38. D. and M. Preston, *Pirate of Exquisite Mind*, pp.348–9.
39. Dampier's correspondence with the Admiralty is in the National Archives: Adm 2/1692 (no folio numbers); the main documents are printed in John Masefield, ed., *Dampier's Voyages* (2 vols, 1906), II, pp.325–30.
40. Dampier to Admiralty, 22 April 1699; printed ibid., II, p.333.
41. Ibid., p.604.
42. See the *London Gazette*, 16 April 1703, reprinted in Masefield, ed., *Dampier's Voyages*, II, p.575.
43. Dampier, *Voyage to New-Holland*, preface.
44. For details of Dampier's route along the Australian coast in 1699, see Marchant, *Island unto Itself*, pp.122–47.
45. Dampier, *Voyage to New-Holland*, p.121.

46. Jonathan Swift, *Gulliver's Travels*, ed. Peter Dixon and John Chalker (Harmondsworth, 1967), p.333.
47. Dampier, *Voyage to New-Holland*, p.122.
48. See Serena K. Marner, 'William Dampier and his Botanical Collection', in Howard Morphy and Elizabeth Edwards, eds, *Australia in Oxford* (Oxford, 1988). The plants are listed by A.S. George and T.E.A. Aplin in John Kenney, *Before the First Fleet: Europeans in Australia 1606–1777* (Kenthurst, NSW, 1995), pp.78, 80.
49. Dampier, *Voyage to New-Holland*, p.108.
50. See Bernard Smith, quoted in Kenny, *Before the First Fleet*, p.146. In *William Dampier in New Holland: Australia's First Natural Historian* (Hawthorn, VIC., 1999), pp.21–98, Alex S. George includes colour photographs of Dampier's specimens in Oxford, set against photographs of the illustrations in *Voyage to New-Holland* and in Plukenet's book. They show the accuracy of Dampier's unknown artist.
51. In 1697, Willem de Vlamingh described some of the vegetation and wildlife of Rottnest Island (including black swans), but apparently brought no live specimens back to Holland.
52. Dampier, *Voyage to New-Holland*, p.81.
53. See John Ray, *Historiae plantarum* (1704), pp.225–6; Robert Huxley, ed., *The Great Naturalists* (2007), p.95; Preston, *Pirate of Exquisite Mind*, p.409.
54. John Ray, *The Wisdom of God* (1691), cited in Huxley, *Great Naturalists*, p.92.
55. Dampier, *Voyage to New-Holland*, p.125.
56. Ibid., pp.108, 110.
57. See ibid., p.108, and Steve Simpson, 'The Peculiar Natural History of New Holland', in Morphy and Edwards, *Australia in Oxford*, p.6.
58. Dampier, *Voyage to New-Holland*, p.111. See p.215 below for François Péron's identification more than ninety years later of the creature as a dugong.
59. Ibid., p.224.
60. See Günter Schilder, *Voyage to the Great South Land: Willem de Vlamingh 1696–1697* (Sydney, 1985).
61. Johannes Heniger, 'Dutch Contributions to the Study of Exotic Natural History in the Seventeenth and Eighteenth Centuries', in William Eisler and Bernard Smith, eds, *Terra Australis: The Furthest Shore* (Sydney, 1988), p.66.
62. A.-F. Frézier, *A Voyage to the South Sea and along the Coasts of Chili and Peru* (1717), pp.151–2.
63. William Funnell, *A Voyage round the World* (1707).
64. John Welbe, *An Answer to Captain Dampier's Vindication* [1708], p.3.
65. Ibid., p.8.
66. Ibid., p.6. It was actually Armstrong's fellow Scot Andrew Barton, killed in a sea action against the English in 1511, whose last words were remembered in the ballad: 'I'll lay me down and bleed awhile/And then I'll rise and fight again.'
67. Woodes Rogers, *A Cruising Voyage round the World*, with introduction by G.E. Mainwaring (1928), pp.190, 195.
68. Robert Harley, *Letters and Papers*, III, in *Manuscripts of the Duke of Portland*, V (Historical Manuscripts Commission, 1899), p.66.
69. Rogers, *Cruising Voyage*, pp.43, 179.
70. B.M.H. Rogers, 'Dampier's Debts', *Mariner's Mirror*, 15 (1924), p.122.
71. Quoted in Peter Earle, *The World of Defoe* (1976), p.47.

Chapter 2 The Alaskan Tribulations of Georg Wilhelm Steller

1. See Raymond H. Fisher, ed., *The Voyage of Semen Dezhnev in 1648* (1981).
2. For a recent biography, see Orcutt Frost, *Bering: The Russian Discovery of America* (New Haven and London, 2003).
3. A full discussion of Bering's first voyage and its aftermath is contained in Raymond H. Fisher, *Bering's Voyages: Whither and Why* (Seattle and London, 1977).

4. The figure is given in James R. Gibson, 'Supplying the Kamchatka Expeditions, 1725–30 and 1733–42', in O.W. Frost, ed., *Bering and Chirikov: The American Voyages and their Impact* (Anchorage, 1992), p.113.

5. A reproduction of the map is contained in Sven Waxell, *The American Expedition*, trans. M.A. Michael (Edinburgh, 1952), between pp.72 and 73.

6. Frank A. Golder, ed., *Bering's Voyages* (2 vols, New York, 1922), I, p.31.

7. Georg Wilhelm Steller, *Journal of a Voyage with Bering 1741–1742*, ed. O.W. Frost, trans. Margritt A. Engel and O.W. Frost (Stanford, 1988), p.49 [hereafter Frost and Engel, *Steller Journal*].

8. Leonard Stejneger, *Georg Wilhelm Steller: The Pioneer of Alaskan Natural History* (Cambridge, MA, 1936), pp.147–8.

9. Fisher, *Bering's Voyages*, p.128.

10. Frost and Engel, *Steller Journal*, p.47.

11. Waxell, *American Expedition*, p.103.

12. Frost and Engel, *Steller Journal*, pp.16–17.

13. Quotations in this paragraph are from ibid., pp. 54, 54, 57.

14. Ibid., p.61.

15. Ibid., pp.194–5.

16. Ibid., p.64.

17. Frost, *Bering*, p.162.

18. Frost and Engel, *Steller Journal*, p.77. In fact, Steller spent only ten hours ('zehn studen') ashore on Kayak Island.

19. Quotations in this paragraph are from ibid., pp.75–6, 21, 77, 78.

20. Golder, *Bering's Voyages*, I, p.120.

21. Waxell, *American Expedition*, p.110.

22. Frost and Engel, *Steller Journal*, p.93.

23. Ibid., pp.93–4.

24. See Kenneth J. Carpenter, *The History of Scurvy and Vitamin C* (Cambridge, 1986).

25. See Vasilii A. Divin, *The Great Russian Navigator, A.I. Chirikov*, trans. and annotated by Raymond H. Fisher (Fairbanks, AK, 1993), p.174.

26. Frost and Engel, *Steller Journal*, p.97.

27. Ibid., p.101.

28. Waxell, *American Expedition*, p.116.

29. Ibid., p.117.

30. Ibid.

31. Frost and Engel, *Steller Journal*, p.105.

32. See Robert Fortune, 'The *St Peter*'s Deadly Voyage Home: Steller, Scurvy and Survival', in Frost, *Bering and Chirikov*, pp.204–28.

33. Frost and Engel, *Steller Journal*, pp.113, 114.

34. Golder, *Bering's Voyages*, I, pp.275, 180–3.

35. Waxell, *American Expedition*, pp.121, 122.

36. Frost and Engel, *Steller Journal*, p.93.

37. Waxell, *American Expedition*, p.123; Golder, *Bering's Voyages*, I, 276.

38. Golder, *Bering's Voyages*, I, pp.188–208.

39. Waxell, *American Expedition*, p.124.

40. Ibid., p.127.

41. Ibid., pp.131–2.

42. See Frost and Engel, *Steller Journal*, pp.215–16 n.13 for an account by a modern physician of the probable causes of Bering's death.

43. Waxell, *American Expedition*, p.135.

44. See Orla Madsen et al., 'Exacavating Bering's Grave', in Frost, *Bering and Chirikov*, pp.229–47.

45. Waxell, *American Expedition*, p.139.

46. See ibid., pp.135–6, and Frost and Engel, *Steller Journal*, pp.143–4.

47. See Frost and Engel, *Steller Journal*, p.17, and Lydia T. Black, *Russians in Alaska 1732–1867* (Fairbanks, AK, 2004), p.56 n.54. Frost, *Bering*, gives the number of pelts as 'nearly 900' (total) and 'nearly 300' (Steller's share).
48. Waxell, *American Journal*, p.199.
49. Ibid., p.205.
50. Frost and Engel, *Steller Journal*, p.155.
51. Ibid., pp.159–60.
52. Stejneger, *Steller*, p.357.
53. Waxell, *American Expedition*, p.142.
54. Stejneger, *Steller*, p.361.
55. Waxell, *American Expedition*, p.152.
56. See Stejneger, *Steller*, p.370. In his journal Steller explained: 'In place of teeth, it has in its mouth two broad bones, one of which is affixed above to the palate, the other on the inside of the lower jaw. Both are furnished with many crooked furrows and raised ridges with which it crunches seaweed as its customary food.' Frost and Engel, *Steller Journal*, p.160.
57. Ibid., p.15.
58. For the report and Steller's reaction, see Carol Urness, ed. and trans., *Bering's Voyages: The Reports from Russia by Gerhard Friedrich Müller* (Fairbanks, AK, 1986), p.155 n.51 and p.39. It should be noted that Steller's family surname was Stöller.
59. For this episode, see Glyn Williams, *Voyages of Delusion: The Search for the Northwest Passage in the Age of Reason* (2002), pp.247–59.
60. Urness, ed. and trans., *Reports from Russia*, p.115.
61. For the convoluted history of Steller's journal, see the editor's comments in Frost and Engel, *Steller Journal*, pp.26–33.
62. Ibid., pp.221–2; for Dampier's description see his *A New Voyage round the World* (1697), ch.III.
63. Waxell, *American Expedition*, pp.194, 196.
64. From Steller's *De bestiis marinis* (St Petersburg, 1751), quoted in Corey Ford, *Where the Sea Breaks its Back* (Portland, OR, 1992), pp.162–3.
65. Stejneger, *Steller*, pp.364–5.
66. Ford, *Where the Sea Breaks its Back*, p.164.
67. For a full study of the subject, see 'The Pictures of the Sea-Cow', in Stejneger, *Steller*, pp.511–23.
68. Frost and Engel, *Steller Journal*, pp.25–6
69. On this, see pp.55–6 below.
70. See the analysis of the list by John F. Thilenius in Frost, *Bering and Chirikov*, pp.413–43.
71. Stejneger, *Steller*, p.537.

Chapter 3 The Fortunes and Misfortunes of Philibert Commerson

1. Robert E. Gallagher, ed., *Byron's Journal of his Circumnavigation 1764–1766* (1964), p.3.
2. William T. Stearn, 'Linnaean Classification, Nomenclature, and Method', in Wilfrid Blunt, *Linnaeus: The Compleat Naturalist* (2004), p.189.
3. Lisbet Koerner, *Linnaeus: Nature and Nation* (Harvard, MA, 1999), pp.39–40, 55.
4. Ibid., p.40.
5. Blunt, *Linnaeus*, p.121.
6. Nicholas Thomas, *Discoveries: The Voyages of Captain Cook* (2003), p.32.
7. Quoted by Janet Browne, 'Botany in the Boudoir and Garden', in David Philip Miller and Peter Hanns Reill, eds, *Visions of Empire: Voyages, Botany, and Representations of Nature* (Cambridge, 1996), p.156.
8. Blunt, *Linnaeus*, p.201.
9. Koerner, *Linnaeus*, p.115.
10. Mary Louise Pratt, *Imperial Eyes: Travel Writing and Transculturation* (1992), p.27.

11. See the petition of François Beau in Etienne Taillemite, 'Hommage à Bougainville', in *Journal de la Société de Océanistes*, Vol. XXIV, No.24 (Dec. 1968), p.38.

12. See Roger L. Williams, *Botanophilia in Eighteenth-Century France* (Dordrecht, 2001), esp. pp.102–40.

13. See 'Sommaire d'observations d'histoire naturelle', in Etienne Taillemite, ed., *Bougainville et ses compagnons autour du monde* (2 vols, Paris, 1977), II, pp.514–22.

14. See Jean-Etienne Martin-Allanic, *Bougainville navigateur et les découvertes de son temps* (2 vols, Paris, 1964), I, p.500.

15. See Taillemite, ed., *Bougainville et ses compagnons*. I, p.87 n.4.

16. John Dunmore, ed. and trans., *The Pacific Journal of Louis-Antoine de Bougainville 1767–1768* (2002), p.xlv.

17. John Dunmore, *Storms and Dreams: The Life of Louis de Bougainville* (Fairbanks, AK, 2007), p.169.

18. Dunmore, ed. and trans., *Bougainville Journal*, pp.xl–xli.

19. Taillemite, ed., *Bougainville et ses compagnons*, I, p.88.

20. Dunmore, ed. and trans., *Bougainville Journal*, pp.14, 19.

21. See Martin-Allanic, *Bougainville navigateur*, I, pp.555–6.

22. Ibid., p.589.

23. Dunmore, ed. and trans., *Bougainville Journal*, p.30.

24. These extracts from Vivez's journal are in ibid., pp.228, 229.

25. Ibid., p.225.

26. Ibid., p.59.

27. Lewis [sic] Antoine de Bougainville, *A Voyage round the World*, trans. J.R. Forster, (1772), pp.218–19.

28. Dunmore, ed. and trans., *Bougainville Journal*, p.60.

29. On this, see ibid., pp.60 n.2, 236, 255, 282.

30. Ibid., pp.72, 74.

31. Ibid., pp.296–7.

32. For a description of Tahiti at the time of the European arrival, see Anne Salmond, *Aphrodite's Island: The European Discovery of Tahiti* (Berkeley, CA, 2010).

33. Dunmore, ed. and trans., *Bougainville Journal*, p.97.

34. The extracts from Vivez's journal are in ibid., pp.229–30.

35. Remarks on Baret are in ibid., pp.97, 228, 293.

36. See Glynis Ridley, *The Discovery of Jeanne Baret* (New York, 2010), esp. chs 6, 7.

37. Dunmore, ed. and trans., *Bougainville Journal*, p.304.

38. L. Davis Hammond, ed., *News from New Cythera: A Report of Bougainville's Voyage 1766–1769* (Minneapolis, MN, 1970), p.26. I say 'presumably' because the note read that the collection was brought back by 'the physician-naturalist aboard the frigate'; and although Commerson was the only 'physician-naturalist' on the expedition, he did not sail on the *Boudeuse* and he did not return to France. Certainly, much of his collection seems to have been disembarked with him in Mauritius.

39. Ibid., p.45.

40. Bougainville, *Voyage round the World*, pp.228–9.

41. Ibid., pp.269, 274.

42. The full text of the 'Post-Scriptum' is printed in Taillemite, ed., *Bougainville et ses compagnons*, II pp.506–10.

43. See Madeleine Ly-Tio-Fane, *Pierre Sonnerat 1748–1814* (Mauritius, 1976), pp.7–9, 54–60.

44. The name has long since disappeared, but in the paperback edition of *The Discovery of Jeanne Baret* (New York, 2011), p.252, Glynis Ridley notes a proposal to name a new species in South America, *Solanum baretiae*, after her.

45. Ly-Tio-Fane, *Sonnerat*, p.61.

46. Paul-Antoine Cap, *Philibert Commerson, naturaliste voyageur* (Paris, 1861), p.164.

47. Taillemite, 'Hommage', p.38.

48. Taillemite, ed., *Bougainville et ses compagnons*, I, p.89.

49. Roger L. Williams, *French Botany in the Enlightenment: The Ill-Fated Voyages of La Pérouse and his Rescuers* (Dordrecht, 2003), p.11.
50. For accusations of plagiarism against Sonnerat, see Ly-Tio-Fane, *Sonnerat*, pp.90–5.
51. Quoted in S. Passfield Oliver, *The Life of Philibert Commerson* (1909), ed. G.F. Scott, pp.236–7.
52. Dunmore, ed. and trans., *Bougainville Journal*, p.lxxi.

Chapter 4 Joseph Banks and Daniel Solander

1. See A.M. Lysaght, ed., *Joseph Banks in Newfoundland and Labrador, 1766* (1971).
2. Banks to Thomas Falconer, April 1768, in Neil Chambers, ed., *The Indian and Pacific Correspodence of Sir Joseph Banks, 1768–1820*, I (2008), p.5.
3. Harold B. Carter, *Sir Joseph Banks 1743–1820* (1988), pp.71–2. Banks admitted that he was taking 'such a Collection of Bottles, Boxes, Baskets bags nets &c &c &c as almost frightens me'.
4. See A.M. Lysaght, 'Banks's Artists and his *Endeavour* Collections', in *Captain Cook and the South Pacific*, The British Museum Yearbook, 3 (1979), pp.9–80.
5. Edward Duyker, *Nature's Argonaut: Daniel Solander 1733–1782* (Melbourne, 1988), p.91.
6. Quoted in John Gascoigne, *Joseph Banks and the English Enlightenment: Useful Knowledge and Polite Culture* (Cambridge, 1994), p.101.
7. Bengt Jonsell, 'Daniel Solander – the Perfect Linnaean . . .', *Archives of Natural History*, 11 (1984), p.448.
8. J.C. Beaglehole, ed., *The Endeavour Journal of Joseph Banks 1768–1771* (2 vols, Sydney, 1962), I, p.22.
9. Banks to William Philip Perrin, 11 August 1768, in Chambers, ed., *Banks's Indian and Pacific Correspondence*, I, p.26.
10. Beaglehole, ed., *Banks Journal*, I, p.23.
11. P. O'Brian, *Joseph Banks: A Life* (1988), p.61. In fact, Linnaeus lived another ten years, dying in 1778.
12. Beaglehole, ed., *Banks Journal*, I, p.30.
13. J.C. Beaglehole, ed., *The Journals of Captain James Cook on his Voyages of Discovery: The Voyage of the Endeavour 1768–1771* (Cambridge, 1955), pp.cclxxxii–cclxxxiii.
14. Chambers, ed., *Banks's Indian and Pacific Correspondence*, I, p.26. At this time Banks would not have been aware of the presence of Commerson and Véron on Bougainville's voyage.
15. Beaglehole, ed., *Banks Journal*, I, p.34.
16. Ibid., p.157.
17. Ibid., p.160.
18. Chambers, ed., *Banks's Indian and Pacific Correspondence*, I, p.33.
19. Joseph Hooker, *Flora Novae-Zelandae*, I (1855), p.iii.
20. Chambers, ed., *Banks's Indian and Pacific Correspondence*, I, p.35.
21. Duyker, *Nature's Argonaut*, p.110.
22. Ibid., p.113.
23. On this episode, see Beaglehole, ed., *Banks Journal*, I, pp.218–23.
24. Beaglehole, ed., *Endeavour Voyage*, p.44.
25. Beaglehole, ed., *Banks Journal*, I, pp.225–6.
26. See ibid., I, pp.243–4, 250–1; II, p.301; Richard Samuel to Banks, 29 July 1768, in Chambers, ed., *Banks's Indian and Pacific Correspondence*, I, p.17.
27. Cook to Sir John Pringle, 7 July 1776, quoted in James Watt, 'Medical Aspects and Consequences of Cook's Voyages', in Robin Fisher and Hugh Johnston, *Captain James Cook and his Times* (Vancouver, 1979), p.135. It is significant that while Cook referred to 'rob' (or concentrate) of citric juice, which lost most of its vitamin content in the processing, Banks used lemon juice from a cask.
28. Beaglehole, ed., *Banks Journal*, I, pp.260–1.
29. Sydney Parkinson, *A Journal of a Voyage to the South Seas* (1773), pp.37–50.
30. Beaglehole, ed., *Banks Journal*, I, p.341.

31. Duyker, *Nature's Argonaut*, p.153.

32. Beaglehole, ed., *Banks Journal*, I, p.255.

33. Ibid., p.40.

34. Beaglehole, ed., *Endeavour Voyage*, p. 501. But see Banks's surprising criticism more than thirty years later of Cook's attitude to the naturalists; p.228 below.

35. Beaglehole, ed., *Banks Journal*, I, pp.312–13. There is a substantial article literature on Tupaia, as well as a full-length biography by Joan Druett, *Tupaia: The Remarkable Story of Captain Cook's Polynesian Navigator* (Auckland, 2011).

36. Beaglehole, ed., *Banks Journal*, I, p.396.

37. For relationships between Cook's crew and the Maori, see Anne Salmond, *Two Worlds: First Meetings between Maori and Europeans 1642–1772* (Auckland, 1991), esp. Part 3.

38. Beaglehole, ed., *Endeavour Voyage*, p.171.

39. Beaglehole, ed., *Banks Journal*, I, p.403.

40. Ibid., p.406; Duyker, *Nature's Argonaut*, p.160.

41. Beaglehole, ed., *Banks Journal*, I, p.443.

42. Ibid., I, pp.435–6; II, p.4.

43. Ibid., I, p.418.

44. Ibid., p.428.

45. Ibid., II, p.10. Among the many uses of flax the most important was for ships' sails, and the Admiralty was continually worried that the country's main source of this essential commodity for the navy was the Baltic region, a supply that might easily be disrupted in time of war.

46. Ibid., I, p.419.

47. Ibid., p.45.

48. *Historical Records of New South Wales*, vol. I, Part 1, p.215.

49. Beaglehole, ed., *Banks Journal*, II, p.58. A quire = four sheets of paper folded to make eight leaves.

50. Ibid., p.62.

51. See Duyker, *Nature's Argonaut*, p.184.

52. See W.T. Stearn, 'The Botanical Results of the *Endeavour* Voyage', *Endeavour*, XXVII (1968), p.9. There are eighteen volumes of Parkinson's drawings in the Natural History Museum, London.

53. Beaglehole, ed., *Banks Journal*, II, p.79.

54. Ibid., p.89.

55. Ibid., p.94.

56. J.C. Beaglehole, *The Life of Captain James Cook* (1974), p.261.

57. For the comments of Banks and Cook on the Aborigines, see my article, '"Far more happier than we Europeans": Reactions to the Australian Aborigines on Cook's Voyage', first published in *Historical Studies*, 19 (1981), pp.499–512.

58. Beaglehole, ed., *Endeavour Voyage*, p.505.

59. These newspaper extracts are printed ibid., pp.642–55.

60. Beaglehole, ed., *Banks Journal*, I, p.53.

61. Duyker, *Nature's Argonaut*, p.222.

62. Ibid., p.224.

63. Chambers, ed., *Banks's Indian and Pacific Correspondence*, I, p.47.

64. J.C. Beaglehole, ed., *The Journals of Captain James Cook on his Voyages of Discovery: The Voyage of the Resolution and Adventure 1772–1775* (Cambridge, 1961), p.913. The James Lind who was approached to sail on Cook's second voyage is not to be confused with the naval surgeon James Lind, who after Anson's voyage investigated possible cures for scurvy.

65. Ibid., p.xxx.

66. Ibid.

67. Chambers, ed., *Banks's Indian and Pacific Correspondence*, I, p.56.

68. Beaglehole, ed., *Resolution and Adventure Voyage*, p.685.

69. Duyker, *Nature's Argonaut*, p.227.

70. Beaglehole, ed., *Resolution and Adventure Voyage*, p.xxix n. In a recent novel, *The Conjuror's Bird* (2005), Martin Davies has speculated that it was the removing of a spare cabin from the *Resolution*'s upper deck, intended by Banks to be the private living quarters of his female companion, that provoked his furious reaction.
71. Quoted in Bernard Smith, *European Vision and South Pacific* (2nd edn, New Haven and London, 1988), p.47.
72. Shown in Beaglehole, ed., *Endeavour Voyage*, p.352.
73. Chambers, ed., *Banks's Indian and Pacific Correspondence*, I, pp.55–6. Banks's biographer gives rather different figures: of plants, more than 3,600 species, 1,400 of them new to science; from the animal kingdom, more than 1,000 species, most of them fishes, anthropods and molluscs. Carter, *Banks*, pp.95–6.
74. O'Brian, *Banks*, pp.168–70.
75. Beaglehole, ed., *Banks Journal*, I, pp.70–1.
76. Ibid., p.120.
77. Duyker, *Nature's Argonaut*, p.128.
78. Beaglehole, ed., *Banks Journal*, I, p.121.
79. Alwyn Wheeler, 'Daniel Solander and the Zoology of Cook's Voyage', *Archives of Natural History*, 11 (1984), p.514.
80. Quoted in Gascoigne, *Banks and the English Enlightenment*, p.32.
81. Carter, *Joseph Banks*, p.142. The *Florilegium* was finally published in a limited edition in thirty-four parts by Electa Historical Editions and the British Museum between 1980 and 1990. This final stage of the saga is described in Brian Adams, *The Flowering of the Pacific* (1986), Part II.

Chapter 5 The Woes of Johann Reinhold Forster

1. Michael E. Hoare, ed., *The Resolution Journal of Johann Reinhold Forster 1772–1775* (4 vols, 1982), I, p.2.
2. Ibid., p.17.
3. Ibid., p.12.
4. Michael E. Hoare, *The Tactless Philosopher: Johann Reinhold Forster (1729–98)* (Melbourne, 1976), p.37.
5. Hoare, ed., *Forster Journal*, I, p.47.
6. Ibid., p.46.
7. J.C. Beaglehole, ed., *The Journals of Captain James Cook on his Voyages of Discovery: The Voyage of the Resolution and Adventure 1772–1775* (Cambridge, 1961), p.8.
8. Ibid., p.717.
9. Hoare, ed., *Forster Journal*, I, p.51.
10. Ibid., p.125.
11. Hoare, *Tactless Philosopher*, p.78.
12. George Forster, *A Letter to the Right Honourable the Earl of Sandwich* (1778), printed as Appendix D in Nicholas Thomas, Oliver Berghof and Jennifer Newell, eds, *George Forster, A Voyage round the World* (2 vols, Honolulu, 2000), pp.788–9.
13. Hoare, ed., *Forster Journal*, I, pp.53–4.
14. The term is explained by Forster in his journal entry for 25 June 1773, where he writes that 'every thing which our Sailors found not to be quite in the common way of a man of war, they call *Experimental*, and gives as an example drinking water distilled from sea water. So, 'Mr. *Wales* the Astronomer, Mr. *Hodges*, the painter, Myself and my Son were comprehended under the name of *Experimental Gentlemen*', ibid., II, p.310.
15. Beaglehole, ed., *Resolution and Adventure Voyage*, p.xlii.
16. Hoare, ed., *Forster Journal*, I, p.148.
17. Ibid., pp.159–60.
18. Anders Sparrman, *A Voyage to the Cape of Good Hope . . . and round the World* (2 vols, 1785), p.81.
19. Quoted in Hoare, ed., *Forster Journal*, I, p.100.

20. Thomas et al., *Forster Voyage*, II, p.790.
21. Ibid., I, p.62.
22. Hoare, ed., *Forster Journal*, II, pp.187–8; Anders Sparrman, *A Voyage around the World* [Stockholm, 1802], trans. Eivor Cormack, in *The Linnaeus Apostles*, 5 (2007), p.381.
23. Hoare, ed., *Forster Journal*, II, p.196.
24. Beaglehole, ed., *Resolution and Adventure Voyage*, pp.98–9.
25. Hoare, ed., *Forster Journal*, II, pp.233, 234.
26. Thomas et al., *Forster Voyage*, I, p.78.
27. Hoare, ed., *Forster Journal*, II, p.241.
28. Sparrman, *Voyage around the World*, pp.391–2.
29. Hoare, ed., *Forster Journal*, II, p.251.
30. George Forster, *Cook, the Discoverer* [German original, Berlin, 1787] (Sydney, 2007), p.204.
31. Hoare, ed., *Forster Journal*, II, p.269.
32. Ibid., p.273.
33. See p.111 below.
34. Thomas et al., *Forster Voyage*, II, p.701.
35. See p.111 below.
36. Hoare, ed., *Forster Journal*, II, p.185 and III, p.387.
37. Ibid., II, p.293.
38. Ibid., III, p.548.
39. Thomas et al., *Forster Voyage*, I, p.280.
40. Hoare, ed., *Forster Journal*, II, p.265.
41. Thomas et al., *Forster Voyage*, II, p.631.
42. Beaglehole, ed., *Resolution and Adventure Voyage*, pp.76, 127, 166–7, 168, 741, 778.
43. Gananath Obeyeskere, *The Apotheosis of Captain Cook: European Mythmaking in the Pacific* (Princeton, NJ, 1992), p.12; but see Nigel Rigby, 'The Politics and Pragmatics of Seaborne Plant Transportation 1769–1805', in Margarette Lincoln, ed., *Science and Exploration in the Pacific: European Voyages to the Southern Oceans in the Eighteenth Century* (Woodbridge, Suffolk, 1998), pp.82–4.
44. Beaglehole, ed., *Resolution and Adventure Voyage*, p.571.
45. Hoare, ed., *Forster Journal*, II, p.289.
46. Ibid., p.284.
47. Ibid., III, p.404.
48. Thomas et al., *Forster Voyage*, II, pp.141–2, 143.
49. Hoare, ed., *Forster Journal*, II, p.331.
50. Ibid., III, p.365. It is not clear from this exchange whether Forster was allowed the use of the great cabin for his work, in the same way that Banks and Solander had been permitted on the *Endeavour*, or whether he is referring to his more occasional social visits to Cook.
51. Ibid., p.551. See also p.123 below.
52. Sparrman, *Voyage around the World*, pp.433–4.
53. Hoare, ed., *Forster Journal*, III, p.396.
54. Ibid., p.432.
55. All quotations in this paragraph are from ibid., pp.438–9.
56. Ibid., p.443.
57. Ibid., p.445.
58. Beaglehole, ed., *Resolution and Adventure Voyage*, p.323; Hoare, ed., *Forster Journal*, III, 451.
59. Hoare, ed., *Forster Journal*, III, p.480.
60. Ibid., pp.500–1.
61. Thomas et al., *Forster Voyage*, II, p.520.
62. Hoare, ed., *Forster Journal*, III, pp.550–1.
63. Ibid., p.552.
64. Ibid., IV, p.578.
65. Ibid., pp.608–9.

66. Quoted in Richard Conniff, *The Species Seekers* (New York, 2011), p.77; see also S.P. Dance, *Shell Collecting: An Illustrated History* (Berkeley, CA, 1966).
67. For this incident, see Hoare, ed., *Forster Journal*, IV, pp.606–7, and Thomas et al., *Forster Voyage*, II, p.774.
68. Hoare, ed., *Forster Journal*, IV, p.647.
69. Beaglehole, ed., *Resolution and Adventure Voyage*, p.543.
70. Thomas et al., *Forster Voyage*, II, pp.585, 586n.
71. Beaglehole, ed., *Resolution and Adventure Voyage*, p.625 and n. Forster's original suggestion was Southern Georgia.
72. Sparrman, *Voyage to Cape of Good Hope*, p.103.
73. Beaglehole, ed., *Resolution and Adventure Voyage*, p.728.
74. Hoare, *Tactless Philosopher*, p.157.
75. Beaglehole, ed., *Resolution and Adventure Voyage*, p.2.
76. Thomas et al., *Forster Voyage*, II, p.806.
77. Ibid., p.797.
78. Ibid., I, p.xxxvi.
79. Ibid., pp.8, xxix.
80. Thomas et al., *Forster Observations*, p.143.
81. Ibid., pp.103, 107–8.
82. For more on this, see P.J. Marshall and Glyndwr Williams, *The Great Map of Mankind: British Perceptions of the World in the Age of Enlightenment* (1982), pp.274–83.
83. J.G. Herder, *Outlines of a Philosophy of the History of Man*, trans. T. Churchill (1800), p.153.
84. Nicholas Thomas, 'Forster, Johann Reinhold, and Georg Forster', in David Buisseret, ed., *The Oxford Companion to World Exploration* (2 vols, Oxford, 2007), I, p. 316.
85. Quoted in John Gascoigne, *Joseph Banks and the English Enlightenment: Useful Knowledge and Polite Culture* (Cambridge, 1994), p.137.
86. Thomas et al., *Forster Voyage*, I, pp.5–6.
87. Hoare, ed., *Forster Journal*, I, p.58.
88. Ibid., p.77. The work, dedicated to Georg Forster, was *Enchiridion histriae naturali inserviens*, published at Halle in 1788.
89. For a recent listing, see Adrienne L. Kaeppler, 'To Attempt New Discoveries in That Vast Unknown Tract', in Michelle Hetherington and Howard Morphy, eds, *Discovering Cook's Collections* (Canberra, 2009), pp.58–77.
90. The information in this paragraph is taken from Adrienne L. Kaeppler, *Holophusicon: The Leverian Museum. An Eighteenth-Century English Institution of Science, Curiosity and Art* (Altenstadt, 2011). The author explains, p.1 n.1, that Lever coined the term Holophusicon or Holophusikon 'to mean that it embraced all of nature (*holo* "whole", *phusikon* "natural")'.
91. Hoare, ed., *Forster Journal*, I, p.69n.
92. Ibid., p.92.
93. Neil Chambers, ed., *The Indian and Pacific Correspondence of Sir Joseph Banks, 1768–1820*, I (2008), p.272.
94. Forster's last years in London are described in Hoare, *Tactless Philosopher*, ch.VIII.
95. Ibid., p.199.
96. Forster, *Cook the Discoverer* [German original, Berlin, 1787] (Sydney, 2007), pp. 165, 171.
97. Robert J. King, 'The Call of the South Seas: George Forster and the Expeditions to the Pacific of Lapérouse, Mulovsky and Malaspina', *Georg-Forster-Studien*, XIII (2008), p.164.
98. Alexander von Humboldt, *Personal Narrative of a Journey to the Equinoctial Regions of the New Continent*, trans. Jason Wilson (1995), p.15.
99. Hoare, *Tactless Philosopher*, p.327. After Johann Reinhold Forster's death, what was left of his ethnographic collection was acquired by the Georg-August University of Göttingen.

Chapter 6 Cook, Vancouver and 'experimental gentlemen'

1. J.C. Beaglehole, ed., *The Journals of Captain James Cook on his Voyages of Discovery: The Voyage of the Endeavour 1768–1771* (Cambridge, 1955), p.289.
2. See Jacob von Stählin, 'Map of the New Northern Archipelago' (1774).
3. For more on this, see Glyn Williams, *Voyages of Delusion: The Northwest Passage in the Age of Reason* (2002), ch.9.
4. For the circumstances of King's appointment, see Steve Ragnall, *Better Conceiv'd Than Describ'd: The Life and Times of Captain James King* (Kibworth Beauchamp, Leics., 2013), chs7, 8.
5. J.C. Beaglehole, *The Journals of Captain James Cook on his Voyages of Discovery: The Voyage of the Resolution and Adventure 1772–1775* (Cambridge, 1961), pp.xlvi–xlvii n., quoting a 1926 translation of Forster's preface. Forster's rendition in German of Cook's remarks as conveyed to him by King was 'Verflucht sind alle Gelehrten und alle Gelehrsamkeit oben drein', which could perhaps be better translated as an attack on philosophers and philosophy since, although 'science' and 'man of science' were in use at this time, the term 'scientist' had to wait until the nineteenth century. See Dan O'Sullivan, *In Search of Captain Cook: Exploring the Man through his Own Words* (2008), p.106.
6. Beaglehole, *Resolution and Adventure Voyage*, p.lxxxiv.
7. Nicholas Thomas, Oliver Bergh and Jennifer Newell, eds, *George Forster: A Voyage round the World* (2 vols, Honolulu, 2000), II, pp.585–6.
8. Beaglehole, ed., *Resolution and Adventure Voyage*, p.959.
9. Neil Chambers, ed., *The Indian and Pacific Correspondence of Sir Joseph Banks, 1768–1820*, I (2008), p.199.
10. Ibid., p.315.
11. James Cook and James King, *A Voyage to the Pacific Ocean* (3 vols, 1784), I, p.5.
12. J.C. Beaglehole, ed., *The Journals of Captain James Cook on his Voyages of Discovery: The Voyage of the Resolution and Discovery 1776–1780* (Cambridge, 1967), II, p.1488.
13. See the National Archives: Adm 2/1332, pp.284–96.
14. See Anne Salmond, *Between Worlds: Early Exchanges between Maori and Europeans 1773–1815* (Honolulu, 1997), p.118.
15. Beaglehole, ed., *Resolution and Discovery Voyage*, I, p.166.
16. Ibid., p.cxci.
17. Cook and King, *Voyage*, I, p.4.
18. The tributes to Anderson are in ibid., p.lxxxiii, and Beaglehole, ed., *Resolution and Discovery Voyage*, I, p.406 and n., and II, p.1430.
19. For these three extracts from Anderson's journal, see Beaglehole, ed., *Resolution and Discovery Voyage.*, II, pp.732, 743, 769.
20. Ibid., I, p.43.
21. Ibid., pp.43, 45–6.
22. For the Anderson references in this paragraph, see ibid., II, pp.1519, 788, 791 and n., 839–40.
23. Cook's description, ibid., I, pp.95–6; Anderson's description, ibid., II, p.851. In his description Anderson was probably influenced by John Ellis's influential *Essay towards a Natural History of the Corallines* (1755). See Rüddiger Joppien and Bernard Smith, eds, *The Art of Captain Cook's Voyages*, III, *The Voyage of the Resolution and Discovery 1776–1780* (New Haven and London, 1988), pp.26–7.
24. Beaglehole, ed., *Resolution and Discovery Voyage*, II, pp.865–6.
25. Ibid., pp.909, 916.
26. Ibid., p.936.
27. Cook and King, *Voyage*, II, ch.XII and pp.334, 375.
28. Beaglehole, ed., *Resolution and Discovery Voyage*, I, p.263n.
29. Cook and King, *Voyage*, II, p.99.
30. J.C. Beaglehole, *The Life of Captain James Cook* (1974), pp.568–9.
31. Now in the Natural History Museum, London: Banks Coll. – And.

32. H.B. Carter, *Sir Joseph Banks 1743–1820* (1988), p.169.
33. Natural History Museum, London: Dawson Turner Copies, Vol. 1, p.304.
34. David Mackay, 'Agents of Empire: The Banksian Collectors and Evaluation of New Lands', in David Philip Miller and Peter Hanns Reill, eds, *Visions of Empire: Voyages, Botany and Representations of Nature* (Cambridge, 1996), pp.39ff.
35. Richard Drayton, *Nature's Government: Science, Imperial Britain, and the 'Improvement' of the World* (New Haven and London, 2000), p. 127.
36. For these, see David Mackay, *In the Wake of Cook: Exploration, Science and Empire 1780–1801* (1985).
37. Beaglehole, *Life of Cook*, p.689.
38. [John Etches], *An Authentic Statement of All the Facts Relative to Nootka Sound* (1790), p.2. King George's Sound was Cook's name for Nootka Sound. It should not be confused with King George the Third's Sound (now King George Sound) in present-day Western Australia, charted and named by George Vancouver in 1791.
39. Neil Chambers, ed., *The Indian and Pacific Correspondence of Sir Joseph Banks, 1768–1820*, II (2009), p.126.
40. John M. Naish, *The Interwoven Lives of George Vancouver, Archibald Menzies, Joseph Whidbey, and Peter Puget* (Lewiston, NY, 1996), p.50.
41. W. Kaye Lamb, ed., *George Vancouver: A Voyage of Discovery to the North Pacific Ocean and round the World 1791–1795* (4 vols, 1984), I, p.30.
42. For the two letters, see Chambers, ed., *Banks's Indian and Pacific Correspondence*, II, pp.136, 157.
43. Robert Galois, ed., *A Voyage to the North West Side of America: The Journals of James Colnett, 1786–89* (Vancouver, 2004), p.100.
44. Ibid., pp.150–1.
45. Neil Chambers, *The Indian and Pacific Correspondence of Sir Joseph Banks, 1768–1820*, III (2010), pp.35–6.
46. See Robert J. King, 'George Vancouver and the Contemplated Settlement at Nootka Sound', *The Great Circle*, 32 (2010), pp.3–30.
47. Chambers, *Banks's Indian and Pacific Correspondence*, III, p.59.
48. John Ellis, *Directions for Bringing over Seeds and Plants from the East-Indies and Other Distant Countries* (London, 1770), p.1. For more on Ellis's designs for plant boxes and those of John Fothergill in the 1790s, see Nigel Rigby, 'The Politics and Pragmatics of Seaborne Plant Transportation, 1769–1805', in Margarette Lincoln, ed., *Science and Exploration in the Pacific: European Voyages to the Southern Oceans in the Eighteenth Century* (Woodbridge, Suffolk, 1998), pp.87–93. For La Péronse's plant boxes see p.152 below.
49. Chambers, ed., *Banks's Indian and Pacific Correspondence*, III, pp.196–8.
50. Ibid., pp.178–9.
51. Ibid., p.294; for similar letters, see pp.234–5, 238, and Neil Chambers, ed., *The Indian and Pacific Correspondence of Sir Joseph Banks, 1768–1820*, IV (2011), pp.122–3.
52. Ibid., IV, pp.199–202.
53. See W. Kaye Lamb, 'Banks and Menzies: Evolution of a Journal', in Robin Fisher and Hugh Johnston, eds, *From Maps to Metaphors: The Pacific World of George Vancouver* (Vancouver, 1993), pp.227–44.
54. Lamb, *Vancouver Voyage*, I, p.227.
55. For valuable detective work on this, see Eric W. Groves, 'Archibald Menzies (1754–1842), an Early Botanist on the Northwestern Seaboard of North America . . .', *Archives of Natural History*, 28 (2001), pp.71–122. Menzies's journal from December 1790 to February 1794 is in the British Library: Add. MS. 36461; that from February 1794 to March 1795 is in the National Library of Australia: MS. 155. Sections of it have been published by C.F. Newcombe, ed., *Menzies' Journal of the Vancouver Voyage April to October 1792*, Archives of British Columbia, V (1923); A. Eastwood, ed., 'Menzies California Journal', *Californian Historical Society*, II (1924), pp.265–340; Wallace M. Olson, ed., *The Alaska Travel Journal of Archibald Menzies, 1793–1794* (Fairbanks, AK,

1993). Because of this multiplicity of sections and editions, the references that follow are usually cited as 'Menzies Journal', with date of the entry.

56. Menzies to Banks, 26 September 1792, Chambers, ed., *Banks's Indian and Pacific Correspondence*, III, p.437.
57. 'Menzies Journal', 2 May 1792.
58. Ibid., 4 May 1792. The modern names of the two plants are given in Newcombe, *Menzies' Journal*, p.20.
59. 'Menzies Journal', 20 May 1792.
60. Lamb, *Vancouver Voyage*, II, 534.
61. 'Menzies Journal', 18 August 1792.
62. Lamb, *Vancouver Voyage*, II, p.587.
63. Newcombe, *Menzies' Journal*, p.128 (no date).
64. José Mariano Moziño, *Noticias de Nutka: An Account of Nootka Sound in 1791*, trans. and ed. by Iris H. Wilson Enstrand (2nd edn, Seattle, 1991), p.x; the list of natural-history specimens is printed on pp.111–23.
65. Menzies to Banks, 14 January 1793, Chambers, ed., *Banks's Indian and Pacific Correspondence*, IV, p.53.
66. The Hawaii and Maui references are in 'Menzies Journal', 26, 27 February and 15 March 1793.
67. Lamb, *Vancouver Voyage*, III, p.990; 'Menzies Journal', 23 July 1793.
68. See Groves, 'Menzies an Early Botanist', p.91.
69. Menzies to Vancouver, 18 November 1793, Chambers, ed., *Banks's Indian and Pacific Correspondence*, IV, p.173.
70. 'Menzies Journal', 21 July 1794.
71. Lamb, *Vancouver Voyage*, IV, p.1382.
72. The plants are listed in James McCarthy, *Monkey Puzzle Man: Archibald Menzies, Plant Hunter* (Dunbeath, 2008), Appendix 5, pp.207–8.
73. 'Menzies Journal', 7 February 1795.
74. The story of how Menzies obtained the seeds has often been regarded as apocryphal, but at a later date Joseph Hooker confirmed that the botanist had told him 'that he took seed from the dessert table of the Governor'. See Groves, 'Menzies an Early Botanist', p.114 n.108, and McCarthy, *Monkey Puzzle Man*, ch.18.
75. Menzies to Banks, 14 September 1795, Chambers, ed., *Banks's Indian and Pacific Correspondence*, IV, p.309.
76. Lamb, *Vancouver Voyage*, I, p.218.
77. James Watt, 'The Voyage of George Vancouver 1791–1795: The Interplay of Physical and Psychological Pressures', *Canadian Bulletin of Medical History/British Columbia History of Medicine*, IV (1987), pp.33–51.
78. Lamb, *Vancouver Voyage*, I, pp.88, 209.
79. See p.139 above.
80. Lamb, 'Banks and Menzies', in Fisher and Johnston, *From Maps to Metaphors*, p.239.
81. Naish, *Interwoven Lives*, p.439.
82. Ibid., p.450.
83. See 'Menzies Journal', 16 June 1792, and Groves, 'Menzies an Early Botanist', pp.83, 112 n.45.
84. Groves, 'Menzies an Early Botanist', p.106.
85. Ibid.
86. Cited in Lamb, *Vancouver Voyage*, I, p.256.

Chapter 7 Naturalists with La Pérouse and d'Entrecasteaux

1. Catherine Gaziello, *L'Expédition de Lapérouse 1785–1788: réplique française aux voyages de Cook* (Paris, 1984).
2. See his *Découvertes des François en 1768 et 1769 dans le sud-est de la Nouvelle-Guinée* (Paris, 1790).

3. The full text of the instructions is printed in John Dunmore, ed., *The Journal of Jean-François de Galaup de la Pérouse* (2 vols, 1994), I, pp.cx–cl, with the king's marginal note on p.cxvii.
4. Ibid., p.xxvii.
5. Set out in Roger L. Williams, *French Botany in the Enlightenment: The Ill-fated Voyages of La Pérouse and his Rescuers* (Dordrecht, 2003), pp.19–28.
6. See ibid., pp.34–49. Thouin's instructions included a comprehensive list of equipment to be supplied to Collignon, ranging from tin chests of various sizes for holding plants and seeds to metal punches for stamping identifying numbers on the specimens collected.
7. See Nigel Rigby, Pieter van der Merwe and Glyn Williams, *Pioneers of the Pacific: Voyages of Exploration, 1787–1810* (2005), pp.46–7.
8. Ibid., p.21.
9. Dunmore, ed., *La Pérouse Journal*, I, p.cxlv.
10. John Dunmore, *Where Fate Beckons: The Life of Jean-François de la Pérouse* (Fairbanks, AK, 2007), pp.195–6.
11. Dunmore, ed., *La Pérouse Journal*, II, p.456.
12. Ibid., p.464.
13. Ibid., I, p.51.
14. Ibid., II, p.469.
15. Ibid., I, p.65.
16. For these voyages, see Glyn Williams, *Voyages of Delusion: The Search for the Northwest Passage in the Age of Reason* (2002), ch.8.
17. The quotations in the preceding two paragraphs are taken from Dunmore, ed., *La Pérouse Journal*, I, pp.95, 97, 129, 132, 133, 165.
18. Ibid., pp.192–3.
19. Gaziello, *L'Expédition de Lapérouse*, p.195.
20. Dunmore, *Where Fate Beckons*, pp.217–18.
21. Dunmore, ed., *La Pérouse Journal*, II, p.486.
22. Ibid., pp.492–3.
23. Ibid., pp.298–9.
24. Ibid., p.316.
25. Ibid., p.513.
26. Ibid., p.525.
27. Ibid., p.349.
28. Edward Duyker, *Père Receveur: Franciscan, Scientist and Voyager with La Pérouse* (Sydney, 2011), p.27.
29. Dunmore, ed., *La Pérouse Journal*, II, p.397.
30. Rigby et al., *Pioneers of the Pacific*, p.53.
31. The events of 11 December 1787 are described in La Pérouse's journal, and in his letter to Fleurieu of 7 February 1788, printed in Dunmore, ed., *La Pérouse Journal*, II, pp.396–411, 536–41.
32. Ibid., p.446.
33. See Duyker, *Père Receveur*, p.21. It is possible that Receveur was killed by Aborigines while out botanising. See 'The Death of Father Receveur' in Dunmore, ed., *La Pérouse Journal*, II, pp.564–9.
34. Paul G. Fidlon and R.J. Ryan, eds, *The Journal of Philip Gidley King: Lieutenant, R.N. 1787–1790* (Sydney, 1980), pp.37–8.
35. John White, *Journal of a Voyage to New South Wales* [1790] (Sydney, 1962), p.115.
36. Dunmore, *Where Fate Beckons*, p.248.
37. David Collins, *An Account of the English Colony in New South Wales* [1798] (Sydney, 1975), p.16.
38. The title of John Dunmore's chapter 22 in *Where Fate Beckons*.
39. Williams, *French Botany*, p.108; Louis-Marie-Antoine Milet-Mureau, *Voyage de la Pérouse autour du monde* (4 vols and atlas, Paris, 1797), I, pp.lv–lx.

40. Williams, *French Botany*, p.109.
41. See Brian Plomley and Josianne Piard-Bernier, *The General: The Visits of the Expedition Led by Bruny d'Entrecasteaux to Tasmanian Waters in 1792 and 1793* (Launceston, Tas., 1993), pp.223–7.
42. Edward Duyker, *Citizen Labillardière: A Naturalist's Life in Revolution and Exploration* (Melbourne, 2003), p.1.
43. The exchange of letters is in Neil Chambers, ed., *The Indian and Pacific Correspondence of Sir Joseph Banks*, III, *1789–1792* (2010), pp.244, 272–3.
44. For more details, see Frank Horner, *Looking for La Pérouse: D'Entrecasteaux in Australia and the South Pacific 1792–1793* (Melbourne, 1995), pp.36–40.
45. Duyker, *Citizen Labillardière*, p.12.
46. Quoted in Horner, *Looking for La Pérouse*, p.52.
47. For details of the expedition's stay at the Cape, see Duyker, *Citizen Labillardière*, pp.87–92.
48. Bruny d'Entrecasteaux, *Voyage to Australia and the Pacific 1791–1793*, ed. and trans. Edward Duyker and Maryse Duyker (Melbourne, 2001), p.34. See also Duyker, *Citizen Labillardière*, p.104.
49. Quoted in S.G.M. and D.J. Carr, 'A Charmed Life: The Collections of Labillardière', in D.J. and S.G.M. Carr, *People and Plants in Australia* (Sydney, 1981), p.87, translating a passage from Jurien's *Souvenirs d'un amiral* of 1860.
50. Quoted in Duyker, *Citizen Labillardière*, p.109. It is interesting to compare this with Linnaeus's clothes and equipment on his Lapland journey on horseback in 1732 (he, of course, had no ship nearby to hold most of his belongings): 'My clothes consisted of a little unpleated coat of West Gothland cloth with facings and a collar of worsted shag, neat leather breeches, a pig-tailed wig, a cap of green fustian, a pair of top boots, and a small leather bag, nearly two feet long and not quite so wide, with hooks on one side so that it can be shut and hung up. In this bag I carried a shirt, two pairs of half-sleeves, two nightcaps, an ink-horn, a pen-case, a magnifying glass, and a small spy-glass, a gauze veil to protect me from midges, this journal and a stock of sheets of paper, stitched together, to press plants between … A short sword hung at my side, and I had a small fowling piece between my thigh and the saddle. I also had a graduated eight-sided rod for taking measurements.' Wilfrid Blunt, *Linnaeus: The Compleat Naturalist* (2004), p.42.
51. E. and M. Duyker, *D'Entrecasteaux Voyage*, p.69.
52. Ibid., pp.79–80.
53. J.J.H. de Labillardière, *Account of a Voyage in Search of La Pérouse* (2 vols, 1800), I, pp.251–2.
54. Ibid., p.351.
55. Duyker, *Citizen Labillardière*, pp.122, 288; William Eisler, *The Furthest Shore: Images of Terra Australis from the Middle Ages to Captain Cook* (Cambridge, 1995), pp.113–15.
56. Horner, *Looking for La Pérouse*, p.103.
57. Riche's account is printed in E. and M. Duyker, *D'Entrecasteaux Voyage*, pp.119–26.
58. For more on this remarkable plant and its distribution, see Duyker, *Citizen Labillardière*, p.141.
59. Labillardière, *Account of a Voyage*, II, p.77.
60. E. and M. Duyker, *D'Entrecasteaux Voyage*, p.141.
61. Plomley and Piard-Bernier, *The General*, p.309.
62. Ibid., p.366.
63. Labillardière, *Account of a Voyage*, II, p.53.
64. Plomley and Piard-Bernier, *The General*, p.378.
65. E. and M. Duyker, *D'Entrecasteaux Voyage*, p.147.
66. Plomley and Piard-Bernier, *The General*, p.343.
67. Labillardière, *Account of a Voyage*, II, pp.86–7.
68. J.C. Beaglehole, ed., *The Journals of Captain James Cook on his Voyages of Discovery: The Voyage of the Resolution and Adventure 1772–1775* (Cambridge, 1961), p.539.

69. S.G.M. and D.J. Carr, 'A Charmed Life', p.90.
70. E. and M. Duyker, *D'Entrecasteaux Voyage*, p.261.
71. Ibid., p.263.
72. Williams, *French Botany*, pp.194–5.
73. This paragraph is a very summary account of the confused and complicated series of events that took place once the expedition arrived in the Dutch East Indies. The episode is covered in detail in Duyker, *Citizen Labillardière*, ch.14, and in Horner, *Looking for La Pérouse*, chs 13, 14.
74. The letters between Banks and Labillardière are printed in Chambers, ed., *Banks's Indian and Pacific Correspondence*, IV, pp.207, 277–8, 386, 408–9, 455; Banks's description of the collection is printed in Duyker, *Citizen Labillardière*, p.208.
75. See Peter Dillon, *Narrative and Successful Result of a Voyage to the South Seas . . . to Ascertain the Actual Fate of La Pérouse's Expedition* (2 vols, 1829).
76. For more details of the shipwreck on Vanikoro, see John Dunmore's comments in *La Pérouse Journal*, I, pp.ccxix–ccxxviii.
77. Duyker, *Citizen Labillardière*, p.232.
78. Chambers, ed., *Banks's Indian and Pacific Correspondence*, III, p.273.
79. E. and M. Duyker, *D'Entrecasteaux Voyage*, pp.xxxix–xl.

Chapter 8 The Scientific and Political Voyage of Malaspina

1. From Malaspina's 'Introducción', intended to form the preface to his published account of the voyage, in Andrew David, Felipe Fernández Armesto, Carlos Novi and Glyn Williams, eds, *The Malaspina Expedition 1789–1794: Journal of the Voyage by Alejandro Malaspina* (3 vols, London and Madrid, 2001–4), I, p.lxxix; hereafter referred to as *Malaspina Journal*.
2. For the Royal Scientific Expedition to New Spain and other major land expeditions by Hipólito Ruiz and José Antonio Pavón (Peru and Chile) and José Celestino Mutis (New Granada), see Iris H.W. Engstrand, *Spanish Scientists in the New World: The Eighteenth-Century Expeditions* (Seattle, 1981), chs 2, 3.
3. For details of Malaspina's life and thought, see John Kendrick, *Alejandro Malaspina: Portrait of a Visionary* (Montreal and Kingston, 1999); for an indication of Malaspina's philosophical views, see the edition and translation by John Black and Oscar Clemotte-Silvero of his *Meditación sobre lo bello en la Naturaleza* (Lewiston, NY, 2007).
4. Neil Chambers, ed., *The Indian and Pacific Correspondence of Sir Joseph Banks, 1768–1820*, IV (2010), p.15 n.6.
5. *Malaspina Journal*, I, pp.312–15 for the full text of the 'Plan'.
6. There are brief biographies of the expedition's officers and supernumeraries in José Ignacio González-Aller Hiero, *Malaspina Journal*, III, pp.333–58.
7. Domingo A. Madulid, 'The Life and Work of Antonio Pineda, Naturalist of the Malaspina Expedition', *Archives of Natural History*, 11 (1982), p.45.
8. Domingo A. Madulid, 'The Life and Work of Louis Née, Botanist of the Malaspina Expedition', ibid., 16 (1989), p.37. It should be noted that the acute accent is now placed on the second 'e' of Neé's name.
9. María Victoria Ibáñez Montoya, ed., *La expedición Malaspina 1789–1794*, IV, *Trabajos científicos y correspondencia de Tadeo Haenke* (Madrid, 1994), p.124.
10. *Malaspina Journal*, I, p.176. For the expedition's botanical work in general, see the catalogue to an exhibition at the Real Jardín Botánico, Pabellón Villanueva, ed., *La Botánica en la expedición Malaspina* (Madrid, 1989).
11. *Malaspina Journal*, I, pp.52, 101, 108.
12. For some of the problems in stuffing and preserving zoological specimens, see Richard Coniff, *The Species Seekers* (New York, 2011), pp.60–6.
13. *Malaspina Journal*, I, p.256.
14. Ibid., p.293.
15. Engstrand, *Spanish Scientists in the New World*, p.50.

16. Eduardo Estrella, 'La Expedición Malaspina en Guayaquil: estudios de historia natural', in Mercedes Palau Baquero and Antonio Oroxco Acuaviva, eds, *Malaspina '92* (Cádiz, 1994), p.71.
17. Ibid., p.70.
18. Engstrand, *Spanish Scientists in the New World*, p.54.
19. *Malaspina Journal*, I, p.258n.
20. Engstrand, *Spanish Scientists in the New World*, pp.63–4.
21. Victoria Ibáñez, 'Botanical and Zoological Investigations on Exploration Voyages to Alaska in the Second Half of the 18th Century', p.12; I am indebted to Robin Inglis for sight of this essay in typescript. Various pages of Haenke's colour codes are illustrated in Ibáñez, *Trabajos científicos de Haenke*, pp.98–101.
22. See p.230 below.
23. See Carmen Sotos Serrano, *Los Pintores de la expedición de Alejandro Malaspina* (2 vols, Madrid, 1982). Much of the information in this paragraph was kindly supplied by Robin Inglis.
24. Engstrand, *Spanish Scientists in the New World*, p.52.
25. Ibid., p.53 and n.
26. Ibid., pp.101–2.
27. *Malaspina Journal*, II, p.53.
28. The substantive part of the new instructions takes up only seven lines in the printed version of Valdés's letter of 22 December 1790 in ibid., II, p.417. Pages 427–84 in the same volume on 'The Ferrer Maldonado Fantasy' include a modern translation of Maldonado's 'Relación' and accompanying documents.
29. See pp.151, 157 above.
30. *Malaspina Journal*, II, pp.431, 465.
31. Ibid., p.104.
32. Ibáñez, *Trabajos científicos de Haenke*, p.137.
33. *Malaspina Journal*, II, p.123. In a gesture to Haenke, Malaspina named a small island near the Hubbard Glacier after him.
34. See María Dolores Higueras, *NW Coast of America: Iconographical Album of the Malaspina Expedition* (Madrid, 1991).
35. *Malaspina Journal*, II, p.477.
36. Ibid., p.176n.
37. Ibid., p.209.
38. Ibid., I, p.lxvii. See Donald C. Cutter, *Malaspina and Galiano: Spanish Voyage to the Northwest Coast 1791 and 1792* (Vancouver, 1992), pp.75–6 for the expedition's report on useful woods, 'Maderas de construcción, de fabricas y muebles'.
39. *Malaspina Journal*, II, p.245.
40. Ibáñez, *Trabajos científicos de Haenke*, p.138.
41. Marjorie G. Driver, ed., *The Guam Diary of Naturalist Antonio de Pineda y Ramirez February 1792* (Mangilao, Guam, 1990), pp.4, 8.
42. *Malaspina Journal*, II, pp.399, 402. A summary account of Haenke's journeyings on Luzon is contained in Ibáñez, *Trabajos científicos de Haenke*, pp.139–40.
43. The quotations in this paragraph are taken from *Malaspina Journal*, II, pp.403–13.
44. Ibid., III, p.75.
45. See Robert J. King, ed., *The Secret History of the Convict Colony: Alexandro Malaspina's Report on the British Settlement of New South Wales* (Sydney, 1990).
46. All references are to Victoria Ibáñez and Robert J. King, 'A Letter from Thaddeus Haenke to Sir Joseph Banks', *Archives of Natural History*, 23 (1996), p.258.
47. The full list is printed ibid., p.257.
48. *Malaspina Journal*, III, p.131.
49. Ibid., p.140.
50. Ibid., pp.156–8.
51. Ibid., p.168.
52. Ibid., p.163.

53. Ibid., p.175.
54. See Pablo Antón Solé, 'Los Padrones de cumplimiento pascual en la expedición Malaspina, 1790–1794', in *La Expedición Malaspina (1789–1794), bicentario de su salida de Cádiz* (Cádiz, 1989), pp.173–238.
55. For Malaspina's praise of Neé and Haenke, see *Malaspina Journal*, III, p.249.
56. See ibid., I, p.xlv.
57. Dolores Higueras Rodríguez, 'The Malaspina Expedition, 1789–1794', in Martínez Shaw, ed., *Spanish Pacific from Magellan to Malaspina* (Brisbane, 1988), p.156. There is an annotated list of the several thousand documents in the Museo Naval, Madrid, relating to the expedition in Dolores Higueras Rodríguez, *Catálogo critic de los documentos de la expeditión Malaspina en el Museo Naval* (3 vols, Madrid, 1985–94); a summary of this and a listing of documents in other archives in Spain and elsewhere are contained in her essay 'The Sources: The Malaspina and Bustamante Expedition: A Spanish State Enterprise', *Malaspina Journal*, III, pp.371–86.
58. See Carlos Novi, 'The Road to San Antón: Malaspina and Godoy', in ibid., pp.313–32.
59. Both quotations are from Malaspina's 'Introducción', ibid., I, pp.lxxxi, lxxxiv.
60. Quoted in Novi, 'Road to San Antón', ibid., III, p.325.
61. See ibid., p.325.
62. Kendrick, *Alejandro Malaspina*, p.139.
63. *Malaspina Journal*, III, p.316.
64. Kendrick, *Alejandro Malaspina*, p.148.
65. *Malaspina Journal*, I, p.xcvi; Engstrand, *Spanish Scientists in the New World*, p.106.
66. See Sandra Knapp, 'Lectotypification of Cavanilles' Names in *Solanum* (Solanaceae)', *Anales del Jardín Botánico de Madrid*, 64 (2007), p.196.
67. Félix Muñoz Garmendia, ed., *La expedición Malaspina 1789–1794*, III, *Diarios y trabajos botánicos de Luis Neé* (Madrid, 1992), p.40.
68. See ibid., p.44; and Madulid, 'Life and Work of Née', 41–3.
69. *Malaspina Journal*, II, p.118n., III, p.385.
70. Josef Opatmý, ed., *La Expedición de Alejandro Malaspina y Tadeo Haenke* (Prague, 2005).
71. *Malaspina Journal*, I, p.xcvi.
72. See Eduardo Estrella, ed., *La Expedición Malaspina 1789–1794*, VIII, *Trabajos . . . de Antonio Pineda Ramírez* (Madrid, 1996); and Madulid, 'Life and Work of Pineda', pp.49–50, 53–6.
73. See *Malaspina Journal*, III, p.385n.
74. O.H.K. Spate, *The Pacific Found and Lost* (Rushcutters Bay, NSW, 1988), p.177.
75. Kendrick, *Alejandro Malaspina*, p.161.

Chapter 9 The Australian Surveys of Baudin and Flinders

1. Alan Frost, *Sir Joseph Banks and the Transfer of Plants to and from the South Pacific 1786–1798* (Melbourne, 1993), p.6.
2. Paul Brunton, ed., *Matthew Flinders: Personal Letters from an Extraordinary Life* (Sydney, 2002), p.51.
3. Quoted in Phyllis I. Edwards, ed., 'The Journal of Peter Good', *Bulletin of the British Museum (Natural History)*, Historical Studies, 9 (1981), p.11.
4. Harold B. Carter, *Sir Joseph Banks* (1988), p.414. There is a copy, in full, of the British passport to Baudin in Jacqueline Bonnemains, Jean-Marc Argentin and Martine Marin, eds, *Mon Voyage aux Terres Australes: journal personnel du commandant Baudin* (Le Havre and Paris, 2001), p.42.
5. Frank Horner, *The French Reconnaissance: Baudin in Australia 1801–1803* (Melbourne, 1987), p.42.
6. For more on this, see Leslie Marchant, *France Australe* (Perth, 1998), and Trevor Lipscombe, 'Two Continents or One? The Baudin Expedition's Unacknowledged

Achievements on the Coast of Victoria', *Victorian Historical Journal*, 78 (2007), pp.23–41.

7. Matthew Flinders, *A Voyage to Terra Autralis . . .* (2 vols, 1814), I, Introduction.
8. David Mackay, *In the Wake of Cook: Exploration, Science and Empire, 1780–1801* (1985), p.5.
9. See Nigel Rigby, '"Not at all a particular ship": Adapting Vessels for British Voyages of Exploration, 1768–1801', in Juliet Wege et al., eds *Matthew Flinders and his Scientific Gentlemen* (Welshpool, WA, 2005), p.21.
10. Carter, *Joseph Banks*, p.415.
11. John White, *Journal of a Voyage to New South Wales* (1790), title-page.
12. Quoted in Bernard Smith, *European Vision and the South Pacific* (2nd edn, New Haven, 1985), p.168.
13. Ibid., p.167.
14. Carter, *Joseph Banks*, p.415.
15. Quoted by Miriam Estensen, 'Matthew Flinders: The Man and his Life', in Wege et al., *Flinders and his Scientific Gentlemen*, p.3.
16. Brunton, ed., *Flinders Letters*, p.39.
17. Nicolas Baudin, *The Journal of Post Captain Nicolas Baudin*, trans. Christine Cornell (Adelaide, 1974), pp.586–90. To distinguish this 'journal de mer' or sea-log from Baudin's incomplete *Journal personnel*, it is referred to in the text as Baudin's log but in the notes, following Cornell's title, as *Baudin Journal*.
18. 'Plan of Itinerary for Citizen Baudin', in ibid., pp.5, 8.
19. The recommendations – in the form of questionnaires – for natural history were surprisingly brief: four questions on zoology and five on botany. See Bonnemains et al., *Baudin journal personnel*, p.50.
20. For a summary of Cuvier's importance, see Philippe Taquet, 'Georges Cuvier: Extinction and the Animal Kingdom', in Robert Huxley, ed., *The Great Naturalists* (2007), pp.202–11.
21. See Joseph-Marie Dégerando, *The Observation of Savage Peoples*, trans. F.C.T. Moore (1969).
22. Quoted in Jacqueline Bonnemains, 'The Artists of the Baudin Expedition', in Sarah Thomas, ed., *The Encounter, 1802. Art of the Flinders and Baudin Voyages* (Adelaide, 2002), p.130. It is assumed that Baudin intended this journal with its illustrations to form the basis of his published account of the voyage, but it finishes at the end of 1801 after the expedition's first visit to Timor. See Margaret Sankey, 'Writing the Voyage of Scientific Exploration: The Logbooks, Journals and Notes of the Baudin Expedition (1800–1804)', *Intellectual History Review*, 20 (2010), p.407. The journal was finally published in 2001.
23. Cornell, trans., *Baudin Journal*, p.21; Bonnemains et al., *Baudin journal personnel*, p.154.
24. For examples, see Ralph Kingston, 'A Not So Pacific Voyage: The "Floating Laboratory" of Nicolas Baudin', *Endeavour*, XXXI (2007), pp.145–51.
25. Cornell, trans., *Baudin Journal*, p.107.
26. For details of the expedition's uncomfortable stay at Isle de France, see ibid., pp.121–36.
27. See Kingston, 'A Not So Pacific Voyage', p.148.
28. Cornell, trans., *Baudin Journal*, p.158.
29. François Péron, *Voyage of Discovery to the Southern Lands*, Books I–III (2nd edn, Paris, 1824), trans. Christine Cornell (Adelaide, 2006), p.56.
30. Edward Duyker, *François Péron: An Impetuous Life* (Melbourne, 2006), pp.52, 105.
31. Jacqueline Bonnemains, Elliott Forsyth and Bernard Smith, *Baudin in Australian Waters: The Artwork of the French Voyage of Discovery to the Southern Lands 1800–1804* (Melbourne, 1988), p.31.
32. Cornell, trans., *Baudin Journal*, p.209. See p.170 above
33. The references in this paragraph are to ibid., pp.200, 217, 239.

34. Péron, *Voyage*, Books I–III, p.102. For a more critical assessment of this incident, see Duyker, *Péron*, pp.91–3.
35. Bonnemains et al., *Baudin journal personnel*, pp.324–5, 328–30, 338–9, 345, 349.
36. Cornell, trans., *Baudin Journal*, p.270.
37. Ibid., pp.313, 315.
38. Ibid., 340.
39. Horner, *French Reconnaissance*, p.230n.
40. Cornell, trans., *Baudin Journal*, pp.344–7, 349; Duyker, *Péron*, p.108.
41. Péron, *Voyage*, Books I–III, p.224. For this incident, see N.J.B. Plomley, *The Baudin Expedition and the Tasmanian Aborigines* (Hobart, 1983), p.195, and, for Baudin's stay generally, Jean Fornasiero, Peter Monteath and John West-Sooby, *Encountering Terra Australis: The Australian Voyages of Nicolas Baudin and Matthew Flinders* (Kent Town, South Australia, 2010), chs 4, 18.
42. See Fornasiero et al., *Encountering Terra Australis*, pp.321–4.
43. Cornell, trans., *Baudin Journal*, p.379.
44. Flinders, *Voyage to Terra Australis*, I, p.103.
45. Cornell, trans., *Baudin Journal*, p.393.
46. For this and other reports of French designs on Port Jackson, see Anthony J. Brown, *Ill-Starred Captains: Flinders and Baudin* (2001), pp.262–8, 274–8.
47. Cornell, trans., *Baudin Journal*, p.430.
48. Ibid., p.442.
49. Duyker, *Péron*, p.156.
50. The above statements by Baudin are in Cornell, trans., *Baudin Journal*, pp.473, 492, 491, 499, 509–11.
51. François Péron, *Voyage of Discovery to the Southern Lands*, Book IV (2nd edn, Paris, 1824), continued by Louis de Freycinet, trans. Christine Cornell (Adelaide, 2003), p.85.
52. See p.27 above.
53. Péron and Freycinet, *Voyage of Discovery*, Book IV pp.151–3.
54. Cornell, trans., *Baudin Journal*, p.525.
55. Ibid., p.560.
56. The words of René Bouvier and E. Maynial, cited in Brown, *Ill-Starred Captains*, p.365.
57. Ibid., p.369.
58. Péron and Freycinet, *Voyage of Discovery*, Book IV, p.14.
59. Formasiero et al., *Encountering Terra Australis*, p.326.
60. François Péron, *Voyage of Discovery to the Southern Lands* (2nd edn, Paris, 1824), *Dissertations on Various Subjects*, trans. Christine Cornell (Adelaide, 2007).
61. Ibid., p.4.
62. Péron, *Voyage of Discovery*, Books I–III, p.xxiv.
63. Brown, *Ill-Starred Captains*, p.421.
64. Geoffrey C. Ingleton, *Matthew Flinders: Navigator and Chartmaker* (Sydney, 1986), p.291.
65. For a summary of recent research on this subject, see Jean Fornasiero and John West-Sooby, 'Naming and Shaming: The Baudin Expedition and the Politics of Nomenclature in the *Terres Australes*', in Anne M. Scott et al., eds, *European Perceptions of Terra Australis* (Farnham, Surrey, 2011), pp.165–84.
66. Ibid., p.178n.
67. Péron, *Voyage*, Books I–III, p.251.
68. Quoted in Anthony J. Brown's Introduction to ibid., p.xxxvi.
69. Quoted in Kingston, 'A Not So Pacific Voyage', p.150.
70. See p.177 above.
71. See Fornasiero et al., *Encountering Terra Australis*, p.348.
72. Horner, *French Reconnaissance*, p.324.
73. Lipscombe, 'Two Continents or One?', p.38.

74. The words are Anthony J. Brown's in his Introduction to Péron, *Voyage*, Books I–III, p.xxix; in South Australia the Baudin Rocks (named by Flinders) lie off the coast at Robe, and there is a Baudin Beach on Kangaroo Island.
75. Brown, *Ill-Starred Captains*, p.159.
76. Edwards, 'Peter Good Journal', p.21.
77. Ibid., p.24.
78. Brown, *Nature's Investigator*, p.102.
79. Ibid., p.127.
80. Flinders, *Voyage to Terra Australis*, I, p.122.
81. Brown, *Ill-Starred Captains*, p.183.
82. T.G. Vallance, D.T. Moore and E.W. Groves, eds, *Nature's Investigator: The Diary of Robert Brown in Australia, 1801–1805* (Canberra, 2001), p.179. [Hereafter Brown, *Nature's Investigator*.]
83. Flinders, *Voyage to Terra Australis*, I, p.220.
84. Brown, *Nature's Investigator*, p.206.
85. Ibid., p.230.
86. See p.213 above.
87. Nigel Rigby, Pieter van der Merwe and Glyn Williams, *Pioneers of the Pacific: Voyages of Exploration, 1787–1810* (2005), p.125.
88. Brown, *Nature's Investigator*, p.205.
89. Brunton, ed., *Flinders Letters*, p.82.
90. The references in this paragraph are to Flinders, *Voyage to Terra Australis*, II, pp.88 (23 November 1802), 189.
91. Quoted by Janette Gathe and Ellen Hickman, 'Ferdinand Bauer: Natural History Artist', in Wege et al., *Flinders and his Scientific Gentlemen*, p.70.
92. Brunton, ed., *Flinders Letters*, p.100.
93. Flinders, *Voyage to Terra Australis*, II, p.322.
94. Ibid., p.328.
95. For details of Flinders's detention, see his record of events in Anthony J. Brown and Gillian Dooley, eds, *Matthew Flinders: Private Journal* (Adelaide, 2005); also Huguette Ly-Tio-Fane Pineo, *In the Grip of the Eagle: Matthew Flinders at Ile de France 1803–1810* (Moka, Mauritius, 1988).
96. Brown, *Ill-Starred Captains*, p.405.
97. Brown, *Nature's Investigator*, p.438.
98. Ibid., pp.418, 420, 423. For Brown's dissatisfaction with the facilities available to him for transporting plants, see Nigel Rigby, 'The Politics and Pragmatics of Seaborne Plant Transportation, 1769–1805', in Margarette Lincoln, ed., *Science and Exploration in the Pacific: European Voyages to Southern Oceans in the Eighteenth Century* (Woodbridge, Suffolk, 1998), pp.88, 96–7.
99. Brown, *Nature's Investigator*, pp.589–90,
100. Brown, *Ill-Starred Captains*, pp.437–8.
101. 'Robert Brown', in the *Oxford Dictionary of National Biography*, 8 (2004), p.108.
102. See D.J. Mabberley, *Jupiter Botanicus: Robert Brown of the British Museum* (1985); also Ann Moyal, 'The Scientific Legacy', in Thomas, *Encounter, 1802*, pp.186–97.
103. Quoted in Smith, *European Vision and the South Pacific*, p.217.
104. See Eric W. Groves, 'Procrastination or Unpredictable Circumstances? The Handling of Robert Brown's Australian Plant Collection(s) in London', in Wege et al., *Flinders and his Scientific Gentlemen*, pp.129–41.
105. See Victoria Ibáñez and H.W. Lack, 'The Early Colour Code of the Bauer Brothers', *Curtis Botanical Magazine* (1995); also M.J. Norst, *Ferdinand Bauer: The Australian Natural History Drawings* (Melboune, 1989).
106. Gathe and Hickman, 'Ferdinand Bauer', in ibid., p.74.
107. Brunton, ed., *Flinders Letters*, p.116.
108. Flinders, *Voyage to Terra Australis*, I, p.ix.

109. Ibid., p.iii. See also Flinders's letter of 17 August 1813 to Banks on the subject: Brunton, ed., *Flinders Letters*, pp.233–5. George Shaw in his *Zoology of New Holland* (1794), I, p.2 seems to have been the first to use the modern name when he wrote of 'This vast Island or rather Continent of Australia, Australasia, or New Holland'.

Chapter 10   Charles Darwin on the *Beagle*

1. For much of the information in this paragraph I am indebted to Randolph Cock. See especially his 'Scientific Sailors and Sailing Scientists: The Pursuit of a Scientific Career in the Royal Navy, 1815–1855', paper read at the International Commission for Maritime History, King's College, London, 4 May 2000; and 'Sir Francis Beaufort and the Co-ordination of British Scientific Activity, 1829–1855', Cambridge PhD thesis, 2002.

2. Iain McCalman, *Darwin's Armada* (2009), p.93.

3. Richard Drayton, *Nature's Government: Science, Imperial Britain, and the 'Improvement' of the World* (New Haven and London, 2000), p.143.

4. As cited by Cock – Hydrographic Office LB13, pp.72–4, 25 June 1845.

5. Louis de Freycinet, *Voyage autour du monde* (5 vols, Paris, 1827–39); see also John Dunmore, *French Explorers in the Pacific: The Nineteenth Century* (Oxford, 1969), pp.63–108.

6. See J.P. Faivre, *L'Expansion française dans le Pacifique de 1800 à 1842* (Paris, 1953).

7. Dunmore, *French Explorers*, p.341.

8. See Simon Werrett, 'Russian Responses to the Voyages of Captain Cook', in Glyn Williams, ed., *Captain Cook: Explorations and Reassessments* (Woodbridge, Suffolk, 2004), pp.184–5.

9. Ibid., p.195.

10. Georg Heinrich von Langsdorff, *Remarks and Observations on a Voyage around the World from 1803 to 1807*, trans. Victoria Joan Moessner, ed. Richard A. Pierce (2 vols, Fairbanks, AK, 1993), I, p.47, II, p.126.

11. Löwenstern's remarks are taken from *The First Russian Voyage around the World: The Journal of Hermann Ludwig von Löwenstern (1803–1806)*, trans. Victoria Joan Moessner (Fairbanks, AK, 2003), pp.39, 90, 13, 28. The reference to 'doctors' in the second quotation is a reminder that Tilesius and Langsdorff, like many shipboard naturalists of the period, had medical training.

12. Glynn Barratt, *Russia in Pacific Waters 1715–1825* (Vancouver, 1981), p.177.

13. Glynn Barratt, *Russia and the South Pacific 1696–1840*, IV, *The Tuamotu Islands and Tahiti* (Vancouver, 1992), pp.3, 5.

14. Frederic Litke, *A Voyage around the World 1826–1829*, I, *To Russian America and Siberia*, ed. Richard A. Pierce (Kingston, Ont., 1987), p.i.

15. Adelbert von Chamisso, *A Voyage around the World . . . In the Years 1815–1818*, ed. and trans. Henry Kratz (Honolulu, 1986), p.8.

16. Ibid., p.23.

17. Ibid., p.21.

18. Ibid., pp.34, 35, 113. It is true that in his mid-thirties Chamisso was one of the oldest men in an exceptionally young ship's company – Kotzebue was only twenty-eight at the outset of the voyage.

19. Ibid., pp.221, 23.

20. Cited by Cock, 'Scientific Sailors and Sailing Scientists', p.14.

21. Both quotations in letter from Henslow to Darwin, 24 August 1831, in Frederick Burkhardt, ed., *Charles Darwin: The Beagle Letters* (Cambridge, 2008), p.19.

22. Ibid., p.30, Darwin to Susan Darwin, 5 September 1831.

23. Ibid., p.32, Darwin to Henslow, 5 September 1831.

24. Ibid., p.66, Darwin to Henslow, 30 October 1831.

25. Darwin Diary, p.8. The manuscript diary, kept at Darwin's home at Down House, Kent, has been made available online at www.darwin-online.org.uk and is cited here. Although it is invaluable as a guide to Darwin's thoughts and actions on the voyage, internal evidence shows that it was not always an immediate record; in places, entries were

written two or three months after the events they describe. After Darwin's return to England the diary formed the basis, in edited form, of his published *Journal and Remarks* (1839).

26. Darwin's Diary, p.13.
27. James Taylor, *The Voyage of the Beagle: Darwin's Extraordinary Adventure aboard Fitzroy's Famous Survey Ship* (2008), p.29.
28. Ibid., p.22.
29. Burkhardt, ed., *Beagle Letters*, p.92.
30. It was then republished as a separate volume as *Journal of Researches into the Geology and Natural History of the Various Countries Visited by H.M.S. Beagle*, best known under its shortened title, *Voyage of the Beagle*.
31. *Narrative of the Surveying Voyages of His Majesty's Ships Adventure and Beagle between the Years 1826 and 1836* (3 vols, 1839): III, Charles Darwin, *Journal and Remarks 1832–1836*, p.599.
32. Darwin's Diary, p.23.
33. Darwin, *Journal and Remarks*, pp.6–7. The octopus in its glass container is still preserved in the storerooms of the Cambridge Zoology Museum. See Rebecca Stott, *Darwin and the Barnacle* (2003), pp.255–7.
34. Darwin Diary, p.49. Bartholomew Sulivan was second lieutenant of the *Beagle*.
35. Taylor, *Voyage of the Beagle*, p.73.
36. Burkhardt, ed., *Beagle Letters*, pp.116–17, Darwin to Caroline Darwin, 25–26 April 1832.
37. Quotations from Richard Keynes, *Fossils, Finches and Fuegians* (2002), pp.161, 245.
38. Burkhardt, ed., *Beagle Letters*, pp.116–17, Darwin to Caroline Darwin, 25–26 April 1832.
39. Ibid., p.122, Darwin to Fox, May 1832; p.142, Darwin to Henslow, July–August 1832; p.169, Darwin to Caroline Darwin, October–November 1832; p.271, Darwin to Catherine Darwin, 6 April 1834.
40. Ibid., p.122, Darwin to Fox, May 1832.
41. Ibid., pp.128, 129, Darwin to Henslow, May–June 1832.
42. Quotations from Michael Dettelbach, 'Global Physics and Aesthetic Empire: Humboldt's Physical Portrait of the Tropics', in David Philip Miller and Peter Hanns Reill, eds, *Visions of Empire: Voyages, Botany and Representations of Nature* (Cambridge, 1996), pp.260, 267.
43. Alexander von Humboldt, *Personal Narrative of Travels to the Equinoctial Regions of America*, trans. Jason Wilson (1995).
44. The two comments on Lyell are from Janet Browne, *Charles Darwin: Voyaging* (1995), p.189.
45. Burkhardt, ed., *Beagle Letters*, p.172, Darwin to Henslow, October–November 1832.
46. Darwin Diary, p.132.
47. Ibid., p.125.
48. Burkhardt, ed., *Beagle Letters*, pp.202, 204, Darwin to Catherine Darwin, 22 May 1833; also p.213, Darwin to Henslow, 18 July 1833.
49. Darwin Diary, p.152.
50. Ibid., p.168.
51. Ibid., p.261.
52. Ibid., pp.249–50.
53. Darwin, *Journal and Remarks*, p.314.
54. Ibid., p.309.
55. Ibid., pp.406–7.
56. Burkhardt, ed., *Beagle Letters*, p.328, Darwin to Henslow, March 1835; p.331, Darwin to Henslow, 18 April 1835.
57. Ibid., p.337, Darwin to Susan Darwin, 23 April 1835.
58. Ibid., p.323, Darwin to Fox, March 1835.
59. Ibid., p.341, Darwin to Catherine Darwin, 31 May 1835.
60. Ibid., p.350, Darwin to Caroline Darwin, July 1835.
61. Ibid., pp.376–7, Darwin to Henslow, January 1836.

62. Keynes, *Fossils, Finches and Fuegians*, p.316.
63. See ibid., p.372, and Peter R. Grant, *Ecology and the Evolution of Darwin's Finches* (Princeton, NJ, 1986).
64. Stephen D. Hopper and Hans Lambers, 'Darwin as a Plant Scientist: A Southern Hemisphere Perspective', *Trends in Plant Science*, 14 (2008), p.427.
65. Darwin Diary, p.694.
66. For a succinct discussion, see 'Islands on his Mind', in McCalman, *Darwin's Armada*, pp.60–81.
67. Keynes, *Fossils, Finches and Fuegians*, p.361.
68. Burkhardt, ed., *Beagle Letters*, pp.386, 387, Darwin to Caroline Darwin, 29 April 1836.
69. Ibid., p.383, Darwin to Fox, 15 February 1836.
70. Darwin Diary, p.447.
71. Browne, *Darwin: Voyaging*, p.358.
72. Darwin, *Journal and Remarks*, pp.454–5.
73. Ibid., pp.461–2.
74. Ibid., pp.474–5.
75. Browne, *Darwin: Voyaging*, p.452.
76. Darwin, *Journal and Researches* (2nd edn, 1845), p.380.
77. Grant, *Ecology and Evolution of Darwin's Finches*, p.8.
78. Richard Keynes, ed., *Charles Darwin's Zoology Notes and Specimen Lists from HMS Beagle* (Cambridge, 2000).
79. 'Arthro' (jointed) and 'balanus' (barnacle) as explained in Stott, *Darwin and the Barnacles*, p.72.
80. Cited in Rebecca Stott, *Darwin's Ghosts: In Search of the First Evolutionists* (2012), p.245.
81. See ibid. for a comprehensive account of Darwin's intellectual predecesors, and pp.299–308 for the text of Darwin's 'Historical Sketch of the Recent Progress of Opinion on *The Origin of Species*' from the book's fourth edition of 1866.
82. The book's full title, even then shorter than Darwin's original choice, was *On the Origin of Species by Means of Natural Selection, or The Preservation of Favoured Races in the Struggle for Life*.
83. Fitzroy, *Narrative*, p.668.
84. Taylor, *Voyage of the Beagle*, p.185.
85. Ibid., pp.177–8.
86. Janet Browne, *Charles Darwin: The Power of Place* (2002), p.94.
87. Ibid., p.123.
88. Quoted in Peter Nichols, *Evolution's Captain* (2003), p.328.
89. Browne, *Darwin: Power of Place*, p.265.
90. Ibid., p.399; Francis Darwin, ed., *The Life and Letters of Charles Darwin* (1887), I, p.61.

Conclusion

1. The quotations from Huxley and Chamisso that follow are taken from Huxley's essay 'Science at Sea', *Westminster Review*, n.s., 61 (1854), pp.98–119.
2. Anthony Pagden, *European Encounters with the New World* (New Haven and London, 1993), p.3.
3. Francis Darwin, ed., *The Life and Letters of Charles Darwin* (1887), I, p.336.
4. Alexander von Humboldt, *Personal Narrative of a Journey to the Equinoctial Regions of the New Continent*, trans. Jason Wilson (1995), p.8.
5. Charles Darwin, *Journal and Remarks 1832–1836*, Vol.III of *Narrative of the Surveying Voyages of His Majesty's Ships Adventure and Beagle* (1839), p.607.

SELECT BIBLIOGRAPHY

This bibliography is limited to works cited in the book's notes. Place of publication is London unless otherwise stated.

A. Journals, Reports, Correspondence

Anon., *Account of Several Late Voyages & Discoveries*, with introduction by Tancred Robinson (1694).

Ayres, Philip, *The Voyages and Adventures of Captain Barth. Sharp and Others in the South Sea* (1684).

Baudin, Nicolas, *The Journal of Post Captain Nicolas Baudin*, trans. Christine Cornell (Adelaide, 1974).

Beaglehole, J.C., ed., *The Journals of Captain James Cook on his Voyages of Discovery: The Voyage of the Endeavour 1768–1771* (Cambridge, 1955); *The Voyage of the Resolution and Adventure 1772–1775* (Cambridge, 1961); *The Voyage of the Resolution and Discovery 1776–1780* (2 vols, Cambridge, 1967).

Beaglehole, J.C., ed., *The Endeavour Journal of Joseph Banks 1768–1771* (Sydney, 1962).

Black, John and Oscar Clemotte-Silvero, trans. and eds, *Meditación sobre lo bello en la Naturaleza* (Lewiston, NY, 2007).

Bonnemains, Jacqueline, Jean-Marc Argentin and Martine Marin, eds, *Mon voyage aux Terres Australes: journal personnel du commandant Baudin* (Le Havre and Paris, 2001).

Bougainville, Lewis [sic] Antoine de, trans. J.R. Forster, *A Voyage round the World* (1772).

Brunton, Paul, ed., *Matthew Flinders: Personal Letters from an Extraordinary Life* (Sydney, 2002).

Bruny d'Entrecasteaux, A.R.J., *Voyage to Australia and the Pacific, 1791–1793*, ed. and trans. Edward Duyker and Maryse Duyker (Melbourne, 2001).

Burkhardt, Frederick, ed., *Charles Darwin: The Beagle Letters* (Cambridge, 2008).

Chambers, Neil, ed., *The Indian and Pacific Correspondence of Sir Joseph Banks, 1768–1820*: Vol. I (2008); Vol. II (2009); Vol. III (2010); Vol. IV (2011).

Chamisso, Adelbert von, *A Voyage around the World . . . In the Years 1815 to 1818*, ed. and trans. Henry Kratz (Honolulu, 1986).

Collins, David, *An Account of the English Colony in New South Wales* [1798] (Sydney, 1975).

Cook, James and James King, *A Voyage to the Pacific Ocean . . .* (3 vols, 1784).

D'Entrecasteaux, see Bruny d'Entrecasteaux, A.R.J.

Dampier, William, *A New Voyage round the World* [1697], with introduction by Albert Gray (1937).

Dampier, William, *A Voyage to New-Holland &c. In the Year 1699* (2 vols, 1703, 1709).

Dampier, William, *Captain Dampier's Vindication* [1707], printed in Masefield, ed., *Dampier's Voyages*, II.

Dampier, William, *Voyages and Discoveries*, II, *Voyages to Campeachy* (1699).

Darwin, Charles, *Journal and Remarks 1832–1836*, Vol. III of *Narrative of the Surveying Voyages of His Majesty's Ships Adventure and Beagle between the Years 1826 and 1836* (1839).

Darwin, Charles, *On the Origin of Species by Means of Natural Selection, or The Preservation of Favoured Races in the Struggle for Life* (1859).

Darwin, Francis, *The Life and Letters of Charles Darwin* (3 vols, 1887).

David, Andrew, Felipe Fernández Armesto, Carlos Novi and Glyn Williams, eds, *The Malaspina Expedition 1789–1794: Journal of the Voyage by Alejandro Malaspina* (3 vols, London and Madrid, 2001–4).

Dillon, Peter, *Narrative and Successful Result of a Voyage to the South Seas . . . to Ascertain the Actual Fate of La Pérouse's Expedition* (2 vols, 1829).

Driver, Marjorie G., ed., *The Guam Diary of Naturalist Antonio de Pineda y Ramirez 1792* (Mangilao, Guam, 1990).

Dunmore, John, ed. and trans., *The Journal of Jean-François de Galaup de la Pérouse* (2 vols, 1994).

Dunmore, John, ed. and trans., *The Pacific Journal of Louis-Antoine de Bougainville 1767–1768* (2002).

Edwards, Phyllis I., ed., 'The Journal of Peter Good', *Bulletin of the British Museum (Natural History)*, Historical Studies, 9 (1981).

Ellis, John, *Essay towards a Natural History of the Corallines* (1755).

Estrella, Eduardo, ed., *La Expedición Malaspina 1789–1794*, VIII, *Trabajos . . . de Antonio Pineda Ramírez* (Madrid, 1996).

Etches, John, *An Authentic Statement of All the Facts Relative to Nootka Sound* (1790).

Fidlon, Paul G. and R.J. Ryan, eds, *The Journal of Philip Gidley King: Lieutenant R.N. 1787–1790* (Sydney, 1980).

Fisher, Raymond H., ed., *The Voyage of Semen Dezhnev in 1648* (1981).

Flinders, Matthew, *A Voyage to Terra Australis . . .* (2 vols, 1814).

Forster, George, *Cook the Discoverer* [German original, Berlin, 1787] (Sydney, 2007).

Freycinet, Louis de, *Voyage autour du monde . . .* (5 vols, Paris, 1827–39).

Frézier, A-F., *A Voyage to the South Sea and along the Coasts of Chili and Peru . . .* (1717).

Funnell, William, *A Voyage round the World* (1707).

Gallagher, Robert E., ed., *Byron's Journal of his Circumnavigation 1764–1766* (1964).

Galois, Robert, ed., *A Voyage to the North West Side of America: The Journals of James Colnett 1786–89* (Vancouver, 2004).

Golder, Frank A., ed., *Bering's Voyages* (2 vols, New York, 1922).

Hammond, L. Davis, *News from New Cythera: A Report of Bougainville's Voyage 1766–1769* (Minneapolis, MN, 1970).

Harley, Robert, *Letters and Papers*, III, in *Manuscripts of the Duke of Portland*, V (Historical Manuscripts Commission, 1899).

Higueras Rodríguez, María Dolores, ed., *NW Coast of America: Iconographical Album of the Malaspina Expedition* (Madrid, 1991).

Hoare, Michael E., ed., *The Resolution Journal of Johann Reinhold Forster 1772–1775* (1982).

Huxley, T.H., 'Science at Sea', *Westminster Review*, n.s., 61 (1854), pp.98–119.

Ibáñez Montoya, María Victoria, ed., *La Expedición Malaspina 1789–1794*, IV, *Trabajos científicos y correspondencia de Tadeo Haenke* (Madrid, 1994).

Joppien, Rüdiger and Bernard Smith, eds, *The Art of Captain Cook's Voyages*, III, *The Voyage of the Resolution and Discovery 1776–1780* (New Haven, 1988).

Keynes, Richard, ed., *Charles Darwin's Zoology Notes and Specimens from HMS Beagle* (Cambridge, 2000).

King, Robert J., ed., *The Secret History of the Convict Colony: Alejandro Malaspina's Report on the British Settlement of New South Wales* (Sydney, 1990).

Krusenstern, Johann von, *A Voyage round the World* . . . (1813).

Labillardière, J.J.H. de, *Account of a Voyage in Search of La Pérouse* (2 vols, 1800).

Lamb, W. Kaye, ed., *George Vancouver: A Voyage of Discovery to the North Pacific Ocean and round the World 1791–1795* (4 vols, 1984).

Langsdorff, George Heinrich von, *Remarks and Observations on a Voyage around the World*, trans. Victoria Joan Moessner, ed. Richard A. Pierce (Kingston, Ont., 1993).

Litke, Frederic, *A Voyage around the World 1826–1829*, I, *To Russian America and Siberia.*, ed. Richard A. Pierce (Kingston, Ont., 1987).

Löwenstern, Hermann Ludwig von, *The First Russian Voyage around the World*, trans. Victoria Joan Moessner (Fairbanks, AK, 2003).

Lysaght, A.M., ed., *Joseph Banks in Newfoundland and Labrador, 1766* (1971).

Masefield, John, ed., *Dampier's Voyages* (2 vols, 1906).

Milet-Mureau, Louis-Marie-Antoine, *Voyage de La Pérouse autour du monde* (4 vols and atlas, Paris, 1797).

Muñoz Garmendia, Felix, ed., *La Expedición Malaspina 1789–1794*, III, *Diarios y trabajos botánicos de Luis Neé* (Mardid, 1992).

Parkinson, Sydney, *A Journal of a Voyage to the South Seas* (1773).

Péron, François, *Voyage of Discovery to the Southern Lands*, Book IV, continued by Louis de Freycinet, trans. Christine Cornell (Adelaide, 2003).

Péron, François, *Voyage of Discovery to the Southern Lands*, Books I–III, trans. Christine Cornell (Adelaide, 2006).

Péron, François, *Voyage of Discovery to the Southern Lands: Dissertations on Various Subjects*, trans. Christine Cornell (Adelaide, 2007).

Ray, John, *Historiae plantarum* (1704).

[Ringrose, Basil], *Bucaniers of America* . . . (1685).

Rogers, Woodes, *A Cruising Voyage round the World* [1712], with introduction by G.E. Mainwaring (1928).

Sloane, Hans, *A Voyage to the Islands Madeira, Barbados, Nieves. S. Christophers and Jamaica, with the Natural History* (2 vols, 1707, 1725).

Sparrman, Anders, *A Voyage to the Cape of Good Hope* . . . *and round the World* (2 vols, 1785).

Sparrman, Anders, *A Voyage around the World* [Stockholm, 1802], trans. Eivor Cormack, in *The Linnaeus Apostles*, 5 (2007).

Steller, Georg Wilhelm, *Journal of a Voyage with Bering 1741–1742*, ed. O.W. Frost, trans. Margrit A. Engel and O.W. Frost (Stanford, 1988).

Swift, Jonathan, *Gulliver's Travels*, ed. Peter Dixon and John Chalker (Harmondsworth, 1967).

Taillemite, Etienne, ed., *Bougainville et ses compagnons autour du monde* (2 vols, Paris, 1977).

Thomas, Nicholas, Harriet Guest and Michael Dettelbach, eds, *Observations Made during a Voyage round the World* [by Johann Reinhold Forster] (Honolulu, 1996).

Thomas, Nicholas, Oliver Berghof and Jennifer Newell, eds, *George Forster, A Voyage round the World* (2 vols, Honolulu, 2000).

Urness, Carol, ed. and trans., *The Reports from Russia by Gerhard Friedrich Müller* (Fairbanks, AK, 1986).

Vallance, T.G., D.T. Moore and E.W. Groves, eds, *Nature's Investigator: The Diary of Robert Brown in Australia 1801–1805* (Canberra, 2001).

Waxell, Sven, *The American Expedition*, trans. M.A. Michael (Edinburgh, 1952).

Welbe, John, *An Answer to Captain Dampier's Vindication* (1708).

White, John, *Journal of a Voyage to New South Wales* [1790] (Sydney, 1962).

Woodward, John, *Brief Instructions for Making Observations in All Parts of the World in Order to Promote Natural History* (1696).

B. Secondary Works

Adams, Brian, *The Flowering of the Pacific* (1986).

Badger, Geoffrey, *The Explorers of the Pacific* (Kenthurst, NSW, 1988).

Barratt, Glynn, *Russia in Pacific Waters 1715–1825* (Vancouver, 1981).
Barratt, Glynn, *Russia and the South Pacific 1696–1840*, III, *Melanesia and the Western Polynesian Fringe* (Vancouver, 1990).
Beaglehole, J.C., *The Life of Captain James Cook* (1974).
Black, Lydia, *Russians in Alaska 1732–1867* (Fairbanks, AK, 2004).
Blunt, Wilfrid, *Linnaeus: The Compleat Naturalist* (2004).
Brown, Anthony J., *Ill-Starred Captains: Flinders and Baudin* (2001).
Browne, Janet, *Charles Darwin: Voyaging* (1995).
Browne, Janet, *Charles Darwin: The Power of Place* (2002).
Carpenter, Kenneth J., *The History of Scurvy and Vitamin C* (Cambridge, 1986).
Carter, Harold B., *Sir Joseph Banks 1743–1820* (1988).
Conniff, Richard, *The Species Seekers* (New York, 2011).
Cutter, Donald C., *Malaspina and Galiano: Spanish Voyages to the Northwest Coast 1791 and 1792* (Vancouver, 1992).
Dance, S.P., *Shell Collectors: An Illustrated History* (Berkeley, CA, 1966).
Davies, Martin, *The Conjuror's Bird* (2005).
Divin, Vasilii A., *The Great Russian Navigator, A.I. Chirikov*, trans. and annotated by Raymond H. Fisher (Fairbanks, AK, 1993).
Drayton, Richard, *Nature's Government: Science, Imperial Britain, and the 'Improvement' of the World* (New Haven and London, 2000).
Druett, Joan, *Tupaia: The Remarkable Story of Captain Cook's Polynesian Navigator* (Auckland, 2011).
Dunmore, John, *French Explorers in the Pacific: The Nineteenth Century* (Oxford, 1969).
Dunmore, John, *Storms and Dreams: The Life of Louis de Bougainville* (Fairbanks, AK, 2007).
Dunmore, John, *Where Fate Beckons: The Life of Jean-François de la Pérouse* (Fairbanks, AK, 2007).
Duyker, Edward, *Nature's Argonaut: Daniel Solander 1733–1782* (Melbourne, 1988).
Duyker, Edward, *Citizen Labillardière: A Naturalist's Life in Revolution and Exploration* (Melbourne, 2003).
Duyker, Edward, *François Péron: An Impetuous Life* (Melbourne, 2006).
Duyker, Edward, *Père Receveur: Franciscan, Scientist and Voyager with La Pérouse* (Sydney, 2011).
Earle, Peter, *The World of Defoe* (1976).
Eisler, William, *The Furthest Shore: Images of Terra Australis from the Middle Ages to Captain Cook* (Cambridge, 1995).
Engstrand, Iris H.W., *Spanish Scientists in the New World: The Eighteenth-Century Expeditions* (Seattle, 1981).
Faivre, J.P., *L'Expansion française dans le Pacifique de 1800 à 1842* (Paris, 1953).
Ford, Corey, *Where the Sea Breaks its Back* (Portland, OR, 1992).
Formasiero, Jean, Peter Monteath and John West-Sooby, *Encountering Terra Australis: The Australian Voyages of Nicolas Baudin and Matthew Flinders* (Kent Town, South Australia, 2010).
Frost, Alan, *Sir Joseph Banks and the Transfer of Plants to and from the South Pacific 1786–1798* (Melbourne, 1993).
Frost, O.W., *Bering and Chirikov: The American Voyages and their Impact* (Anchorage, 1992).
Frost, O.W., *Bering: The Russian Discovery of America* (New Haven and London, 2003).
Gaziello, Catherine, *L'Expédition de Lapérouse: réplique française aux voyages de Cook* (Paris, 1984).
Gascoigne, John, *Joseph Banks and the English Enlightenment: Useful Knowledge and Polite Culture* (Cambridge, 1994).
George, Alex S., *William Dampier in Australia: Australia's First Natural Historian* (Hawthorn, VIC, 1999).
Gill, Anton, *The Devil's Mariner: A Life of William Dampier, Pirate and Explorer* (1997).
Grant, Peter R., *Ecology and the Evolution of Darwin's Finches* (Princeton, NJ, 1986).
Greenblatt, Stephen, *Marvelous Possessions: The Wonder of the New World* (Chicago, 1991).

Herder, J.G., *Outlines of a Philosophy of the History of Man*, trans. T. Churchill (1800).
Hoare, Michael E., *The Tactless Philosopher: Johann Reinhold Forster (1729–98)* (Melbourne, 1976).
Hooker, Joseph, *Flora Novae-Zelandae*, I (1855).
Horner, Frank, *The French Reconnaissance: Baudin in Australia 1801–1803* (Melbourne, 1987).
Horner, Frank, *Looking for La Pérouse: D'Entrecasteaux in Australia and the South Pacific 1792–1793* (Melbourne, 1995).
Huxley, Robert, ed., *The Great Naturalists* (2007).
Ingleton, Geoffrey C., *Matthew Flinders: Navigator and Chartmaker* (Sydney, 1986).
Kaeppler, Adrienne L., *Holophusicon: The Leverian Museum. An Eighteenth-Century English Institution of Science, Curiosity and Art* (Altenstadt, 2011).
Kendrick, John, *Alejandro Malaspina: Portrait of a Visionary* (Montreal and Kingston, 1999).
Kenney, John, *Before the First Fleet: Europeans in Australia 1606–1777* (Kenthurst, NSW, 1995).
Keynes, Richard, *Fossils, Finches and Fuegians* (2002).
Koerner, Lisbet, *Linnaeus: Nature and Natives* (Harvard, MA, 1999).
Ly-Tio-Fane, Madeleine, *Pierre Sonnerat 1748–1814* (Mauritius, 1976).
Ly-Tio-Fane Pineo, Huguette, *In the Grip of the Eagle: Matthew Flinders at Ile de France 1803–1810* (Mauritius, 1988).
Mabberley, D.J., *Jupiter Botanicus: Robert Brown of the British Museum* (1985).
McCalman, Iain, *Darwin's Armada* (2009).
Marchant, Leslie, *An Island unto Itself: William Dampier and New Holland* (Carlisle, WA, 1988).
Marchant, Leslie, *France Australe* (Perth, 1998).
Marshall, P.J. and Glyndwr Williams, *The Great Map of Mankind: British Perceptions of the World in the Age of Enlightenment* (1982).
Martin-Allanic, Jean-Etienne, *Bougainville navigateur et les découvertes de son temps* (2 vols, Paris, 1964).
Naish, John N., *The Interwoven Lives of George Vancouver, Archibald Menzies, Joseph Whidbey, and Peter Puget* (Lewiston, 1996).
Nicolas, Peter, *Evolution's Captain* (2003).
Norst, M.J., *Ferdinand Bauer: The Australian Natural History Drawings* (Melbourne, 1989).
Obeyeskere, Gananath, *The Apotheosis of Captain Cook: European Mythmaking in the Pacific* (Princeton, NJ, 1992).
Oliver, S. Passfield, *The Life of Philibert Commerson*, ed. G.F. Scott (1909).
Opatmý, Josef, *La Expedición de Alexjandro Malaspina y Tadeo Haenke* (Prague, 2005).
O'Sullivan, Dan, *In Search of Captain Cook: Explaining the Man through his Own Words* (2008).
Plomley, Brian and Josiane Piard-Bernier, *The General: The Visits of the Expedition Led by Bruny d'Entrecasteaux to Tasmanian Waters in 1792 and 1793* (Launceston, TAS, 1993).
Plomley, N.J.B., *The Baudin Expedition and the Tasmanian Aborigines* (Hobart, 1983).
Pratt, Mary Louise, *Imperial Eyes: Travel Writing and Transculturation* (1992).
Preston, Diana and Michael, *A Pirate of Exquisite Mind: A Life of William Dampier* (1997).
Real Jardín Botánico de Madrid, catalogue, *La Botánica en la expedición Malaspina* (Madrid, 1989).
Ridley, Glynis, *The Discovery of Jeanne Baret* (New York, 2010).
Rigby, Nigel, Pieter van der Merwe and Glyn Williams, *Pioneers of the Pacific: Voyages of Exploration, 1787–1810* (2005).
Salmond, Anne, *Two Worlds: First Meetings between Maori and Europeans 1642–1772* (Auckland, 1991).
Salmond, Anne, *Aphrodites's Island: The European Discovery of Tahiti* (Berkeley, CA, 2010).
Schilder, Günter, *Voyage to the Great South Land: Willem de Vlamingh 1696–1697* (Sydney, 1985).
Shaw, George, *Zoology of New Holland* (2 vols, 1794).

Shipman, Joseph C., *William Dampier: Seaman-Scientist* (Lawrence, KS, 1962).
Smith, Bernard, *European Vision and the South Pacific* (2nd edn, New Haven and London, 1988).
Sotos Serrano, Carmen, *Los Pintores de la expedición de Alejandro Malaspina* (2 vols, Madrid, 1982).
Spate, O.H.K., *The Pacific since Magellan*, II, *Monopolists and Freebooters* (Canberra, 1983).
Spate, O.H.K., *The Pacific Found and Lost* (Rushcutters Bay, NSW, 1988).
Sprat, Thomas, *History of the Royal Society* (1967).
Stejneger, Leonard, *Georg Wilhelm Steller: The Pioneer of Alaskan Natural History* (Cambridge, MA, 1936).
Stott, Rebecca, *Darwin and the Barnacles* (2003).
Stott, Rebecca, *Darwin's Ghosts: In Search of the First Evolutionists* (2012).
Taylor, James, *The Voyage of the Beagle: Darwin's Extraordinary Adventure aboard Fitzroy's Famous Survey Ship* (2008).
Thomas, Nicholas, *Discoveries: The Voyages of Captain Cook* (2003).
Williams, Glyn, *Voyages of Delusion: The Search for the Northwest Passage in the Age of Reason* (2002).
Williams, Roger L., *Botanophilia in Eighteenth-Century France* (Dordrecht, 2001).
Williams, Roger L., *French Botany in the Enlightenment: The Ill-Fated Voyages of La Pérouse and his Rescuers* (Dordrecht, 2003).

C. Essays and Articles

Anton Solé, Pablo, 'Los Padrones de cumplimiento pascual en la expedición Malaspina, 1790–1794', in *La Expedición Malaspina (1789–1794): bicentario de su salida de Cádiz* (Cádiz, 1989), pp.173–238.
Barr, Joel H., 'William Dampier at the Crossroads: New Light on the "Missing Years", 1691–1697', *International Journal of Maritime History*, 8 (1996), pp.97–117.
Bonnemains, Jacqueline, 'The Artists of the Baudin Expedition', in Sarah Thomas, ed., *The Encounter, 1802: Art of the Flinders and Baudin Voyages* (Adelaide, 2002), pp.126–47.
Browne, Janet, 'Botany in the Boudoir and Garden', in David Philip Miller and Peter Hanns Reill, eds, *Visions of Empire: Voyages, Botany and Representations of Nature* (Cambridge, 1996), pp.153–72.
Carr, S.G.M. and D.J., 'A Charmed Life: The Collections of Labillardière', in D.J. and S.G.M. Carr, *People and Plants in Australia* (Sydney, 1981), pp.79–115.
Dettelbach, Michael, 'Global Physics and Aesthetic Empire: Humboldt's Physical Portrait of the Tropics', in David Philip Miller and Peter Hanns Reill, eds, *Visions of Empire: Voyages, Botany and Representations of Nature* (Cambridge, 1996), pp.258–92.
Eastwood, A., ed., 'Menzies California Journal', *Californian Historical Society*, II (1924), pp.265–340.
Estensen, Miriam, 'Matthew Flinders: The Man and his Life', in Juliet Wege et al., *Matthew Flinders and his Scientific Gentlemen* (Welshpool, WA, 2005), pp.1–11.
Estrella, Eduardo, 'La Expedición Malaspina en Guayaquil: Estudios de Historia Natural', in Mercedes Palau Baquero and Antonio Oroxco Acuaviva, eds, *Malaspina '92* (Cádiz, 1994), pp.67–78.
Fornasiero, Jena and John West-Sooby, 'Naming and Shaming: The Baudin Expedition and the Politics of Nomenclature in the *Terres Australes*', in Anne M. Scott et al., eds, *European Perceptions of Terra Australis* (Farnham, Surrey, 2011), pp.165–84.
Fortune, Robert, 'The *St Peter*'s Deadly Voyage Home: Steller, Scurvy and Survival', in O.W. Frost, ed., *Bering and Chirikov: The American Voyages and their Impact* (Anchorage, 1992), pp.204–28.
Gathe, Janette and Ellen Hickman, 'Ferdinand Bauer: Natural History Artist', in Juliet Wege et al., eds *Matthew Flinders and his Scientific Gentlemen* (Welshpool, WA, 2005), pp.67–75.

Gibson, James R., 'Supplying the Kamchatka Expeditions, 1725–30 and 1733–42', in O.W. Frost, ed., *Bering and Chirikov: The American Voyages and their Impact* (Anchorage, 1992), pp.90–116.

Groves, Eric W., 'Archibald Menzies (1754–1842), an Early Botanist on the Northwestern Seaboard of North America . . .', *Archives of Natural History*, 28 (2001), pp.71–122.

Groves, Eric W., 'Procrastination or Unpredictable Circumstances? The Handling of Robert Brown's Australian Plant Collection(s) in London', in Juliet Wege et al., eds *Matthew Flinders and his Scientific Gentlemen* (Welshpool, WA, 2005), pp.129–41.

Higueras Rodriguez, María Dolores, 'The Malaspina Expedition (1789–1794): A Venture of the Spanish Enlightenment', in Carlos Martínez Shaw, ed., *Spanish Pacific from Magellan to Malaspina* (Brisbane, 1988), pp.147–63.

Higueras Rodriguez, María Dolores, 'The Sources: The Malaspina and Bustamante Expedition: A Spanish State Enterprise', in Andrew David et al., eds, *The Malaspina Expedition 1789–1794*, III (2004), pp.371–86.

Hopper, Stephen D. and Hans Lambers, 'Darwin as a Plant Scientist: A Southern Hemisphere Perspective', *Trends in Plant Science*, 14 (2008), pp.421–35.

Ibáñez, Victoria and H.W. Lack, 'The Early Colour Code of the Bauer Brothers', *Curtis Botanical Magazine* (1995).

Ibáñez, Victoria and Robert J. King, 'A Letter from Thaddeus Haenke to Sir Joseph Banks', *Archives of Natural History*, 23 (1996), pp.255–60.

Jonsell, Bengt, 'Daniel Solander – the Perfect Linnaean', *Archives of Natural History*, 11 (1984), pp.443–50.

Kaeppler, Adrienne L., 'To Attempt New Discoveries in That Vast Unknown Tract', in Michelle Hetherington and Howard Morphy, eds, *Discovering Cook's Collections* (Canberra, 2009), pp.58–77.

King, Robert J., 'The Call of the South Seas: George Forster and the Expeditions to the Pacific of Lapérouse, Mulovsky and Malaspina', *Georg-Forster-Studien*, XIII (2008), pp.149–74.

King, Robert J., 'George Vancouver and the Contemplated Settlement at Nootka Sound', *The Great Circle*, 32 (2010), pp.3–30.

Kingston, Ralph, 'A Not So Pacific Voyage: The "Floating Laboratory" of Nicolas Baudin', *Endeavour*, XXXI (2007), pp.145–51.

Knapp, Sandra, 'Lectotypification of Cavanilles' Names in *Solanum* (Solanaceae)', *Anales del Jardín Botánico de Madrid*, 64 (2007), pp.195–203.

Lamb, W. Kaye, 'Menzies and Banks: Evolution of a Journal', in Robin Fisher and Hugh Johnston, eds, *From Maps to Metaphors: The Pacific World of George Vancouver* (Vancouver, 1993), pp.227–44.

Lipscombe, Trevor, 'Two Continents or One? The Baudin Expedition's Unacknowledged Achievements on the Coast of Victoria', *Victorian Historical Journal*, 78 (2007), pp.23–41.

Lysaght, A.M., 'Banks's Artists and his *Endeavour* Collections', *Captain Cook and the South Pacific*, The British Museum Yearbook, 3 (1979), pp.9–80.

Mackay, David, 'Agents of Empire: The Banksian Collectors and Evaluation of New Lands', in David Philip Miller and Peter Hanns Reill, eds, *Visions of Empire: Voyages, Botany and Representations of Nature* (Cambridge, 1996), pp.38–57.

Madsen, Orla et al., 'Excavating Bering's Grave', in O.W. Frost, ed., *Bering and Chirikov: The American Voyages and their Impact* (Anchorage, 1992), pp.229–47.

Madulid, Domingo A., 'The Life and Work of Antonio Pineda, Naturalist of the Malaspina Expedition', *Archives of Natural History*, 11 (1982), pp.43–59.

Madulid, Domingo A., 'The Life and Work of Louis Née, Botanist of the Malaspina Expedition', *Archives of Natural History*, 16 (1989), pp.33–48.

Marner, Serena K., 'William Dampier and his Botanical Collection', in Howard Morphy and Elizabeth Edwards, eds, *Australia in Oxford* (Oxford, 1988), pp.1–3.

Moyal, Ann, 'The Scientific Legacy', in Sarah Thomas, ed., *The Encounter, 1802: Art of the Flinders and Baudin Voyages* (Adelaide, 2002), pp.186–97.

Novi, Carlos, 'The Road to San Antón: Malaspina and Godoy', in Andrew David et al., *The Malaspina Expedition 1789–1794*, III, pp.313–32.

Rigby, Nigel, 'The Politics and Pragmatics of Seaborne Plant Transportation, 1769–1805', in Margarette Lincoln, ed., *Science and Exploration in the Pacific: European Voyages to the Southern Oceans in the Eighteenth Century* (Woodbridge, Suffolk, 1998), pp.81–100.

Rigby, Nigel, '"Not at all a particular ship": Adapting Vessels for British Voyages of Exploration, 1768–1801', in Juliet Wege et al., *Matthew Flinders and his Scientific Gentlemen* (Welshpool, WA, 2005), pp.13–23.

Rogers, B.M.H., 'Dampier's Debts', *Mariner's Mirror*, 15 (1924), pp.322–4.

Sankey, Margaret, 'Writing the Voyage of Scientific Exploration: The Logbooks, Journals and Notes of the Baudin Expedition (1800–1804)', *Intellectual History Review*, 20 (2010), pp.401–13.

Simpson, Steve, 'The Peculiar Natural History of New Holland', in Howard Morphy and Elizabeth Edwards, eds, *Australia in Oxford* (Oxford, 1988).

Stearn, W.T., 'The Botanical Results of the *Endeavour* Voyage', *Endeavour*, XXVII (1968).

Stearn, W.T., 'Linnean Classification, Nomenclature, and Method', in Wilfrid Blunt, *Linnaeus: The Compleat Naturalist* (2004), pp.256–69.

Thomas, Nicholas, 'Forster, Johann Reinhold, and Georg Forster', in David Buisseret, ed., *The Oxford Companion to World Exploration*, 1 (Oxford, 2007), pp.314–17.

Watt, James, 'Medical Aspects and Consequences of Cook's Voyages', in Robin Fisher and Hugh Johnston, eds, *Captain James Cook and his Times* (Vancouver, 1979), pp.129–57.

Watt, James, 'The Voyage of George Vancouver 1791–1795: The Interplay of Physical and Psychological Pressures', *Canadian Bulletin of Medical History/British Columbia History of Medicine*, IV (1987), pp.33–51.

Werrett, Simon, 'Russian Responses to the Voyages of Captain Cook', in Glyndwr Williams, ed., *Captain Cook: Explorations and Reassessments* (Woodbridge, Suffolk, 2004), pp.179–97.

Wheeler, Alwyn, 'Daniel Solander and the Zoology of Cook's Voyage', *Archives of Natural History*, 11 (1984), pp.505–15.

Williams, Glyndwr, '"Far more happier than we Europeans": Reactions to the Australian Aborigines on Cook's Voyage', *Historical Studies*, 19 (1981), pp.499–512.

D. Miscellaneous

Chapter 1. William Dampier's manuscript journal, 'The Adventures of William Dampier ... in the South Seas', is in the British Library: Sloane MSS. 3236. His correspondence with the Admiralty is in the National Archives: Adm 2/1692.

Chapter 6. William Anderson's manuscript botanical notes from Cook's third voyage are in the Natural History Museum: Banks Coll. – And. The museum also holds letters to Joseph Banks from James King and others about the third voyage in the Dawson Turner Copies, Vol. 1. Archibald Menzies's journal, 1790–4, is in the British Library: Add. MS. 36461; 1794–5 is in the National Library of Australia: MS 155.

Chapter 8. Victoria Ibáñez, 'Botanical and Zoological Investigations on Exploration Voyages to Alaska in the Second Half of the 18th Century', unpublished essay.

Chapter 10. For Randolph Cock's research into scientists in the Royal Navy, see 'Sir Francis Beaufort and the Co-ordination of British Scientific Activity, 1829–1855', Cambridge PhD thesis, 1902; and in 'Scientific Sailors and Sailing Scientists: The Pursuit of a Scientific Career in the Royal Navy, 1815–1855', paper read at the International Commission for Maritime History seminar, King's College, London, 4 May 2000. Darwin's diary kept on the voyage of the *Beagle* is available at www.darwin-online.org.uk.

INDEX

Aborigines, Dampier describes (1688), 12–14, 27, 87; skirmish with (1699), 23; bodymarkings, 24; at Botany Bay, 85, 87, 163; at Endeavour River, 87; Cook and Banks describe, 87; Tasmanian, 167, 171–2, 221; artefacts lost, 220; cave paintings, 225; treatment of corpse, 225
Académie des Sciences (Paris), 3, 32, 49, 68, 70, 152, 177
Academy of Sciences (St Petersburg), 32–3, 34, 35, 50, 237
Acapulco, 8, 29, 185, 186
Achin, 11, 14,
Adanson, Michel, his *Familles naturelles des plantes*, 56
Admiralty Islands, 166, 168
Adventure Bay (Tasmania), 128, 170
Ahutoru (Tahitian), 65, 68, 69
Alaska, 33, 36–43, 53, 133, 145, 186–8
Albemarle Island (Galapagos), 145–6, 250
Alcalá Galiano Dionisio, 197
Aleuts, encountered by Steller, 42–3; described by Waxell, 42–3; appearance, 42–3; kayaks, 43
Allen, John, 204
alligators (Campeche), 18
Amboina, 3, 169
Amiens, Treaty of, 213, 223
Anaura Bay, 83
Anderson, William, on Cook's second voyage, 103, 111, 124, 128; on Cook's third voyage, 124, 126–31, 153, 171; death, 126–7; tributes to, 127;

criticism of Cook, 127, 129–30; frustration, 128, illness, 130–1; specimens and lists, 131
Andes, 182, 195, 248, 249
Anne, queen, 22, 26
Anson, George, voyage, 17
Antarctic Circle, 101
anteater, giant (Campeche), 17
Archipel de la Recherche, 221
Arica, 28
Armstrong, Johnny, 29, 266 n.66
Ascension Island, 22
Ashmolean Museum, Oxford, 91
Avacha Bay, 33, 35, 37, 40, 44, 46

Bahia, 23
Balade, 111
banded hare-wallaby (*Lagostrophus fasciatus fasciatus*), 26–7
Banks, Joseph (later Sir), 3, 5, 16, 57, 124; and Aborigines, 13, 85, 87; early years, 73–4; on Cook's first voyage, 74–88, 228; ignores predecessors, 76, 94, 270 n.14; botanical work, 76, 77, 78, 79, 80, 82, 83–4, 85–6, 103, 192; narrow escape, 78; relations with Cook, 78–9, 81–2, 131–2; praises breadfruit, 80; activities on Tahiti, 81, 113, 139; regret at Maori deaths. 83; escape from Great Barrier Reef, 86; describes kangaroo, 86; illness, 87; praised, 88–9; prepares for second voyage, 88, 89; and scurvy, 89; withdraws from second voyage, 90, 95, 100, 272 n.70;

sexual reputation, 90; natural-history collection, 91–2, 92–3, 262, 272 n.73; his florilegium, 91, 92, 93–4, 272 n.81; elected president of Royal Society, 93; and J.R. Forster, 96, 97, 98, 105–6, 107, 108–9; given shells, 110; and Anderson, 128, 131; role as promoter, 132, 201; interest in northwest coast, 133–4, 135, 136, 137; contact with Menzies, 133–4, 135, 137–9; relations with Vancouver, 137–8; helps Labillardière, 165, 175–6, 177–8; contact with Malaspina, 180; supervises Flinders expedition, 201–5, 231; contact with Brown, 221–2, 227–8; future of botany, 229, 262

Banks Island, 135

Baret, Jean (Jeanne), joins *Etoile*, 59; hard-working collector, 60, 61–2, 66, 70; gender revealed, 64–6; early years, 65; praised, 65, 66, 70; assaulted, 66; plant named after, 70; later years, 70

barnacles, Darwin's work on, 256

Barrington, Daines, 96, 97, 114

Barrow, John, 218, 232

Bass, George, 201

Bass Strait, 201, 221, 223

Batavia (Jakarta), 67, 87

Baudin, Nicolas, 5; voyage, 202, 205–16; equipment, 205; instructions, 205–6; early years, 206; journal, 206, 283 n.17; desertions, 207; friction with scientists, 207–8, 208–9, 209, 210, 211, 214, 215; friction with crew, 213, 214, 215, 220; encounters Flinders, 212–13; criticises cartographers, 215; death, 216, 218, 219; official account, 218–19; place names discarded, 219; character, 220–1

Bauer, Ferdinand, 184, 204, 225; complaints, 225–6; artistic output, 228–9, 229; assessment, 230

Bay of Islands, 83

Bayly, William, 104, 123, 124, 153

Beautemps-Beaupré, Charles-François, 168, 173, 210

Bellingshausen, Fabian Gottlieb von, 235, 237

Bencooly (Benkulen), 12

Bering, Vitus, 4, 33, and First Kamchatka Expedition, 33, and Second Kamchatka Expedition, 33–45, 133, 234, 263; illness, 37; death and burial, 45–6

Bering Island, shipwreck on, 47–8, 50

Bering Strait, 33, 133, 236

Berlin Academy, 32, 245

Bernier, Pierre-François, 208; death, 216

Bernier Island, 208, 209

Bernizet, Gérault-Sébastien, 159

Betge, Matthias, 35, 40, 47

Billings, Joseph, voyage, 234, 235

Bird Island, 42

Bissy, Frédéric, 208

Bligh, William, 5, 17, 139, 201; criticised, 170–1

Bodega y Quadra, Juan Francisco de la, 141, 142

Bolts, Williams, 120

Bonaparte, Mme (later Empress Josephine), 176, 207, 217, 218, 220

Bongaree (Aborigine), 224

Borabora, 131

Bory de Saint Vincent, Jean-Baptiste, 207–8

Boswell, James, 126

Botany Bay, 6; Cook visits, 85–6; La Pérouse visits, 160, 162–4; Malaspina visits, 180–1; climate, 201

Bougainville, Louis-Antoine de, 54; instructions, 59; relations with Commerson, 60–1, 66; reaches Tahiti, 62; *Voyage*, 62–3, 68–9, 96; enthusiasm for Tahitian life, 62–4, 68, 193; reservations about Tahiti, 68–9; and Baret, 64, 65; visits Samoa, 160

bougainvillea (*Bougainvillea spectabilis*), 60, 71, 77

Bourbon (Réunion), 69, 79

Brambila Ferrari, Fernando, 184

Brazil, 2, 31; plants from, 25

breadfruit (*Artocarpus altilis*), 5; Dampier's description, 16; and Anson's voyage, 17; praised by Banks, 80; sent to West Indies, 132, 135; in Samoa, 162

Brett, Peircy, 17

British Museum, 16, 74, 75, 91, 96, 117, 125, 148, 229

Brosses, Charles de, 58–9, 74

Brown, Robert, 203, 221, 253; contact with Banks, 221–2, 224, 227, 228; loses way, 222; specimens, 223, 224, 225, 229; relations with Flinders, 228; brings collection home, 228; his *Pordromus florae Novae Hollandiae*, 229; his 'General Remarks', 231

Buache, Philippe, 49

Buache de Neuville, Jean-Nicolas, 151, 186

buccaneers, in South Sea, 7–8, 9–11

Buchan, Alexander, 74; death, 78
Buffon, George-Louis Leclerc, 4, 13, 56, 71, 119, 132, 151, 152, 154; his *Histoire naturelle*, 57, 71
Burnett, 'Mr', 90
Burney, James, 130
Bustamante, José, 179–80, 186, 193, 195, 196; journal published, 200
Bynoe, Benjamin, 243
Byron, John, 54, 73, 76

Caley, George, 228, 229
Californian redwood, (*Sequoia semperverens*), 189
Callao, 184, 194
Campeachy (Campeche), 9, 14, 17–18
cannibalism, Maori, 84; in New Caledonia, 173
Cape Blanco, 32
Cape Deshneva, 32
Cape Horn, 7, 34, 79, 154, 181, 185, 228, 238
Cape Leeuwin, 208, 221
Cape of Good Hope, 3, 99, 112, 128, 202, 238
Cape Verde Islands, 23, 134
Cardero y Meléndez, José de, 183, 187, 188
Carlos III, king, 179, 180
Carlos IV, king, 180, 197
Caro, Jean-Louis, 60
Carteret, Philip, 54, 61
Carvajal, Ciriaco González, collections, 189
Castries, Charles-Eugène-Gabriel de, 151, 158
Catesby, Mark, 40
Catherine II, empress, 95
Cavanilles, Antonio José, 198
celery, wild (*Apium prostractum*), 79
Chamisso, Adelbert von Chamisso, specimens, 236; influence, 236; published accounts, 236–7; complaints, 237–8, 262; criticises Kotzebue, 237, 260–1; specimens spoilt, 238
Charles Island (Galapagos), 250
Charles X, king, 177
Charlotte, queen, 112, 132, 175
Chatham Island (Galapagos), 251
Chatham Strait, 145
Chilean strawberry (*Fragaria chilensis*), 152
Chirikov, Alexsei, voyage, 36, 39, 41–2, 133
Chukchi, 43
Churchill, Awnsham and John, 31

Clerke, Charles, quarrels with J.R. Forster, 103, 111; illness, 130–1; death, 131; grave, 159
Cleveley, John, Jnr, 93
cockroaches, destructive nature and identity, 130, 184
coconut, uses of, 16
Cocos Islands, 251
Collignon, Jean-Nicolas, 152, 157, 278 n.6; sows seeds, 156; injured, 158–9; at Tutuila, 160; wounded, 162
Collins, David, 163
Colnett, James, 133; relations with northwest coast peoples, 135
Columbus, Christopher, 1
Commerson, Philibert, 55, 94; early years 55, 58; as doctor, 58, 61; appointed to Bougainville expedition, 58–9; partnership with Baret, 59, 60, 61–2, 65–6, 69–70; incomplete journal, 59; complaints, 59–60, 67; relationship with Bougainville, 60–1; praise for Tahiti, 64, 68, 69, 72; collection, 67–8, 70–2, 269 n.38; stay on Isle de France, 69–70; death, 70; elected to Académie, 70
Concepción, 155, 194; earthquake, 249
Cook Inlet, 126, 145
Cook Islands, 128–9
Cook, James, 4–5, 13, 57; first Pacific voyage, 73–88; instructions, 76; relations with Banks, 78–9, 81–2; and lemon juice, 79–80; his journals, 79, 80; praises naturalists, 81; and Aborigines, 85, 87, 171; achievements, 87–8; ignored on return, 88; promoted, 88; second voyage, 98–112; relations with J.R. Forster, 99, 103, 106, 108, 109–10, 111; illness, 103; motivation, 108; journal, 113; published account, 113–14; voyage collections, 116, 118; third voyage, 122–31, 133, 234; and speculative cartography, 122–3; and science, 123–4, 125–6, 153, 275 n.5; instructions, 126, 127; behaviour, 129–30; death, 132–3, 150; safe conduct, 150, 175; and flax, 152; model for Flinders, 202, 230; Russian tributes to, 235
Cooper, Joseph, 100, 103
corals, 115, described by Cook, 128–9, by Anderson, 129, by Flinders, 224, by Darwin, 251–2
cormorant, spectacled (*Phalacrocorax perspicillatus*), 48

Covington, Syms, 247
Cowley, Ambrose, 18–19
Coxe, William, 123
Cranstoun, Alexander, 136, 139
crocodile, specimen, 182
Croyère, Louis Delisle de la, 34, 50; death, 36; grave, 159
Cuiver, Georges, 71, 205–6, 217

D'Auribeau, Alexandre d'Hesmivy, 164, 167, 173, 174, 175
D'Entrecasteaux Channel, 168, 210
D'Entrecasteaux, Joseph-Antoine Bruny, voyage, 164–73, 221; relation with scientists, 166, 167, 169, 170; describes Tasmanian Aborigines, 171–2; decline, 172; death, 173; journal published, 174
D'Urville, Dumont, voyages, 234
Dagelet, Joseph Lepaute, 158
Dampier Archipelago, 23
Dampier, William, 4; early years, 8; as logwood cutter, 9; *Voyages to Campeachy*, 9, 17–18; as buccaneer, 9–11; *New Voyage round the World*, 10, 11–14, 15, 20; visits New Holland (1688), 10–11, 12–13; manuscript journals, 11–14; *Voyages and Descriptions*, 14–15; discourse on trade winds, 15; visits Galapagos Islands, 18–20; on authorship, 20; financial distress, 19; commands discovery voyage (1699–1701), 21–7; visits New Holland (1699), 21, 23–7; court-martialled, 22; *Voyage to New-Holland*, 22–7; skirmish with Aborigines, 23–4; collects plants, 25–6; description of New Holland, 26–7; *St George* voyage, 29–30; accused of cowardice, 29; pilot with Woodes Rogers, 30–1; death, 31; importance of journals, 31; description of southern manatee, 51
Darwin Mount, 247
Darwin Sound, 247
Darwin, Caroline, 244, 245, 252
Darwin, Catherine, 244, 245, 250
Darwin, Charles, 6, 115; joins Fitzroy, 239; early years, 239, 241; seasickness, 241; diary, 241, 286 n.25; published account, 241; specimens, 241, 243, 245, 246, 253; nicknames, 243; relationship with Fitzroy, 243–4, 252–3, 258–9; and Humboldt, 245, 246; shipments to England, 245, 246, 247; inland travels, 247–8, 250;

challenges Lyell's theories, 249, 252; visits Galapagos, 250–1; and Galapagos finches, 250, 251, 253, 254, 255; and Galapagos tortoises, 250–1, 254; on coral reefs, 251–2; hatred of sea, 252; returns home, 252; and natural selection, 253, 254–5, 256–7; essay on transmutation, 254, 255–6; published works, 255, 257; moves to Down House, 255–8; *Origin of Species*, 255, 257; studies barnacles, 256; later years, 257–9; on ocean voyaging, 263
Darwin, Emma, 255–6, 257
Darwin, Robert Waring, 240, 247
Darwin, Susan, 244, 249
Decaen, Charles-Mathieu-Isadore, 227, 230
Defoe, Daniel, 14, 31
Dégerando, Joseph-Marie, 206
Delisle, Joseph-Nicolas, 34, 36 50; inaccurate report of Second Kamchatka Expedition, 49–50
Deschamps, Louis-Auguste, 165, 166; praised, 169; criticises Aborigines, 171; collection, 176
Deshnev, Semen, 32
Diderot, Denis, 262
Dillon, Peter, 176, 177
dingoes, 26
Dixon, George, 133, 187
Doubtful Sound, charted, 191
Dougals, David, 149
Douglas, Dr John, 123, 130
Douglas-fir (*Pseudotsuga menziesii*), 148
Down House, Darwin's life at, 255–9
Drake, Francis, 2, 7
Dryander, Jonas, 223
Duclos-Guyot, Pierre, 60
Duncan, Charles, 133
Dusky Sound, 102, 103 104, 139, 191
dysentery, among discovery crews, 87, 210, 216, 225

East Falkland, 195
East India Company (British), 132, 203
East India Company (Dutch), 2–3, 110, 169, 174
Easter Island, visited by Cook, 155–6; visited by La Pérouse, 155–6
Echeverria, Atanasio, 142
Edwards, Edward, 168
elephant seals, southern (*Mirounga leonine*), 214
Ellis, John, 75–6, 92, 137
Ellis, William, 125

emus (*Dromaius novaehollandiae*), 213, 214, 216
Endeavour River, 86
Eromanga, 110
Escholtz, Johann Friedrich, 236
Eskimos, Chugach, 38
Esperance Bay, 170
Espinosa y Tello, José Maria de, 199–200
Etches, John, 133; and Richard Cadman, 133, 135
Eua, 107
eucalyps, 128, 167; *see also* Tasmanian bluegum
Evelyn, John, 20
Every, Henry, 11
Exquemelin, Olivier, 8, 10
'experimental gentlemen', 98, 124, 272 n.14

Falck, Johan Petter, 57
Falkland Islands, 59, 233
False Ladyslipper *(Calypso bulbosa)*, 140
Fernández de Oviedo y Valdés, Gonzalo, 1
Ferrer Maldonado, Lorenzo, and Northwest Passage, 186–8
Fesche, Charles-Félix-Pierre, 60
Feuillée, Louis, 28
finches, Galapagos, 250, 251, 253, 254, 255
First Fleet, 135, 162–3
Fitzroy, Robert, and Darwin, 239; his instruments, 240; character, 243–4; relationship with Darwin, 244, 246, 258, 259; depression, 248–9; his *Narrative*, 258; protest at *Origin of Species*, 258; death, 258; work as hydrographer, 258–9
flax, New Zealand *(Phormium tenax)*, 84, 152, 172
Fleurieu, Charles Claret de, 151, 154, 155, 158, 162, 164, 169, 178, 205
Fleuriot de Langle, Paul-Antoine-Marie, 153, 156, 160; praises naturalists, 155; at Tutuila, 160–2; killed, 161
Flinders, Ann, 204–5, 223–4, 230
Flinders, Matthew, 137; proposes survey voyage, 201–3; detained on Isle de France, 218, 227; voyage, 221–7; meets Baudin, 221–3; crew deaths, 222, 226; circumnavigates Australia, 226; published account, 226–7, 230–1; chart of Australia, 227, 231; place names discarded, 227; disappointments, 231; death, 231

Flinders, Samuel, 204, 226
Forster, George, boyhood, 95–6; accompanies father on *Resolution*, 97, 98; plant drawings, 99, 102, 118, 125; supports father's complaints, 100, 111; on hardships, 102, 103; reaction to Tahiti, 106; attacked, 109; publishes *Voyage*, 114–15; praises father's *Observations*, 116; returns to Germany, 119; gratitude to Banks, 119; praises Cook 120; plans further Pacific voyage, 120; influence on Humboldt, 120; death, 120
Forster, Johann Reinhold, 6, early years, 95–6; visits Volga, 95–6; relations with son, 95–6; in London, 96–8; appointed to Cook's second voyage, 97–8; character, 98; endurance, 99, 102, 108; relations with Cook, 99, 106, 108, 109–10, 111, 273 n.50; relations with shipmates, 99, 100–1, 103, 110, 111; complaints about conditions, 100, 101–3, 106, 107, 108, 110, 184, 262; praises Banks, 105–6, 107; reaction to Tahiti, 106; criticises predecessors, 108; injured, 108; attacked, 109; collecting in New Caledonia, 111, 178; fears for future, 112; received by king, 112; honorary degree, 112; praised by Gilbert White, 113; dispute over published account, 97, 113–14; his *Observations*, 115–16; his collections dispersed, 116–17, 274 n.99; praises son, 116–17; financial difficulties, 118, 119, 120; his *Characteres generum plantarum*, 118; belated publication, 118–19; relies on Banks, 119; post at Halle, 119; challenges Buffon, 119; further publications, 121; death, 121; praised by Herder, 121; on Cook and science, 123–4
Forsters Passage, 112
Fox, William Darwin, 244, 249–50, 252
foxes, Arctic, depredations, 45, 51
Franklin, John, 204
Franz I, Emperor of Austria, 118
Freycinet, Henri de Saules de, 213, 223
Freycinet, Louis-Claude de Saules de, 214; account of Baudin voyage, 218, 219; atlas, 218–19, 227; expedition, 233, 238; specimens, 233
Frézier, A.-F., 28, voyage, 28–9, 152
Fuegians, described by Bougainville, 61, by G. Forster, 104; on *Beagle*, 240; and Darwin, 247, 257

Funnell, William, 29
fur seal, northern (*Callorhinus ursinus*), 48, 51
Furneaux, Tobias, 98, 104–5, 171

Galapagos Islands, vi; visited by Dampier, 18–20, 30, 250; Spanish descriptions, 18–19; visited by Darwin, 19, 20, 250–1; visited by Menzies, 145–6
Gama Land, 34, 36
gardens, planted on Cook's second voyage, 104–5, 107
Gennes, J.-B., de, 28
Geographe Bay, 209
George III, king, 97, 103, 112, 132, 144
George, prince of Denmark, 22
Gilbert, Joseph, 100, 103–4
Giraudais, Chesnard de la, 59, 67
Gmelin, Johann Georg, 34, 35, 52, 53
Godoy y Alverez de Faria, Manuel de, 197
González Guitérrez, Pedro María, 194
Good, Peter, 204, 221, 224, 225; death, 226
Gore, John, 131, 155
Gould, John, and Galapagos specimens, 253–4
Gravière, Jurien de la, 167
Great Australian Bight, 139, 221
Great Barrier Reef, 67, 86, 224
Great Northern Expedition, *see* Second Kamchatka Expedition
Green, Charles, 74; death, 82
Greppi, Paulo, 196
Gronovius, Johan the Younger, 56
Guam, 10, 16, 17, 189–90
Guayaquil, 182
Guio y Sánchez, José, 183, 186
Gulf of Carpentaria, 222, 224
Gulf St Vincent, 214, 222
Gulliver's Travels, 14, 23–4
Gvosdev, Mikhail, 33

Haenke, Tadeo, 5; joins Malaspina, 181–2; collections, 182, 187, 188–9, 190, 192, 193; painting methods, 184, 230; disappointment, 188; on Guam, 189; dangers on Luzon, 190; at Port Jackson, 192; writes to Banks, 192; at Vava'u, 193; leaves Malaspina, 194; later years, 195, 196; publication of notes, 199, 200
Haida (northwest coast), 141
Halle, University, 35, 95, 119, 120
Hamelin, Emmanuel, 207, 211, 212, 213
Harley, Robert, 1st earl of Oxford, 30
Hartlib, Samuel, 2

Hawaiian Islands, visited by Cook, 130, Colnett, 134, Menzies, 142–3, La Pérouse, 155, 156; intended visit by Malaspina, 181, 186, 187
Hawke, Edward, Lord, 75
Hawkesworth, John, *Voyages*, 80, 90, 91, 105, 113, 193
Henslow, Professor John Stevens, 239, 244, 246, 248, 249, 250, 253
Herder, Johann Gottfried, 116, 121
Hermann, Paul, 3
Hernández, Francisco, 1–2
Herschel, John, 253
hippopotamus, supposed, 27, 215
Hitia'a (Tahiti), 62
Hodges, William, 21, 98, 104, 112, 114
Hooker, Joseph Dalton (later Sir), 77, 229, 253, 254, 263
Hooker, Sir William, 148, 239
Hope, Professor John, 133–4
Hornsby, Professor Thomas, 123
Huahine, 106, 131
Hubbard Glacier, 188
Hudson's Bay Company, 136, 149, 150
Hulme, Nathaniel, 79
Humboldt, Alexander von, 6, 115, 180, 233, 240; criticises naturalists, 245–6; hardships, 263
hummingbird (Campeche), 17–18
Hunter, John, report of La Pérouse, 166, 167, 168
Huon de Kermadec, Jean-Michel, 164, 172; death, 173
hurricane, described, 9
Huxley, T.H., 236, 253; experiences at sea, 260–1

ice, on Cook's second voyage, 101, 107, 108; on Cook's third voyage, 123, 133
Institut National (later Institut Impérial), 202, 205, 207, 208, 218
Irkutsk, 49
Isle de France (Mauritius), 67, 163, 151, 164, 205, 207, 216, 221; Commerson's stay, 69–70

Jacquin, Nikolaus, 230
Jamaica, 3, 8, 9
Japan, 2, 33, 35, 151
Jardin du Roi (later Jardin des Plantes), 2, 56, 71, 151, 152, 175, 206
Jeoly (Filopino), 11, 14
Johnson, Dr Samuel, 115, 126
Joseph II of Austria, collections, 206

Jossigny, Paul-Philippe Sauguin de, 70, 71
Juan Fernandez, 29, 30
Jussieu, Antoine-Laurent de, 205, 217; his
    Genera plantarum, 56, 60, 71;
    classification method, 178, 229, 261

Kaempfer, Engelbert, 3
Kajeli (Buru Island), 174
Kalm, Pehr, 57
Kamchadals, 40, 43, 46, 49, 52, 53
Kamchatka, 32–3, 35–6, 45, 52, 53, 159
Kamehameha (of Hawai'i), 144
Kangaroo Island, 214, 221, 222
kangaroo, described, 86–7; drawn, 91; dried
    specimens, 192; classification, 204; live
    specimens, 213, 214, 216, 217
Kastri Bay, 158
Kauai, 130
Kayak Island, 38–40, 53
Kealakekua Bay, 122, 125
Kerguelen cabbage (*Pringlea antiscorbutica*),
    128
Kerguelen Island, 127
King George Sound (WA), 139, 214
King George's Sound Company, 135,
King Island (Bass Strait), 214
King Sound (WA), 11
King, James, 123–4, 132
King, Philip Gidley (governor NSW),
    213, 226
King, Philip Gidley (midshipman), 240,
    242
Khitrov, Sofron, 36, 38, 43–4
Knapton, James, 10, 14, 16; and John, 22
Kotzebue, Otto von, voyages, 235, 236–7;
    published account, 237
Krasheninnikov, Stepan P., 52, 159
Krusenstern, Adam Johann von, voyage,
    236; atlas, 237
Kupang, 210
Kuril Islands, 33, 151
Kwakiutl (northwest coast), 141

La Coruña, 11
La Motte du Portail, Jacques-Malo de, 172
La Pérouse, Jean-François de Galaup de,
    posthumous *Voyage*, 121, 176; voyage,
    137, 150–64; instructions, 151;
    relations with scientists, 154, 157, 158,
    158–9; journal, 155, 163; instructions
    changed, 155, 160; searches for
    Northwest Passage, 156–7, 186; loses
    boat crews, 157; criticises
    philosophers, 157, 162; strait named
    after, 159; hardships, 159, 163; visits

Botany Bay, 162–4; expedition
    disappears, 163, 263; search for,
    164–72, 224, 233; relics found, 176–7;
    possible survivors, 177
Labillardière, Jacques-Julien Houton de,
    165, 166; advice from Banks, 165;
    collections, 166–7, 167, 173, 175,
    175–6; criticises sailors, 168–9;
    observations, 169; contact with
    Aborigines, 171; criticises
    d'Entrecasteaux, 172; taken prisoner,
    174; publishes account, 177; plants of
    New Holland, 177; plants of New
    Caledonia, 173, 178; Baudin
    specimens, 219–20
Lacépède, Bernard-Germaine-Etienne, 71
Lagrange Bay, 23
Lahaye, Félix, 165, 176
Lahontan, Lom d'Arce de, 43
Lalande, Lefrançais de, 68
Lamanon, Jean-Honoré-Robert de Paul,
    153; criticised by La Pérouse, 154,
    158; collects seeds, 154; searches for
    specimens, 157; observations sent
    home, 158; killed, 162
Lamarck, Jean-Baptiste, 206
Lamartinière, Joseph-Hughes de, 153, 157;
    praised, 155; disappointed, 156–7; at
    Tutuila, 160; collection, 163
Landais, Pierre, 60, 67
Langsdorff, George Heinrich von, 235;
    specimens destroyed, 235; disputes,
    235–6
Lassenius, Peter, and scurvy, 40
Lavaux, Simon-Pierre, 153
Lawson, Nicolas, 250
Lee, James, 125
Lemonnier, Louis-Guillaume, 71
Lepekhin, Thomas, 38
Lesseps, Jean-Baptiste Barthélémy de, 154,
    160, 176
Lesueur, Charles-Alexandre, 206–7, 208,
    214, 220; artistic output, 217, 218
Lever, Ashton, his Holophusicon, 117–18,
    274 n.90; collections dispersed, 118
Lind, James (naval surgeon), and Anson's
    voyage, 89, 271 n.64
Lind, James (scientist), and Cook's second
    voyage, 97
Linnaeus, Carl, 4, 25, 74, 75, 188; his
    Species plantarum, 52, 55, 134; praises
    Steller, 53; develops binomial system,
    55, 77, 134, 178; his *Systema naturae*,
    55–6, 96; opposition to his system, 56,
    229; his *Philosophia botanica*, 57; his

'apostles', 57; interest in national economy, 57; praises Banks, 89; criticises Solander, 89; worries about Cook's second voyage, 92; and J.R. Forster, 96, 98; clothing, 168 n.50; criticised, 185, 204

Linnean Society, 135, 148, 201

Lisianskii, Yuri, 235

Litke, Fedor Petrovich, and scientific research, 237

lizard, bobtail (*Tiliqua rugosa*), 26

Louis XVI, king, 151, 152, 154, 156; executed, 174

Louis XVIII, king, 175

Löwenstern, Hermann Ludwig von, 235–6

Lyell, Charles, 246, 248, 253; influence on Darwin, 246, 249, 252

Macau, 158, 190

McCormick, Robert, 243

MacGillivray, John, 260

Macquarie, Lachlan, 231

Madagascar, 69, 70, 207

Madeira, 77, 99

madrona tree (*Arbutus meziesii*), 139–40

'Mai (Society Islander), 113

Malabar, 3

Malacca (Melaka), 11, 14

Malaspina, Ajandro, 5, 141; motives for voyage, 179–81; programme, 181; voyage, 181–96; shipments, 184, 185, 189; relations with naturalists, 184–5; crew losses, 185, 187, 195; lack of instructions, 186, 194; at Port Mulgrave, 187–8; acts of possession, 188, 193; tribute to Pineda, 191; investigates Port Jackson, 192; problems with crews, 193–4; ill health, 193–4; journal, 196; planned publication, 196, 197; promoted, 197; arrested, 197; imprisoned, 197, 200; journal published, 200; last years, 200

Malmaison, 218

Manby, Thomas, 147

Manila galleon, 29

Maori, on Cook's first voyage, 82–4; deaths, 83; and cannibalism, 84; on second voyage, 104, 105; judged by J.R. Forster, 104

marijuana, and Dampier, 11

Marion Dufresne, Marc-Joseph, 171

Marquesas, 98, 108

Matavai Bay, Tahiti, 64, 109

Maugé, René, 210; death, 211

Maui, 143, 156

Maurelle, Francisco, 193

Maurits, Count Johan of Nassau-Siegen, 2

Mendaña, Alvaro de, 67, 173

Menzies, Archibald, with Colnett, 133–5; sends specimens to Banks, 134, 135; joins Vancouver expedition, 136; relations with Vancouver, 136–7, 137–8, 139, 142, 144, 146–7, 149; his journal, 138–9, 147–8; his botanical work, 139–41, 142–3, 145–6; describes salmon, 143–4; climbs Mauna Loa, 144; neglects collection, 148; remembered in Hawaii, 149; later life, 149

Mercury Bay (Te Whanganui-o-Hei), 84

Mertens, Karl-Heinrich, specimens, 237

Mexico City, 185

Milet-Mureau, Louis-Marie-Antoine Destouff, 162, 176

Miller, James and John Frederick, 93

Mindanao, 10, 11, 191

Moll, Herman, 10, 23, 27

Moluccas, 69, 151

Mongez, Jean-André, 153, 157, 159

monkey puzzle tree or Chile Pine (*Araucaria araucana*), 146, 277 n.74

Monkhouse, William Broughan, 82, 83; death, 87

Montagu, Charles, Lord, 16, 21

Monterey pine (*Pinus radiate*), 145

Monterey, 142, 145, 157–8, 188–9

Montevideo, 247

Mount Avachinskaya, 159

Mount St Elias, 37, 156

Moziño, José Mariano, 142

Müller, Gerhard Friedrich, 34, 35; accounts of Bering's expeditions, 50

Mulovsky, Grigory Ivanovich, 120

Murchison, Roderick, 253

Murray, Thomas, 16

Muséum d'Histoire naturelle (Le Havre), 220

Muséum National d'Histoire naturelle (Paris), 205, 217, 220, 233

Nagai Island, 40

Napoleon Bonaparte, 200, 202–3, 206, 218

Nass River, 143

Nassau-Siegen, Charles-Othon d'Orange et de, 66

Neé, Louis, 181, 182, 184, 185, 196; lists useful trees, 183; travels to Mexico City, 186, 187, 189; specimens misattributed, 188, 198; on Guam, 189–90; collections on Luzon, 190; at

Botany Bay, 192, 198; leaves
Malaspina expedition, 194; rejoins
expedition, 195; collections, 195, 198;
publications, 198; notes lost, 199;
death, 199; notes published, 200
Nelson, David, 125, 128
Nepean, Evan, 137–8, 146, 203, 204
New Britain, 22, 27, 173
New Caledonia, 98, 111, 124, 128, 163,
168, 172–3; published flora, 173, 178
New Guinea, 22, 25, 27, 67, 163, 173;
plants from, 25
New Hebrides (Vanuatu), 67, 98
New Holland, 10–11,12–14; plants from,
24–6; east coast surveyed by Cook's,
84–7, 151
New Ireland, 27, 65, 168
New South Wales, 87, 137, 237
New Zealand, on Cook's first voyage, 82–4;
on Cook's second voyage, 98, 101–3,
104–5
Nicobar Islands, 11, 12
Nieuhof, Jan, 31
Nihau, 130
Niue, 109
Nomuka, 129
Nootka Sound, 130, 133, 141–2, 188;
crisis, 136
Norfolk Island, 162
North West Cape (WA), 23
Northwest Passage, 122, 125–6, 139, 145,
151, 156–7, 186–8, 234, 236
Novo y Colson, Pedro, 200

Okhotsk, 33, 34, 35, 160
Orford, earl of, 21
Owen, Richard, 253

Palapag (Samar), 190
Pallas, Peter Simon, 50, 52
Panama Isthmus, 7, 9, 12,
paper mulberry (*Morus papyrifera*), 152
Parkinson, James, 118
Parkinson, Stanfield, 91
Parkinson, Sydney, 21, work on Cook's first
voyage, 74, 76, 77, 103; painting
difficulties, 80; natural-history notes,
80; drawing techniques, 82, 86; death,
87, 91; kangaroo representation, 91;
drawings, 92, 93, 125
Patagonia, 60
Patten, James, 104
Pembroke, 8th earl, 22
Pennant, Thomas, 96, 99
pepper, Guinea, 28

Pepys, Samuel, 20
Péron, François, personality, 208; loses way,
209, 215; narrow escape, 209–10;
investigates Aborigines, 212; at Port
Jackson, 213–14; collection, 214, 217,
220; identifies dugong, 215; writes
official account, 218–19, 227; death,
219; tribute, 219
Peru, viceroy of, 184
Peter I (the Great), 32, 34
Petit, Nicolas, 206–7, 212, 218
Philadelphus coronarius, later *Philadelphus
lewisii*, 148–9
Philip, Governor Arthur, 162
Philippines, 158, 190–1
Pickersgill, Richard, 85
Pico de Teide (Tenerife), 154, 166
Picquet, Antoine, 209, 210
Pineda y Ramírez, Antonio de, 181, 182,
184, 185, 196; collection spoilt, 182;
finds valuable specimens, 182–3, 185;
travels to Mexico City, 186, 187, 189;
investigates Guam, 189–90; journeys
across Luzon, 190–1; death, 191;
notebooks, 191, 194, 199; notes
published, 200
Pineda y Ramírez, Arcadio de, 191,
194, 199
Piron, Jean, 171, 177
pitcher plant, insect-trapping (*Cephalotus
follicularis*)
Pitt, Hon. William (Lord Camelford), 147
Pitt's Island (Vanikoro), 168, 173, 176–7,
233
plant boxes, 137, 152, 223, 228
plant frames, on Vancouver's voyage, 137,
142, 144, 145, 146, 152; on Flinders'
voyage, 204, 222, 224
Plenisner, Friedrich, 45; draws sea
mammals, 52
Plukenet, Leonard, 26
Poivre, Pierre, 67, 69, 70
Port de Français (Lituya Bay), 156–7
Port Egmont, 182
Port Jackson (Sydney), 162; Malaspina
visits, 191–2; and Baudin expedition,
212, 213–14; and Flinders expedition,
223–6
Port Mulgrave (Yakutat Bay), 187–8
Port Phillip, 214
Portland, 3rd duke, 148
Portlock, Nathaniel, 133, 138
Portuguese man-of-war, 84–5
Postnikov, Alexsey, 200
Poverty Bay (Tuuranga-nui), 82–3

Pozo Ximémez, José de, 183
Presl, K.B., 199
Prévost, Guillaume (the elder), 153; criticised, 155, 158
Prévost, Jean-Louis-Robert (the younger), 153
Prince William Sound, 130
Pringle, Sir John, 79–80, 124, 128
proas (Guam), 17
Puget Sound, 139, 140, 141
Puget, Peter, 140, 141
Pulteney, Richard, 75

Queen Charlotte Islands, 142
Queen Charlotte Sound (Totara-nui), 84, 102, 104–5, 107, 128
Queens Cup (*Clintonia uniflora*), 144

Raiatea, 82, 106, 131
Ravenet y Bunel, Juan Francisco, 184, 190
Ray, John, 3; plant classification, 25
Real Gabinete de Ciencias Naturales, 189, 198
Real Jardín Botánico, 5, 180, 198
Realejo, 183
Receveur, Claude-François-Joseph, 153, 157, 159–60; praised, 155, 159; wounded, 162; death, 163
Recherche Bay, visited by d'Entrecasteaux, 167–8
Resanoff, Nicolas Petrovich, 235
Revilla Gogedo, Conde de, 185
rhea (South American), 253
Rhododendron, Large-flowered (*Rhododendron californicum*), 140
Rhododendron, Pacific (*Rhododendron macrophyllum*), 140
Riche, Claude, 165, 174; clothing, 168; ccomplaints, 169; narrow escape, 170, 209
Riédlé, Anselm, death, 210, 211
Ringrose, Basil, 8, 10
Rio de Janeiro, 59, 60, 77
River Plate, 60, 61
Roberts, Henry, 136
Robertson, William, 123
Robinson, Tancred, 15
Rogers, Woodes, 28; voyage to South Sea (1708–11), 30; account of voyage, 30–1
Rollins, Claude Nicolas, 153, 159
Ross, James Clark, 234, 253
Rossel, Elisabeth-Paul-Edouard de, 171, 174; edits d'Entrecasteaux's journal, 178

Rottnest Island, 24
Royal Botanic Gardens (Kew), 132, 146, 148, 224, 239
Royal Horticultural Society, 149
Royal Philippines Company, 180, 194
Royal Scientific Expedition to New Spain, 179, 180
Royal Society (London), 3, 15, 16, 25, 32, 91, 96; and Cook's first voyage, 73, 74, 75, 76, 78; and Cook's second voyage, 89
Rumiantsev, Nicolai, 236
Rumphius, Georg Eberhard, 3, 169
Russian American Company, 236
Russian fur traders, 123, 133

Saint-Hilaire, Geoffroy, 206
Sakhalin, 158
Salish (northwest coast), 141
salmonberry (*Rubus americanus*, later *Rubus spectabilis*), 39
Samoa, 66, 160
Samuel, Richard, 79
San Blas, 185
Sandwich, 4th earl, 97, 100, 113, 114, 132–3
Santa Barbara, 144
Santa Catarina, 154, 235
Santa Maria, 29
Scholient, Ernst, 99
Scouler, John, 149
scurvy grass (*Cardamine glacialis*), gathered by Steller, 40, 47; gathered by Banks, 78
scurvy, 7, 185; on Second Kamchatka Expedition, 36, 40–7 *passim*; on Anson's voyage, 41; investigated by Lind, 41; on Bougainville's voyage, 62; affects Banks, 79; on Cook's second voyage, 79–80; on *Prince of Wales*, 134; on Vancouver's voyage, 141, 145; on La Pérouse's voyage, 161; on d'Entrecasteaux's voyage, 169, 173, 174; on Baudin's voyage, 210, 213; on Flinders' voyage, 225
sea lion (*Eumetopias jubatus*), described by Steller, 50
Sea of Okhotsk, 33
sea otter (*Enhydra lutris*), 45; as food, 47; pelts prized, 46, 268 n.47; described by Steller, 50; pelts collected, 133, 135, 156
Second Kamchatka Expedition, 33–49; instructions, 34; course of, 35–49; inaccurate reports of, 49

Selkirk, Alexander, 30
Sessé y Lacasta, Martínde, 179
Shark Bay, 23, 24, 27, 209, 215
Sharpe, Bartholomew, 8, 10, 18
Sheffield, Rev.W., 91–2
shells, collected, 26, 27, 105, 110, 111, 124,
    169, 190, 211, 215
Shelvocke, George, 28
Sherardian Herbarium, 24
Ships:
    Adventure, 98, 101, 102, 107, 113
    Astrolabe (La Pérouse), 153–64
    Astrolabe (d'Urville), 233
    Atrevida, 181–96
    Bachelor's Delight, 19–19
    Beagle, 239–52, influence on Darwin,
        257, 259, cabin reconstructed, 257
    Boudeuse, 55–68
    Bounty, 5, 17, 135
    Boussole, 153–64; wrecked, 177
    Casuarina, 213
    Cato, wrecked, 226
    Chatham, 136, 139–47
    Cinque Ports, 29
    Cumberland, 227
    Cygnet, 10, 12
    Descubierta, 181–96
    Discovery (1776–80), 123–32
    Discovery (1790–5), 136–47
    Dolphin, 61, 64
    Endeavour, 73–87
    Erebus, 253
    Espérance, 164–74
    Etoile, 55–68
    Géographe, 215–17; cargo of specimens,
        217
    Jardinière, 206
    King George, 133
    Lady Nelson, 224
    Nadezhda, 235, 236
    Niger, 93
    Pandora, 168, 177
    Porpoise, wrecked, 226
    Predpriyatie, 237
    Prince of Wales, 133
    Princess Royal, 133
    Queen Charlotte, 133
    Rattlesnake, 164–74
    Resolution (1772–5), altered, 89–90,
        98–113; (1776–80), 121–30
    Roebuck, 121–30
    Ryurk, 236, 237–8
    Seniavin, 237
    St Gabriel, 33
    St George, 22, 29

    St Paul, 35, 36, 39, 42, 50
    St Peter, 35–41; wrecked, 44–5, 46;
        rebuilt, 48–9
    Swallow, 61
    Uranie, 231, 238
    Vostok, 237
Shumagin Islands, 40, 42
Siberia, 49, 53
Sloane, Hans, 3, 12, 16
Smith, Christopher, 171
Smith, James Edward, 135, 203–4
Société des Observateurs de l'Homme, 206
Society for the Encouragement of Arts,
    Manufactures and Commerce, 96
Society Islands, 79, 80, 102, 113, 131
Society of Antiquaries, 96
Solander, Daniel, 57, role on Cook's first
    voyage, 74–91 *passim*; compiles
    Tahitian vocabulary, 81; illness, 87;
    praised, 88–9; relationship with
    Banks, 90, 91, 124; death, 93; and J.R.
    Forster, 96, 97, 103, 107, 108–9
Solomon Islands, 67, 163, 173
Sonnerat, Pierre, 70, 71, 270 n.50
South Georgia, 112
South Pole, 90, 98, 108, 234
South Sandwich Islands, 112
southern continent, great (*Terra Australis
    Incognita*), 76, 82, 108
Spanberg, Martin, 35
Sparrman, Anders, 57, 99, 100, 102, 116,
    118; attacked, 106, 109; described
    provisions, 112
Spencer Gulf, 214, 221, 222
spiders (Campeche), 18
Spöring, Herman Dietrich, 74, 80, 84, 87
Sprat, Thomas, 15
St Peter and St Paul (Petropavlovsk),
    35–6, 159
Stel, Simon and Adrian van der, 3
Steller, Georg Wilhelm, 4, 94; early years,
    34–5; joins Bering, 35; lifestyle, 35;
    relationship with Bering, 36–7;
    criticism of naval officers, 37, 46;
    explores Kayak Island, 38–40; his
    frustration, 38–9, 267 n.18; fights
    scurvy, 40–1, 45, 47; leads survivors,
    46; praised by Waxell, 48; travels in
    Kamchatka, 49; death, 49; praised by
    Müller, 50; his journal published, 50;
    his *De bestiis marinis* published, 50;
    describes and dissects sea cow, 50–2;
    notes and specimens assessed, 52–3,
    159; achievements, 53; praised by
    Linnaeus, 53

Steller's eider (*Anas stelleri*), 48
Steller's jay (*Cyanocitta stelleri*), 40
Steller's sea cow (*Hydrodamalis gigas*),
    attempts to catch, 47–8; value as food,
    48, 51; skeleton abandoned, 48; palatal
    plates, 48, 268 n.56; exterminated, 50;
    described, 50–1, 53; dissected, 51–2
Stephens, Philip, 75
Stokes, John Lort, 240, 241, 242
Stokes, Pringle, 239
Stone, Sarah, 117
Strait of Anian, 186
Strait of Georgia, 140, 148
Strait of Juan de Fuca, 139, 142, 197
Strait of Le Maire, 78
Strait of Magellan, 7, 60, 61, 62, 248
Strait of Tartary, 158
Stubbs, George, 91
Sturt's desert pea (*Clianthus dampieri* later
    Swainsona formosa), 25
Sullivan, Bartholomew, 243, 259
Sumatra, 11
sundew plant (*Drosera* species), 170
Surabaya (Java), 174
Suria Lozano, Tomás de, 183, 187, 188;
    journal, 183–4
swans, black (*Cygnus atratus*), 192, 213, 228
Swift, Jonathan, 14, 23

Tahiti, 17; reached by Bougainville, 62;
    sensuous images of, 62–3, 64, 69; 'an
    enchanted island', 63–4, 66, 68, 69;
    disillusionment, 68–9; named
    Nouvelle-Cythère, 66; first printed
    descriptions, 68; and Cook's first
    voyage, 73, 74, 80–2, 195; plant life,
    80, 139, 152; further voyage to, 88,
    237; and Cook's second voyage, 98,
    105–6, 108–9; and Cook's third
    voyage, 130, 131; and Darwin,
    250, 252
Tana, 109, 111
Tasman Sea, 84
Tasman, Abel, 7, 23
Tasmania, *see* Van Diemen's Land
Tasmanian bluegum (*Eucalyptus globulus*),
    167
Tasmanians, described by Anderson,128;
    by Baudin, 210, 211–12
Tenerife, 134, 154, 166
'Terre Napoléon', 218, 227
Thouin, André, 152, 278 n.6
Thunberg, Carl Peter, 57
Tierra del Fuego, 78, 93, 104, 134, 247, 248
Tilesius, Wilhelm, 235

Timor, 2, 10, 22, 24, 27, 210, 216, 225;
    plants from, 25
Tlingit (northwest coast), 141, 145, 157,
    187, 188
tobacco, on northwest coast, 145
Tofiño de San Miguel, Vicente, 180
Tolaga Bay (Uawa), 84
Tonga, 106–7, 126, 129–30, 163, 172
Tonkin (Vietnam), 11, 14
Torres Strait, 87, 224
tortoises, Galapagos (*Chelonoidis nigra*),
    19–20; 146
transit of Venus, 73, 82
Tres Marias Islands, 145
Trial Rocks, 226
Tuamotus, 62, 108
Tupaia (Society Islander), 82; death, 87
Tutuila (Samoa), visited by La Pérouse,
    160–2; boat crews attacked, 161–2

Uppsala, university of, 55, 57

Vaillant, Sébastien, 55
Valdés y Bazán, Antonio, 180, 186, 196, 197
Valdés y Flores, Cayetano, 197
Van Diemen's Land (Tasmania), 131;
    visited by d'Entrecasteaux, 167–8,
    170–2; charted by Beautemps-
    Beaupré, 168; proved to be island, 201
Vancouver Island, named, 141, 142
Vancouver, George, published account, 121;
    voyage, 136–47, 202, 204; relations
    with Menzies, 137–8, 139, 140, 142,
    144, 146–7; hardships of survey,
    140–1, 143; ill health, 147; abuses
    crew, 147; published account, 121,
    148, 197
Vancouver, John, 148
Vatoa (Fiji), 109–10
Vava'u, visited by Malaspina, 193
Ventenat, Louis, 171
Véron, Pierre-Antoine, 55, 59, 60
Vestiges of the Natural History of Creation,
    influence on Darwin, 255
Vivez, François, 60, 63; dislikes
    Commerson, 61; suspects Baret's
    gender, 61, 64–5; possible rapist, 66
Vlamingh, Willem de, 23, 27
volcanoes, 115

Wales, William, 98, 100, 103, 112, 115
Walker, John, 126
Wallace, Alfred Russel, impact on Darwin,
    257; hardships, 263
Wallis, Samuel, 54, 61, 62, 64, 69, 73, 76

Wardian container, 5
Warrington Academy, 96, 97
Waxell, Sven, account of Second
    Kamchatka Expedition, 36–48 *passim*,
    49, 51, 52
Webber, John, 125, 133, 153
Welbe, John, 29
Wesley, Rev. John, 117
West Falkland, 182, 195
West India Company (Dutch), 2
Westall, William, 204, 225
Wheeler, Alwyn, 93
White, John (governor Roanoke colony), 2

White, John (surgeon-general NSW), 203
Wickham, John Clements, 243
Wiles, James, 171
Wilkes, Charles, 234
William IV, 146
Witsen, Nicolas, 3, 27
Woodward, John, 15, 25

Yezo, 34
Yushin, Kharlam, 36, 44

Zoological Society (London), 253,
    257, 262